Alternative Fuels for Transportation

Mechanical Engineering Series

Frank Kreith, Series Editor

Alternative Fuels for Transportation
Edited by Arumugam S. Ramadhas

Computer Techniques in Vibration
Edited by Clarence W. de Silva

Distributed Generation: The Power Paradigm for the New Millennium
Edited by Anne-Marie Borbely and Jan F. Kreider

Elastic Waves in Composite Media and Structures: With Applications to Ultrasonic Nondestructive Evaluation
Subhendu K. Datta and Arvind H. Shah

Elastoplasticity Theory
Vlado A. Lubarda

Energy Audit of Building Systems: An Engineering Approach
Moncef Krarti

Energy Conversion
Edited by D. Yogi Goswami and Frank Kreith

Energy Management and Conservation Handbook
Edited by Frank Kreith and D. Yogi Goswami

The Finite Element Method Using MATLAB, Second Edition
Young W. Kwon and Hyochoong Bang

Fluid Power Circuits and Controls: Fundamentals and Applications
John S. Cundiff

Fundamentals of Environmental Discharge Modeling
Lorin R. Davis

Handbook of Energy Efficiency and Renewable Energy
Edited by Frank Kreith and D. Yogi Goswami

Heat Transfer in Single and Multiphase Systems
Greg F. Naterer

Heating and Cooling of Buildings: Design for Efficiency, Revised Second Edition
Jan F. Kreider, Peter S. Curtiss, and Ari Rabl

Intelligent Transportation Systems: Smart and Green Infrastructure Design, Second Edition
Sumit Ghosh and Tony S. Lee

Introduction to Biofuels
David M. Mousdale

Introduction to Precision Machine Design and Error Assessment
Edited by Samir Mekid

Introductory Finite Element Method
Chandrakant S. Desai and Tribikram Kundu

Machine Elements: Life and Design
Boris M. Klebanov, David M. Barlam, and Frederic E. Nystrom

Mathematical and Physical Modeling of Materials Processing Operations
Olusegun Johnson Ilegbusi, Manabu Iguchi, and Walter E. Wahnsiedler

Mechanics of Composite Materials
Autar K. Kaw

Mechanics of Fatigue
Vladimir V. Bolotin

Mechanism Design: Enumeration of Kinematic Structures According to Function
Lung-Wen Tsai

Mechatronic Systems: Devices, Design, Control, Operation and Monitoring
Edited by Clarence W. de Silva

The MEMS Handbook, Second Edition (3 volumes)
Edited by Mohamed Gad-el-Hak

MEMS: Introduction and Fundamentals
MEMS: Applications
MEMS: Design and Fabrication

Multiphase Flow Handbook
Edited by Clayton T. Crowe

Nanotechnology: Understanding Small Systems
Ben Rogers, Sumita Pennathur, and Jesse Adams

Nuclear Engineering Handbook
Edited by Kenneth D. Kok

Optomechatronics: Fusion of Optical and Mechatronic Engineering
Hyungsuck Cho

Practical Inverse Analysis in Engineering
David M. Trujillo and Henry R. Busby

Pressure Vessels: Design and Practice
Somnath Chattopadhyay

Principles of Solid Mechanics
Rowland Richards, Jr.

Thermodynamics for Engineers
Kau-Fui Vincent Wong

Vibration Damping, Control, and Design
Edited by Clarence W. de Silva

Vibration and Shock Handbook
Edited by Clarence W. de Silva

Viscoelastic Solids
Roderic S. Lakes

Alternative Fuels for Transportation

Edited by

Arumugam S. Ramadhas

CRC Press is an imprint of the
Taylor & Francis Group, an **informa** business

CRC Press
Taylor & Francis Group
6000 Broken Sound Parkway NW, Suite 300
Boca Raton, FL 33487-2742

Printed in the United States of America on acid-free paper
10 9 8 7 6 5 4 3 2 1

International Standard Book Number: 978-1-4398-1957-9 (Hardback)

Visit the Taylor & Francis Web site at
http://www.taylorandfrancis.com

and the CRC Press Web site at
http://www.crcpress.com

Dedicated to my parents, in-laws, and teachers

who are great sources of inspiration and support

Contents

Preface

Two problems currently confronting the world are the energy crisis and environmental pollution. Energy consumption in the world is increasing faster than its generation. It is estimated that the existing petroleum oil and natural gas reserves will be sufficient for only another few decades. The transportation sector and decentralized power generation completely depend upon petroleum products, particularly gasoline and diesel. The transportation sector is growing at a faster rate than reserves due to rapid technological development in the automotive industry and an increase in the use of personal vehicles in developed and developing countries. Thus the demand for petroleum accelerates the crude oil petroleum production peaks as well as its cost.

Vehicle pollution has been a serious concern for the past few decades all over the world. Combustion products of petroleum fuels, such as carbon dioxide, are a major contributor to greenhouse gases. An increase in the concentration of greenhouse gases in the atmosphere will lead to global climate change.

In the early 1960s, the United States initiated laws to control vehicle emissions and most other counties in the world have followed its lead. Moreover, the increasing operating cost of refineries to meet the latest emissions norms has put pressure on refining margins, and remains a problem converting refinery streams into products with acceptable fuel specifications.

The combination of a short supply of fossil fuel reserves, environmental pollution, and volatility in crude oil prices has generated interest for using alternative fuels. In the long term, the role and interrelationship of alternative energy sources with other markets demand further attention and consideration. Renewable alternative fuels (i.e., biofuels) have low-net greenhouse gases; hence, many countries are promoting them as a part of their national plan to reduce greenhouse gases and to improve their energy security. Stringent emission standards have been enforced in many countries which has led to improvement in automotive technology and the use of alternative fuels like biofuels and gaseous fuels at an optimum blend.

This book addresses the need for energy researchers, engineers (mechanical, chemical, and automobile), doctoral students, automotive power train researchers, vehicle manufacturers, oil industry researchers, and others interested in working on alternative fuel sources. Each chapter in the book contains the potential of the alternative fuels, production methods, properties, and vehicle tests, as well as their merits and demerits.

The structure of this book is such that each chapter—describing a particular fuel—is completely self-contained. The book is intended for anyone in need of comprehensive understanding of alternative fuels. Anyone would be able to reference a particular fuel that is of interest without having to read other chapters to gain a full understanding of fuel technology.

Chapter 1 highlights the importance of moving the focus toward alternative fuels and the problems resulting from dependence on petroleum products and their environmental impacts. It outlines the parameters for the selection of alternative fuels and the scenario of alternative fuel vehicles. Chapter 2 describes the characterization of vegetable oils and various methods for use in diesel engines. The performance, emissions, and durability studies on diesel engines using vegetable oil–diesel blends and challenges posed by vegetable oils in engines are discussed. The necessity of the transesterification process and the various methods to produce biodiesel from vegetable oils are reviewed in Chapter 3. Biodiesel properties, testing methods, the effect of properties on engines and emissions, engine performance, emissions and durability, and the challenges posed by biodiesel are discussed in detail.

Chapters 4 and 5 discuss the use of alcohols in internal combustion engines. The methods to produce alcohols, properties, ways to use alcohols in gasoline and diesel engines, materials compatibility, safety aspects, and challenges are reviewed. In Chapter 6, dimethyl ether production methods, economics, fuel properties, its applications in the engine as well as domestic applications, safety aspects, and its future scope are described.

Chapter 7 discusses the potential of liquefied petroleum gas, its production, properties, its use in gasoline and diesel engines, conversion kits, material compatibility, and safety aspects. The natural gas production, utilization, properties, its storage, distribution, safety aspects, its usage in gasoline and diesel engines, and natural gas advantages and disadvantages are discussed in Chapter 8. In Chapter 9, clean fuel hydrogen production from various sources; its properties; well-to-wheel analysis; safety; economics; hydrogen-fueled spark ignition engines and compression ignition engines; hythane engine; its storage in various forms such as liquid, solid, and gaseous; its benefits; and challenges are all elaborated.

Zero emission vehicles are divided into three chapters: electric vehicles, fuel cells, and hybrid vehicles. Chapter 10 details the construction and working principle of the motor, battery, and controllers; solar panel–operated electric vehicles, and the current scenario. In Chapter 11 the working principle, various types of fuel cells, fuel cell arrangement, fuel cell fueling systems, fuel processing systems, technical issues, fuel cell vehicle configuration, and the current scenario of the fuel cell market are discussed. Based on the available energy sources, the battery electric vehicle and fuel cell electric vehicle cannot compete with the internal combustion engine vehicle in terms of driving range or initial cost. In the future, the hybrid electric vehicle (commonly called the hybrid vehicle) is not only a major class of alternative fuel vehicles, but also a practical solution for commercialization of super ultralow emission vehicles. Chapter 12 presents an overview of the latest hybrid vehicles, with emphasis on their configurations, classification, and operations as well as energy efficiency and environmental benefits.

The first generation of biofuels may face problems in the future if food crops are used for fuel purposes. Chapter 13 gives an overall picture of future fuels that can be produced from nonfood crops. Biodiesel produced from algae, ethanol from cellulosic materials, biohydrogen, Fisher–Tropsch diesel, and dimethyl ether by gasification processes are discussed.

This book can be considered a textbook for college curricula that deal with fuel technology, internal combustion engines, alternative fuels, renewable energy, and sustainable development. Expert readers might include mechanical engineers, automobile engineers, chemical engineers, fuel engineers, fuel processors, policy makers, environmental engineers, and graduate/postgraduate/doctoral students.

I express my sincere thanks to my teachers and my friends who helped in the preparation of the book.

Authors wish to acknowledge Elsevier publications, the American Society of Agricultural and Biological Engineers, the Advanced Fuels and the Advanced Vehicles Research Center (U.S. Department of Energy) for their kind permission to reproduce figures and tables from their publications/data bank.

I express my sincere thanks to my mentors, particularly Dr. S. Jayaraj (National Institute of Technology Calicut), Dr. C. Muraleedharan (National Institute of Technology Calicut), Dr. G. Lakshmi Naryana Rao (QIS College of Engg, Ongole), Dr. N. Nallusamy (Sri Venkateswara College of Engg, Chennai) and my friends who helped in the preparation of the book. Special thanks to Ashish Kachhawa, of Indian Oil (R&D) for designing the theme of the cover of the book.

I must, especially, express my indebtedness to my wife Gomathy, my daughter Shivani, and my son Arvind for their patience and support all the way.

Dr. Arumugam Sakunthalai Ramadhas

Editor

Dr. Arumugam Sakunthalai Ramadhas received his BE (1996) from Madurai Kamaraj University, Madurai; his ME (1999) from Annamalai University, Chidambaram; and his PhD (2007) from the National Institute of Technology Calicut in mechanical engineering. He has published more than 25 research papers in peer-reviewed international journals and spoken at conferences/seminars, and contributed two chapters on biodiesel in the *Handbook of Plant-Based Biofuels* (2008). He is a reviewer of reputed international journals in the area of fuels and engines and a member of the Society of Automotive Engineers, India, and the Combustion Institute of India. He has a decade of experience in academia and industry. Presently he holds a position of Senior Research Officer in the Engine Testing of Fuels and Emissions Department, Research and Development Center of Indian Oil Corporation Ltd., India. His research focuses on conventional and alternative fuels, gasoline additives, engine technologies, and emissions.

Contributors

Mustafa Canakci
Department of Automotive Engineering Technology
Alternative Fuels R&D Center
Kocaeli University
Izmit, Turkey

K. T. Chau
Department of Electrical and Electronic Engineering
International Research Centre for Electric Vehicles
The University of Hong Kong
Hong Kong, People's Republic of China

Abhijeet Chausalkar
Engine Testing Laboratory
Indian Oil Corporation Ltd. (R&D Center)
Faridabad, India

Carroll E. Goering
Department of Agricultural and Biological Engineering
University of Illinois at Urbana-Champaign
Champaign, Illinois

Alan C. Hansen
Department of Agricultural and Biological Engineering
University of Illinois at Urbana-Champaign
Champaign, Illinois

Oguzhan Ilgen
Department of Chemical Engineering
Alternative Fuels R&D Center
Kocaeli University
Izmit, Turkey

Simon Jayaraj
Department of Mechanical Engineering
National Institute of Technology Calicut
Calicut, India

Chandrasekaran Muraleedharan
Department of Mechanical Engineering
National Institute of Technology Calicut
Calicut, India

Nallusamy Nallusamy
Department of Mechanical Engineering
Sri Venkateswara College of Engineering
Chennai, India

Fernando Ortenzi
CTL Centre for Transport and Logistics
Sapienza University of Rome
Rome, Italy

Giovanni Pede
Department of Energy Technologies
Italian National Agency for
New Technologies
Rome, Italy

Sethuraman Pitchumani
Central Electro Chemical Research
Institute-Madras Unit
CSIR Madras Complex
Chennai, India

Arumugam Sakunthalai Ramadhas
Engine Testing Laboratory
Indian Oil Corporation Ltd.
(R&D Center)
Faridabad, India

Gattamaneni Lakshmi Narayana Rao
QIS Institute of Technology
Ongole, India

Mohamed Younes El-Saghir Selim
Mechanical Engineering
Department
College of Engineering
United Arab Emirates University
Al Ain, United Arab Emirates

Paramasivam Sakthivel
Engine Testing Lab
Indian Oil Corporation Ltd.
(R&D Center)
Faridabad, India

Ashok K. Shukla
Solid State and Structural
Chemistry Unit
Indian Institute of Science
Bangalore, India

Spencer C. Sorenson
Department of Mechanical
Engineering
Technical University
of Denmark
Lyngby, Denmark

Parthasarathy Sridhar
Central Electro Chemical
Research Institute-Madras
Unit
CSIR Madras Complex
Chennai, India

1

Fuels and Trends

Arumugam Sakunthalai Ramadhas

CONTENTS

1.1 Introduction

Energy is the prime mover for economic growth of any country and is vital to the sustenance of modern economy. Future economic growth crucially depends on the long-term availability of energy from sources that are affordable, accessible, and environmentally friendly as well. The major sources of energy in the world are fossil fuels (petroleum oil, coal, and natural gas), renewable energy (hydro, wind, solar, geothermal, marine energy, and combustible wastes), and nuclear energy. These primary energy sources are converted into secondary energy sources; that is, coal and crude oil are converted into electricity and steam.

Combustible wastes include animal products, biomass, and industrial wastes. Coal is the major source of energy for electric power generation where as petroleum products are for the transport sector. Energy consumption is unevenly divided across various sectors of industrialized economies, as it is unevenly divided across geographic regions. Energy consumption continues to grow throughout the world, with most of the growth occurring by use of petroleum, coal, and natural gas as is shown in Figure 1.1 (World Energy Outlook 2007).

Petroleum oil, which is the most important and abundantly available energy source, is largely consumed in the world. The price of crude oil is very volatile and supply is driven by market price. While developed industrialized countries consume around 43 million barrels daily on an average, developing countries consume only 22 million barrels per day (MBD). Coal is the second most abundant source of energy in the world and is mainly used in power generation. Natural gas has been the energy source with highest rates of growth in recent years. The high end-use efficiency of natural gas has made it a popular choice for power generation projects. Hydroelectricity has been a major use of hydro sources of energy around the globe. Renewable sources of energy are gaining popularity. However, fuel prices and regulatory policies of different countries play important roles in the development of renewals (www.economywatch.com/energy-economy/scenario.html). The hydrocarbon industry across the world has been a major driver of economic growth and development for developed as well as developing countries. The commercial transportation sector uses a major share on diesel vehicles where as the public transportation—two wheelers and light commercial vehicles—auto market completely depend on gasoline-fueled engines. It is predicted that the global crude oil demand will rise by about 1.6% from 75 MBD in the year 2000 to 120 MBD in 2030 according to the IPCC assessment

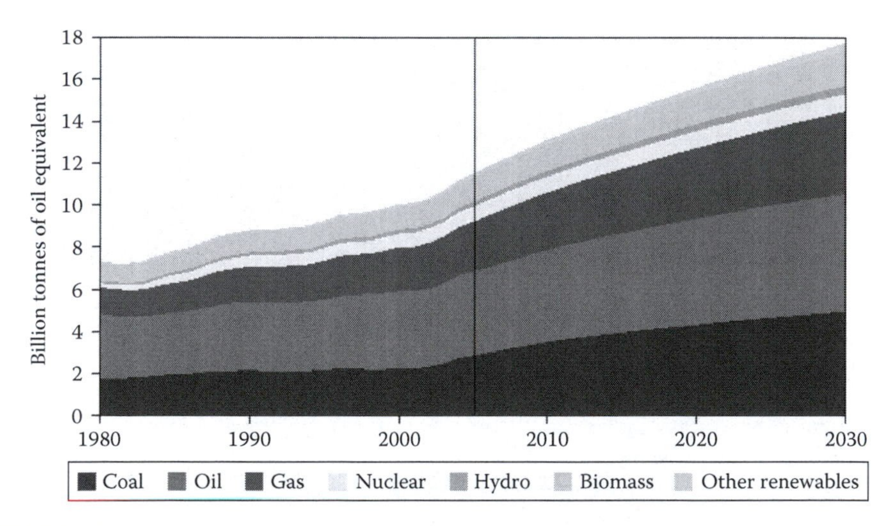

FIGURE 1.1
World energy consumption trend. (From World Energy Outlook 2007: China and India Insights, International Energy Agency (IEA) © OECD/IEA, 2007, Figure 1.1, page 76, www.iea.org)

TABLE 1.1

World's Primary Oil Demand (Million Barrels Per Day)

	1980	2000	2006	2010	2015	2030	2006–2030 (%) p.a.
OECD	41.8	46.0	47.3	49.0	50.8	52.9	0.5%
North America	20.9	23.4	24.9	26.2	27.7	30.0	0.8%
Europe	14.7	14.2	14.3	14.5	14.7	14.7	0.1%
Pacific	6.3	8.4	8.1	8.3	8.3	8.1	0.0%
Transition economies	9.4	4.2	4.5	4.7	5.1	5.6	0.9%
Russia	—	2.6	2.6	2.8	3.0	3.3	0.9%
Developing countries	11.3	23.1	28.8	33.7	38.7	53.3	2.6%
China	1.9	4.7	7.1	9.0	11.1	16.5	3.6%
India	0.7	2.3	2.6	3.1	3.7	6.5	3.9%
Other Asia	1.8	4.5	5.5	6.2	6.9	8.9	2.0%
Middle East	2.0	4.6	6.0	7.0	7.9	9.5	1.9%
Africa	1.3	2.3	2.8	3.1	3.4	4.8	2.2%
Latin America	3.5	4.7	4.8	5.2	5.6	7.1	1.6%
Int. marine bunkers and stock changes	2.2	3.6	4.1	3.7	3.9	4.5	—
World	64.8	77.0	84.7	91.1	98.5	116.3	1.3%
European Union	—	13.6	13.8	13.8	14.0	13.8	0.0%

Source: World Energy Outlook 2007: China and India Insights, International Energy Agency (IEA) © OECD/IEA, 2007, Table 1.2, page 80, www.iea.org.

on climate change (Biofuel 2003). The primary oil demand of the world is shown in Table 1.1.

1.2 Energy Security

Energy is one of the important inputs for any country's economic growth and development. Although 80% of the world's population is in the developing countries, their energy consumption amounts to only 40% of the total energy consumption. The high living standards in the developed countries are attributable to high-energy consumption levels. Also, the rapid population growth in the developing countries has kept the per capita energy consumption low as compared with that of developed countries. Continuous availability of energy in various forms in sufficient quantity and at reasonable prices is required. Though the crude reserves are spread all over the world, the major petroleum resources are available only in selected regions particularly in Middle East countries (about 63%). The international trade in oil is subject to violent fluctuations and has often led to war-like situations in the past, especially involving supply from the Gulf countries. The amount of crude oil is finite being the product of burial and transformation of biomass over 200 million years. Even now fossil fuels are being created under pressure and temperature; however, they are more rapidly consumed than they are produced. So there is a chance of a fuel shortage in the near future. The petroleum resources around the world are continuously utilized and thus these resources are fast depleting. This fast depletion of the resources leads to one another global factor (i.e., the ever-increasing cost of fuels). Hence, there is an urgent need to understand the energy crisis and transition from conventional to nonconventional sustainable energy sources. Figure 1.2 shows the breakup of energy consumption in the world (Deimbras 2009).

Countries having insufficient crude oil resources and depending upon the import of crude oil may seriously affect its economic development and sovereignty in international political relations. Therefore, as a matter of necessity and national self-reliance, such countries strive to achieve self-sufficiency in fuel availability. In general, the world depends on the limited oil reserves (particularly Middle East countries) and if there is a reduction in the production of crude oil, it will affect the world economy. Table 1.2 shows the crude oil production projection (IEA 2007).

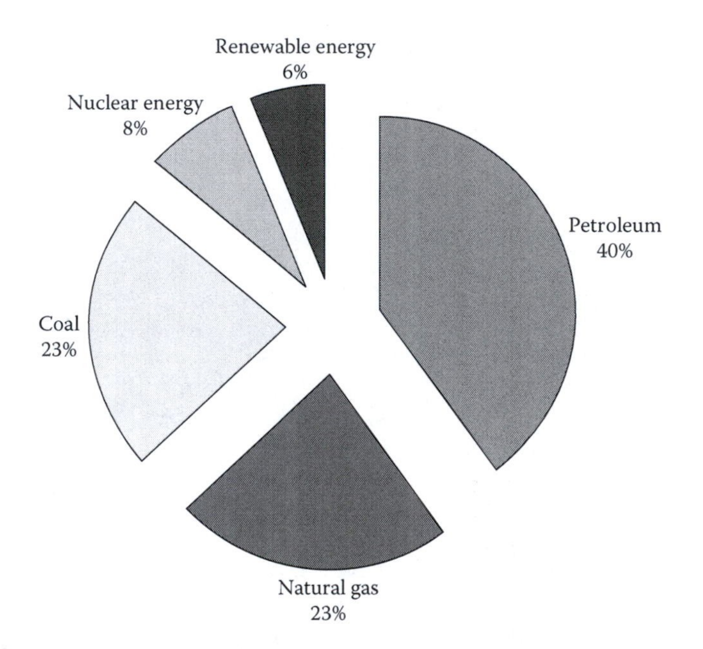

FIGURE 1.2
Energy consumption in the world. (From World energy outlook 2007: China and India Insights, International Energy Agency (IEA) © OECD/IEA, 2007, Fig 1.1, page 76, www.iea.org)

TABLE 1.2

Crude Oil Production Projection (Million Barrels Per Day)

		2030		
Projection	**2007**	**Reference**	**Low Oil Price**	**High Oil Price**
Conventional liquids				
Conventional crude oil and lease condensate	71.0	77.3	93.6	57.7
Natural gas plant liquids	8.0	12.4	11.2	12.1
Refinery gain	2.1	2.7	3.2	2.1
Subtotal	81.1	92.4	108.1	71.9
Unconventional liquids				
Oil sands, extra-heavy crude oil, shale oil	2.0	5.6	6.7	6.1
Coal-to-liquids, gas-to-liquids	0.2	1.6	0.8	2.8
Biofuels	1.2	5.4	3.3	7.7
Other	0.3	0.4	0.4	0.4
Subtotal	3.7	13.0	11.2	17.0
Total	84.8	105.4	119.3	88.9

Source: Annual Energy Outlook 2009, with projection to 2030, World Energy Outlook © DOE/EIA-0383, 2009, Table 5, page 30, www.iea.org.

Arab–Israeli wars in 1967 and 1973, the oil embargo in 1973, the Islamic revolution in Iran in 1979, the Iran–Iraq war in 1980, the Iraqi attack on Kuwait in 1990, and the latest Iraq war resulted in awareness about energy insecurity in the world. During the Iranian revolution in 1979–1980 and with the beginning of Iran–Iraq war in the 1980s, there was an increase in oil prices with the drop in oil consumption; that is, 63 MBD in 1980 to 59 MBD in 1983. However, oil consumption of the world was about 87 MBD in 2008. The consumption of petroleum products depends on economic growth, government policies, and vehicle population. Rapid economic growth of China and India has significant increase in crude oil consumption. Being fifth largest energy consumer in the world, India imports about 70% of crude oil requirement during 2003 and 2004 (Pandey 2008).

During the year 2007–2008, the price of crude of oil went up to $140 per barrel. If the oil price goes up, the import cost also goes up immediately, apart from the demand-pull price increase due to the continuing need for oil in transport, power, and industry sectors. Hence, the oil import bill has serious consequences on the nations' economy. Substitution of fuels for petroleum products therefore has a positive effect on reducing imports. Efforts are being made to reduce the consumption of fossil fuels and maximize the utilization of environment-friendly energy sources and fuels for meeting energy needs.

1.3 Environmental Pollution

Presently the world is confronted with the twin crisis of fossil fuel depletion and environmental concern. Environmental pollution is an important public health problem in most cities of the developing world. Epidemiological studies show that air pollution in developing countries accounts for tens of thousands of excess deaths and billions of dollars in medical costs and lost productivity every year. These losses, and the associated degradation in quality of life, impose a significant burden on people in all sectors of society, but especially the poor (Faiz 1996).

Burning of any hydrocarbon fuel produces carbon dioxide, which is accumulated in the atmosphere. CO_2 is a greenhouse gas that is slowly building in quantity in the atmosphere and it is believed this will raise the temperature of the planet, causing dramatic climate changes. The majority of developed countries have committed to targeted emission reduction through the Kyota Protocol of 1997, which entered into force in February 2005.

Combustion of hydrocarbon fuel generates both "direct" and "indirect" greenhouse gases. Direct gases emitted by transport include carbon dioxide, methane, nitrous oxide, and carbonate fuel cells are radiatively active. The indirect greenhouse gases include carbon monoxide, other oxides of nitrogen, and nonmethanic volatile organic carbons. These do not have a strong radiative effect in themselves, but influence atmospheric concentrations of the direct greenhouse gases by, for example, oxidizing to form CO_2 or contributing to the formation of ozone—a potent direct greenhouse gas.

The Human Development Report of 2007 says that developed countries should cut their carbon emissions by 20–30% before 2030 and at least by 80% by the year 2050. If emissions continue to rise following current trends, then stocks of the greenhouse emissions will be increasing at 4–5 ppm per year; by the year 2035, it may almost double the current rate. Accumulated stock will have risen to 550 ppm. Even without further increases in the rate of emissions, stocks would reach over 600 ppm by 2050 and 800 ppm by the end of the twenty-first century. India accounts for 5.5% of CO_2 emissions with the population of 17.2% world share. The report also says that the developing countries like India should cut their emissions by 20%. It would stabilize CO_2 equivalent concentration at 450 ppm in the atmosphere, which is currently at 379 ppm (Human Development Report 2007/2008).

It is expected that burning hydrocarbon fuel in the presence of air produces carbon dioxide and water. But, internal combustion engines do not completely burn petroleum products. Hence, these engines release unburned or partially burned/oxidized gases and nitrogen oxides into the atmosphere. The transport sector is a major contributor of air pollution particularly in cities and the vehicular pollution is the primary cause of air pollution in the urban areas (60%), followed by industries (20–30%), and fossil fuels. The pollutants released in the atmosphere interact with other pollutants (like photochemical reactions) and disturb the ecological balance. Faced with the growing air pollution related problems, a number of countries have established ambient air quality standards to protect their environment.

The main pollutants of internal combustion engines include particulate matter, unburned hydro carbons, CO, SO_2, and NO_x. These pollutants have serious effects on human health. Research is being carried out to produce alternative fuels, mainly directed toward reducing these harmful pollutant emissions. At this point, it would be appropriate to discuss the related health effects of various diesel pollutants (Ramadhas 2005).

The principal source of nitrogen oxides—nitric oxide (NO) and nitrogen dioxide (NO_2) are collectively known as NO_x. Nitric oxide (NO) emissions from engine exhaust is oxidized to nitrogen dioxides (NO_2), which can react with unburned hydrocarbon emissions and other sources allowing the concentration of ozone to increase around dense traffic. The NO and NO_2 concentrations are greatest in urban areas where traffic is heaviest. Exposure to NO_2 is linked with increased susceptibility to respiratory infection and viral infections such as influenza, lung irritation, edema of lungs, increased airway resistance in asthmatics, bronchitis and pneumonia, decreased pulmonary function, and increase in sensitivity to allergens. Nitrogen oxides combine with water vapor to form nitric acid and are removed from the atmosphere by direct deposition to the ground, or transferred to cloud or rainwater, thereby contributing to acid deposition.

Major sources of sulfur dioxide (SO_2) emissions are diesel automobiles, powerhouses, and petroleum industries. The SO_2 can impair lung function by constricting airways and damaging lung tissue, aggravate asthma and emphysema, and lead to suffocation and irritation to the throat. The SO_2 in ambient air is also associated with asthma and chronic bronchitis. Sulfur dioxide is a corrosive acid gas, which combines with water vapor in the atmosphere to produce acid rain. It causes acidification of lakes and streams and can damage trees, crops, historic buildings, and statues.

Concern about the potential health impacts of PM10 has increased very rapidly over recent years. Airborne particulate matter varies widely in its physical and chemical composition, source, and particle size. The particulates in air of a very small size ($< 10\ \mu m$) are of major current concern as compared to larger particulates. The smaller particulates are small enough to penetrate deep into the lungs and so potentially pose significant health risks.

There are two main groups of hydrocarbons of concern: volatile organic compounds (VOCs) and polycyclic aromatic hydrocarbons (PAHs). Volatile organic compounds are released in vehicle exhaust gases either as unburned fuels or as combustion products, and are also emitted by the evaporation of solvents and motor fuels. The VOCs effects include eye irritation, respiratory irritation, and cancer. The VOC condense together with SO_2 to create particulates, including smoke, soot, and dust. Benzene content in atmosphere is the main sources of emissions emitted in vehicle exhaust.

Carbon monoxide (CO) is a colorless, odorless, and poisonous gas produced by incomplete combustion of hydrocarbon fuels. The CO interferes with the body's ability to absorb oxygen, impair perception and thinking, slows reflexes, cause drowsiness, and can even cause unconsciousness and death at high levels. Inhalation of CO by pregnant women may threaten growth and mental development of the fetus.

Carbon dioxide, methane, and nitrous oxide and CFCs are called greenhouse gases. The global temperature is increased because of the greenhouse gases. Carbon dioxide pollution building up in the atmosphere is the single biggest contributor to global warming. The reduction in CO_2 is possible only if there is a reduction in combustion of fuel or use of nonhydrocarbon fuels like hydrogen.

1.4 Alternative Fuels

Fast dwindling reserves of fossil fuels, particularly petroleum products, cause hazards of environmental pollution by their combustion. Attempts were made to develop the future energy technologies that are energy efficient, environmentally friendly, and economically viable. Small incremental improvements in current energy technologies are sufficient to tackle the energy crisis and environmental problems. Alternative fuels are derived from other than crude oil resources. In general, alternative fuels include all the fuel used in vehicles other than gasoline and diesel. There are various alternative fuels that can be used with the current petrol or diesel internal combustion engine with little or no modification. The advantages with these fuels include cleaner burning than petroleum-derived fuels, producing lower emissions, and if it is derived from renewable biomass sources it will decrease the dependency on nonrenewable petroleum. However, alternative fuels, need not necessarily refer to a source of renewable energy. Each fuel has its own distinct advantages and disadvantages associated with cost, availability, environmental impact, vehicle/engine modification, safety and customer acceptance, and legislation. Alternative fuels are receiving attention because of the following reasons:

i. Alternative fuels are mostly produced from domestic resources that reduce the energy dependence.

   Use of locally available resources for fuel purposes can reduce crude oil import bill. Most of the alternative fuels, for example, alcohols, biodiesel can be produced from biomass resources and agricultural wastes and electricity for battery operated vehicles can be produced from solar and fuel cells. Hydrogen can be produced from biomass gasification or electrolysis of water. Hence, even a small percentage substitution of different alternative fuels reduces the crude oil import significantly.

ii. Alternative fuels generally reduce the vehicle exhaust emission and hence improve the environmental air quality.

Alternative fuels are capable of reducing the engine emissions as compared to petroleum products. The molecular structure of alternative fuels (CH_3OH, C_2H_5OH, and CH_4, etc.) is much simpler than gasoline/diesel (mixture of different molecules). Moreover, a low C:H ratio of alternative fuels generates less hydrocarbon emissions on combustion. Hydrogen is the clean fuel and generates nil hydrocarbon emissions. Emissions released from a centralized plant can be more easily controlled than vehicular emissions.

iii. Some alternative fuels have the potential to operate at a lower cost compared to petroleum products.

Success of any new product or fuel depends on its cost. Currently, cost of the most alternative fuels is a little bit higher than conventional fuel. However, the cost of biodiesel and compressed natural gas (CNG) are cost competitive with petroleum. For the development of alternative fuels, government legislations and incentives are required to a certain extent. The large-scale production of alternative fuel could make these alternative fuels cost competitive (Maxwell and Jones 1995).

The following parameters are to be considered while deciding the alternative fuel (Agarwal 1999):

- The fuel should have high volumetric and mass energy density.
- Ease of transportation from production site to delivery points.
- Long-storage life of fuel, minimum handling, and distribution problems.
- Environmental compatibility: While using alternative fuel, the engine performance is expected to improve significantly with regard to regulated emissions and unregulated emissions.
- Manufacturer's warranty: The alternative fuel must guarantee the lifetime of the equipment; its reliability and operational capability are not modified.
- Investment cost: Additional investment on an existing engine must be small to ensure that the operation is competitive with petroleum fuel.
- Modification of existing engines: Engine modification should be simple, inexpensive, and easily reversible. Such modification should not affect the use of traditional fuel in order to preserve engine

compatibility with the use of two fuels. Switch over of operation from alternative fuel mode to conventional fuel mode should be done easily.

The search for an alternative fuel has produced a long list of new candidates and a series of arguments, which support and project their characteristics. Suitability of each of these fuels for internal combustion engines has been under investigation throughout the world. Some of the important fuels are listed here:

i. Alcohols (methanol and ethanol)
ii. Vegetable oils and biodiesel
iii. Gaseous fuels (natural gas, hydrogen, and liquefied petroleum gas)
iv. Ethers
v. Electric/fuel cell/hybrid vehicles
vi. Future fuels

1.4.1 Alcohols

Alcohols are considered as a substitute or additive component for gasoline as they possess a higher octane number. The move toward unleaded fuels produced excessive exhaust valve wear in gasoline engines. This problem was solved by incorporating hardened valve seats and satellite coating of valves. These lead to an increase in the use of alternative fuels, particularly methyl tertiary butyl ether (MTBE), methanol/ethanol blends with gasoline because of their higher octane number.

1.4.1.1 Methanol

Methanol was used as an automobile fuel during the 1930s to replace/supplement gasoline supplies in high-performance engines. With the U.S. Government's mandatory phase-out of lead as a gasoline octane additive, a low concentration of methanol is found to be a good nonmetallic substitute. MTBE replaced methanol and tertiary butyl alcohol for a short while. However, the use of MTBE phased out due to the environmental concern.

With the advent of flexi-fuel vehicles (FFV) in the 1990s, the methanol became a prime candidate for use in vehicles because of its lower emission characteristics and high-octane rating. To solve the cold-start problems in methanol, a percentage of gasoline is added. Blends of 15% gasoline and 85% of methanol are called M85. The FFV vehicles are also able to operate on gasoline, methanol, and their blends. For the adjustment of air–fuel ratio and spark timing there is a sensor in the fuel line that measures the percentage of methanol in the gasoline and gives the information to the engine computer. Though it is a commercial success, methanol has not become the competitor of gasoline because of its higher cost. Currently, methanol is again considered for fuel cell vehicles; that is, methanol is used as fuel to derive the hydrogen for the operation of fuel cells.

1.4.1.2 Ethanol

Ethyl alcohol, commercially known as ethanol, possesses a number of characteristics favoring its use as an automobile fuel. Nikolaus August Otto, the German inventor of the internal combustion engine, conceived his invention to run on ethanol. Ethanol is a by-product in the production of sugar. It can be considered a renewable fuel as it is produced from the sources where there is a potential for greenhouse gas emissions abatement. It depends on the production process, especially the nature of energy inputs used in distillation and other phases. The use of ethanol in spark ignition engines started from the 1950s in countries like United States, Germany, and France. During World War I and World War II, ethanol was used as a substitute fuel for commercial as well as military vehicles. Currently, it is blended with gasoline in most of the countries and it is a major fuel in Brazil.

A close observation on its properties shows that it can be considered as a suitable automobile fuel. The autoignition temperature of ethanol is significantly higher than gasoline and this makes ethanol less susceptible to ignition. Presently, the focus is on ethanol for the replacement of MTBE and oxygenates in the gasoline. At present, ethanol is blended with gasoline for improving the octane number of gasoline for eliminating the blending of MTBE. Ethanol molecules contain oxygen and therefore, it allows the engine to completely combust the fuel, resulting in less emission of carcinogenic gases like carbon monoxide, NO_x, and so on. Moreover, it reduces the particulate concentration and greenhouse gas effect.

Though most of the work has been done on ethanol as a substitute for gasoline fuel, ethanol is also a suitable alternative fuel for diesel engines. Vaporizing ethanol in a naturally aspirated engine is an inexpensive way of using ethanol in the diesel engine and replacing about 60% diesel consumption. For higher blends of ethanol (> 20%), high concentrations of additives are needed to stabilize the mixture or attain the required cetane number. The higher percentage of ethanol in diesel requires a double-injection device or so-called dual fuel or fumigation systems. Gaseous pollutants and combustion noise levels could be significantly reduced, but the complexity to control these devices is restricted. However, use of 100% ethanol in diesel engines requires some modification in the engine fuel system.

1.4.2 Vegetable Oils and Biodiesel

Dr. Rudolf Diesel, inventor of the diesel engine, has demonstrated that his engine ran on peanut oil. With the advent of low-cost crude oil, appropriate crude oil fractions were refined to serve as fuel and hence diesel engine and diesel fuel evolved together. When the first energy crisis arose in the 1970s, the research on vegetable oil for fuel purposes picked up again. With increases in crude oil prices, limited resources of fossil oil, and environmental concerns, there has been a renewed focus on vegetable oils and animal fats to make biodiesel fuels. Vegetable oil fuels are not now petroleum competitive fuels because they are more expensive than petroleum fuels. However, vegetable oils and derivatives have the potential to substitute for a fraction of the petroleum distillates and petroleum-based petrochemicals in the near future. Vegetable oil fueled engines require frequent maintenance (like injector and combustion chamber cleaning) and hence these are suitable for stationary engines that are used for power generation and generator-motor pump sets in rural areas. These could reduce a significant amount of energy savings in terms of diesel or electricity consumption.

Biodiesel is methyl or ethyl esters of fatty acids derived from edible and nonedible type vegetable oils (used or fresh) and animal fats. The major sources for biodiesel production can be jatropha, karanji, palm, soy bean, and sun flower. During the years 1970 and 1980, a significant amount of research work was conducted with neat vegetable oil and its blends, partially esterified oils in blends with diesel. Use of neat oils causes varieties of engine problems. Based on the results obtained, research focused on esters due to their superior fuel properties than neat vegetable oils.

Feedstock accounts for 60–80% of the production cost of biodiesel. The biodiesel processing technologies—feed stock and by-product (glycerol) utilization—play an important role in success of biodiesel against petroleum diesel. Biodiesel can be blended with diesel at any ratio (however currently the blend is limited to 20%) and operates the diesel engine, which requires little or no modification. The use of biodiesel in conventional diesel engines substantially reduces the exhaust emissions.

Well-to-wheel energy consumption of biodiesel is higher than for fossil diesel but generally lower than for gasoline. Well-to-wheel emissions of biodiesel are very similar to diesel emissions; that is, higher NO_x and particulates but relatively low for CO and hydrocarbons and very low net CO_2 emissions because it is derived from biomass. Jatropha in India, palm oil in Malaysia, and soybeans in the United States are the major feed stocks for their biodiesel production. The annual production of biodiesel is increasing rapidly worldwide, from 10,000 tons in the year 2000 to 3.5 million tons in 2006 (Pandey 2008). In this scenario, it is necessary to go for second-generation biofuels (i.e., choose nonfood crops for biodiesel production) so the quantity of biodiesel can be increased to sustain the market.

1.4.3 Gaseous Fuels

1.4.3.1 Natural Gas

Natural gas has received a great deal of attention and has been successfully implemented in various parts of the world, such as Argentina, Russia, Italy, and India and gaining importance for transport vehicles. Natural gas occurs as gas under pressure in rocks beneath the earth's surface or more often in solution with crude oil as the volatile fraction of petroleum is composed of mainly methane with varying amounts of the paraffinic hydrocarbon family, ethane, propane, butane, (methane hydrate). Natural gas commercially has been using as a fuel for centuries in China. However, for the past few decades natural gas has been receiving more attention due to the increase in price of petroleum products.

Natural gas reserves are evenly distributed on a global basis and therefore provide better security of supply. Natural gas is being used in domestic and industrial sectors. Natural gas essentially consists of methane (80–98%) depending upon geographic origin and its properties are closer to that of methane. Natural gas has been one of the most widely experimented fuels proposed to replace liquid petroleum fuels. Research Octane Number (RON) of CNG is approximately 130. The excellent antiknocking property of CNG allows use of a higher compression ratio for increased power output and fuel economy compared to petrol.

The CNG has an excellent lean flammability limit and has the potential to reduce regulated and nonmethane hydrocarbon emissions compared to conventionally fueled engines. Natural gas can be stored in gaseous form at ambient temperatures and under high pressure (approx. 200 bar) as CNG and in liquid form at cryogenic temperatures (–161°C) and at atmosphere pressure as liquefied natural gas (Nine et al. 1997).

1.4.3.2 Liquefied Petroleum Gas

Liquefied petroleum gas (LPG), a mixture of propane (C_3H_8) and butane (C_4H_{10}) gas, is a popular fuel for internal combustion engines. It is a nonrenewable fossil fuel that is prepared in a liquid state under certain conditions. This popularity comes from many features of the fuel such as its high octane number for spark ignited engines, comparable to gasoline heating value that ensures similar power output. The LPG is stored as a liquefied gas under pressure at ambient temperature. The percentage composition of the mixture depends upon the season, as a higher percentage of propane is kept in winter and the same for butane in summer.

1.4.3.3 Hydrogen

Worldwide preparation to meet internationally enforceable greenhouse gas emission limits from vehicles with the use of hydrogen in engines and fuel cells may cause sustainable contribution to reduction in environmental pollution. Hydrogen is one of the clean fuels in the world, as it does not contain carbon compounds. Hydrogen is a clean and efficient energy carrier with the potential to replace liquid and gaseous fossil fuels. Significant work on hydrogen (i.e., demonstration of hydrogen) for automobiles and power generation has been carried out all over the world.

Hydrogen can be combusted directly in the IC engines or it can be used in the fuel cell to produce electricity, which can operate the vehicle. Hydrogen can be introduced into the engines by manifold induction, direct injection to the cylinder, and hydrogen–diesel duel fuel mode. On combustion of hydrogen, only water vapor is emitted. Therefore, the use of hydrogen as a transportation fuel would result in few or no emissions that affect air quality.

Hydrogen is manufactured from water using energy from either fossil or nonfossil fuel sources. The use of hydrogen has the potential to improve the air quality and climate change. The methods to produce hydrogen include electrolysis, photolysis, thermochemical water splitting, and thermal water splitting. In near term, hydrogen can also be produced from coal gasification and from petroleum and natural gases. Hydrogen can also be produced from various biomass sources. Gases generated by gasification can be steam reformed to produce hydrogen and followed by a water–gas shift reaction to further enhance hydrogen production. Biomass thermochemical conversion is one of the economical ways of producing renewable hydrogen in large-scale. Anaerobic fermentation enables the mass production of hydrogen from relatively simple processes using a wide spectrum of potentially utilizable substrates, including refuse and waste products. Moreover, fermentative hydrogen production generally proceeds at a higher rate and does not rely on the availability of light sources.

Hydrogen is a low-density gas. At ambient temperature and atmospheric pressure, 1 kg of the gas has a volume of about 11 m^3. The storage of hydrogen in a compact and efficient manner is a major technological challenge and becomes an important area of research for the promotion in the use of hydrogen as an automotive vehicle fuel. Hydrogen storage implies the reduction of an enormous volume of hydrogen gas. The hydrogen may be compressed to store in a cylinder, the temperature of gas to be decreased below the critical temperature (i.e., stored as liquid or in solid-state storage like metal hydrides). A broad research and development on different aspects comprising hydrogen production, storage, and use in vehicular and fuel cells will improve its practicability and acceptance.

1.4.4 Ethers

Ethers are oxygenating fuel that improves the combustion efficiency. Dimethyl ether is the commonly used blending component in gasoline fuel. Moreover, DME is a potential alternative fuel that can be used in engines as well as onboard hydrogen generation fuel cells. DME can be produced from natural gas and gasification of coal or biomass and synthesis. DME is the simplest ether, consisting of two methyl groups bonded to central oxygen and is expressed by its chemical formula CH_3OCH_3. Approximately 1 lakh to 1.5 lakh tons of DME is produced per annum worldwide. DME is a colorless gas at room temperature with an ethereal odor, has a vapor pressure of 5.93 bar at 25°C, and is highly flammable in air. DME has no corrosive effects on metals but it is a good solvent. DME has high cetane number (55–60). However, the calorific value of DME is 28.5 MJ/kg whereas for diesel 42.5 MJ/kg. DME contains oxygen but no C–C bonds and has the advantage of reducing carbon particulate matter emissions formation. DME is a gas at ambient conditions but it can be liquefied under pressure like LPG (Sorenson and Mikkelsen 1995).

Diethyl ether (DEE) can be used as a renewable fuel additive. DEE has long been known as a cold-start aid for engines, but little is known about using DEE as a significant component in a blend or as a complete replacement for fuel. It is an excellent engine fuel with higher energy content than ethanol. DEE is liquid at ambient conditions that make it attractive for fuel handling and infrastructure requirements. Storage stability of DEE and its blends are of concern because of its tendency to oxidize and form peroxides in storage. DEE has several favorable properties, including an outstanding cetane number and reasonable energy density for onboard storage. (Bailey, Eberhardt, and Goguen 1997).

1.4.5 Electric/Fuel Cell/Hybrid Vehicles

The control of tailpipe emissions and maximum possible reduction in fuel consumption are the top priority objectives for any automobile designer and manufacturer. Electricity stored in the battery can be used to operate the vehicles by means of an electric motor. These electric vehicles are noiseless. Electricity normally is generated from coal, natural, solar, fuel cell, on board diesel engine, gas, and nuclear energy. If the electricity produced from nonfossil fuel, then electric vehicles become zero emission vehicles. Thermal energy is converted to electricity and stored in a battery, and again it is converted into work. So, while calculating overall efficiency of vehicles, these conversion efficiencies have to be considered. The storage capability of the battery plays an important role in the success of electric vehicles.

Pure electric vehicles currently do not have adequate range when powered by batteries alone, and hence it cannot be used for driving long distances. Electric vehicles require recharging of the battery. Hybrid electric vehicles (HEV) combine the alternative energy source (engine/fuel cell) to run the vehicle and charge the battery. Hence, HEV is gaining acceptance and is overcoming some of the problems of pure electric vehicles.

Hybrid electric vehicles combine the internal combustion engine of a conventional vehicle with the battery and electric motor of an electric vehicle resulting in better fuel economy. The power of the hybrid vehicle's internal combustion engine generally ranges from 1/10th to 1/4th of the conventional automobile. In a HEV, when the driver applies the brakes, the motor becomes a generator, using the kinetic energy of the vehicle to generate electricity that can be stored in the battery for later use. The high efficiency and fuel economy are achieved by carefully adjusting the operating parameters of power-plant (IC-engine or motor) always around optimum efficiency conditions. A hybrid can achieve the cruising range and performance advantages of conventional vehicles with the low-noise, low-exhaust emissions, and energy independence benefits of electric vehicles.

The fuel cell can be used in hybrid vehicles to charge the battery instead of the internal combustion engine. Fuel cells are electrochemical devices that convert the chemical energy of a reaction directly into electrical energy. The basic physical structure or building block of a fuel cell consists of an electrolyte layer in contact with a porous anode and cathode on either side. Gaseous fuels are fed continuously to the anode (negative electrode) compartment and an oxidant (i.e., oxygen from air) is fed continuously to the cathode (positive electrode) compartment; the electrochemical reactions take place at the electrodes to produce an electric current. Biomass generated gaseous fuels such as hydrogen can also be used as fuel for the fuel cells.

1.4.6 Future Fuels

Next generation fuels include biodiesel produced from algae or nonfood products, cellulosic ethanol and fuels produced from gasification of biomass. Synthetic fuel is a liquid fuel obtained from natural gas, coal, oil shale, and biomass sources through the Fischer–Tropsch synthesis process. The production of synthetic fuel is generally the hydrogen addition process. The source of hydrogen can be intramolecular in which a carbonaceous low–hydrogen residue is produced. Hydrogenation can be either direct or indirect. Direct hydrogenation involves exposing raw material to hydrogen at high pressure. Indirect hydrogenation involves reaction of the feedstock with steam, and hydrogen is generated within the system.

The Fischer–Tropsch diesel is manufactured by the liquefaction of synthesis gas (mixture of CO and H_2) produced from gasification of biomass and fossil fuels. Depending upon the synthesis process, several types of fuels can be produced. German scientists Franz Fischer and Hans Tropsch established Fischer–Tropsch synthesis in 1923. The Fischer–Tropsch synthesis process, which produces diesel from gasification of biomass is called green diesel.

Fischer–Tropsch (FT) diesel can be substituted directly for conventional (petroleum-derived) diesel to fuel diesel-powered vehicles, without modification to the vehicle engine or fueling infrastructure. Shell markets FT diesel as a premium diesel blend in Europe and Thailand. The output of the Fischer–Tropsch process is typically 50% high-quality, sulfur-free, high cetane synthetic diesel; 30% naphta; and 20% other products. Some of the production plants in the world is listed below (Rehnlund 2007; Nylund, Aakko-Saksa, and Sipilä 2008).

Sasol, South Africa:	156,000 bpd, mainly coal
Shell, Malaysia:	15,000 bpd, natural gas
Qatar:	2 × 34,000 bpd, natural gas
Sasol Shell, United States:	5,000 bpd, waste coal
Shell, Qatar:	140,000 bpd

Currently, the following three technologies are used to produce liquid fuel viz. (CTL: coal to liquid, BTL: biomass to liquid, and GTL: gas to liquid. Coal to liquid process converts coal directly into liquids (direct liquefaction). Liquefaction is a thermal conversion process of biomass or other organic matters into primarily liquid oil in the presence of a reducing reagent. Liquefaction generally carried out at the moderate temperatures (from 550 to 675 K) and at high pressures. Biomass to liquid is a term describing processes for converting biomass into a range of liquid fuels such as gasoline, diesel, and petroleum refinery feedstocks. BTL can convert types of biomass, such as wood and agricultural residues that are difficult to handle using other biofuel production processes. In the gas to liquid process, biomass is turned into synthesis gas by gasification. Syngas can be burned directly to produce electricity or converted into hydrocarbons (such as gasoline and diesel), alcohols, ethers, or chemical products. In the long run, FT fuels compatible with existing vehicles are predicted to take the lead as alternative fuels for transport. The VTT Technical Research Centre of Finland reported that as the quality of FT diesel fuels is high, FT diesel might also be used to upgrade lower quality diesel fuels (Nylund, Aakko-Saksa, and Sipilä 2008).

The number of alternative fuels on the road are increasing every year. The growth of vehicles run on alternative fuel is depicted in Table 1.3. Ever increasing population of vehicles aggravates the demand of fuels. The alternative fuels can partially substitute the petroleum demand. The research on alternative fuels for its use in commercial vehicles is at peak in this scenario. Automotive manufacturers, oil industry, and academic institutions are doing research on production and storage of alternative fuels as well as to reduce the exhaust emissions of vehicles and increase the percentage of alternative fuels in petroleum fuel. However, the success of alternative fuels depends upon government polices and cost of the fuel.

TABLE 1.3

On-Road Alternative Fuel Vehicles Available

Fuel Type/Configuration	1998	1999	2000	2001	2002	2003	2004	2005	2006	Total
E85 flex fuel vehicle	216,165	426,724	600,832	581,774	834,976	859,261	674,678	743,948	1,011,399	5,949,757
Compressed natural gas (CNG)	10,221	13,425	9,501	11,121	8,988	6,122	7,752	3,304	3,128	73,562
Dedicated	4,143	4,891	3,997	5,506	5,397	3,397	4,398	2,276	2,066	36,071
Nondedicated	6,078	8,534	5,504	5,615	3,591	2,725	3,354	1,028	1,062	37,491
Electric	1,844	1,957	6,215	6,682	15,484	12,395	2,200	2,281	2,715	51,773
Liquefied petroleum gas (LPG)	5,620	5,955	4,435	3,201	1,667	2,111	2,150	700	473	26,312
Dedicated	3,525	2,273	1,056	633	532	287	164	241	277	8,988
Nondedicated	2,095	3,682	3,379	2,568	1,135	1,824	1,986	459	196	17,324
Liquefied natural gas (LNG)	380	40	411	393	147	111	136	68	92	1,778
Hydrogen	0	0	0	0	2	6	31	74	40	153
TOTAL	234,230	448,101	621,394	603,171	861,264	880,006	686,947	750,375	1,017,847	6,103,335

Source: www.afdc.energy.gov/afdc/data

References

Agarwal, A. K. 1999. Performance evaluation and tribological studies on biodiesel fueled compression ignition engine. PhD Thesis, IIT Delhi, India.

Bailey, B., J. Eberhardt, and S. Goguen. 1997. Diethyl ether as a renewable diesel fuel. *SAE* 972978.

Biofuel, 2003. Planning commission, Government of India. New Delhi.

Deimbras, A. 2009. *Biofuels*. London: Springer.

Faiz, A., S. Christopher, and M. P. Weaver. 1996. *Air pollution from motor vehicles*. Washington, DC: The World Bank.

Human Development Report. 2007/2008. Fighting climate change: Human solidarity in a divided world. United Nations Development Programme, Palgrave Macmillan, New York.

Maxwell, T. T., and J. C. Jones. 1995. Alternative fuels: Emissions, economics and performance. *SAE*.

Nine, R. D., N. N. Clark, B. E. Mace, and L. E. Igazzar. 1997. Hydrocarbon speciation of a lean burn spark ignited engine. *SAE* 972971.

Nylund, N.-O., P. Aakko-Saksa, and K. Sipilä. 2008. Status and outlook for biofuels, other alternative fuels and new vehicles, Espoo 2008. VTT Tiedotteita, Research Notes, 2426.

Pandey, A. 2008. *Handbook of plant-based biofuels*, 3–13. Boca Raton, FL: CRC Press.

Ramadhas, A. S. 2005. Theoretical modeling and experimentation of different alternative fuels for compression ignition engines. PhD Thesis, National Institute of Technology, Calicut, India.

Rehnlund, B. 2007. Synthetic gasoline and diesel oil produced by Fischer–Tropsch Technology. A possibility for the future? IEA/AMF Annex XXXI report.

Sorenson, S. C., and S. E. Mikkelsen. 1995. Performance and emissions of a 0.273 liter direct injection diesel engine fuelled with neat dimethyl ether. *SAE* 950564.

World Energy Outlook. 2007. China and India insights. International Energy Agency (IEA).

2

Vegetable Oils

Arumugam Sakunthalai Ramadhas, Chandrasekaran Muraleedharan, and Simon Jayaraj

CONTENTS

2.1 Introduction

Fossil fuels (i.e., coal and petroleum products) are the major sources of energy in the world. Coal is considered the primary energy source for electric power generation and petroleum products for the transportation sector. Various alternatives are being tried for the substitution of petroleum products, which include biomass derived sources like vegetable oils, biodiesel, alcohols, other gaseous fuels like natural gas and LPG, and engine power trains such as fuel cells and hybrid systems. These efforts were made to reduce the emissions and to improve the fuel security. Significant research work is being carried out around the world on production as well as the application of biomass energy for fuel purposes.

The concept of using vegetable oil as fuel dates back to 1895 when Dr. Rudolf Diesel developed the first diesel engine to run on vegetable oil. Dr. Diesel had demonstrated his engine at the World Exhibition in Paris in 1900 using peanut oil. With the arrival of petroleum and its appropriate fractions, low-cost petroleum products replaced the vegetable oils for use in the engines. However, during the energy crisis periods (1970s), the vegetable oils and alcohols were widely used as engine fuel. Due to the ever rising crude oil prices and environmental concern, there has been a renewed focus on the vegetable oils and their derivatives for use in engines.

Vegetable oils have two broad classifications: edible oils (sunflower, soy bean, palm oil, etc.) and nonedible oils (jatropha, karanji, rubber seed oil, etc.). Edible type oils are mainly used for food purposes whereas nonedible oils are used for food purposes. The nonedible vegetable oils serve as an important raw material for the manufacture of soaps, paints, varnishes, hair oil, lubricants, textile auxiliaries, and various sophisticated products. After extraction of oil from oil seeds, the oil cakes can be used as cattle feed and fertilizer. Moreover, these oil cakes can be used as biomass feed stock for gasification process. Vegetable oils are derived mainly from four sources (Agarwal 1999). These are

1. Cultivated oil seeds (i.e., groundnut, rape-mustard, soybean, sesame, sunflower, safflower, castor, and linseed)
2. Perennial oil-bearing materials (i.e., coconut and palm)
3. Derived oil-bearing material (i.e., cottonseed and rice bran)
4. Oil seeds of forest and tree origin (i.e., karanji and rubber seed oil)

Most of the countries are encouraging efforts to cultivate the oil yielding trees for vegetable oil production. The annual oil seed production of the world is 390 million metric tonnes in the year 2007. The break up of various oil seeds is shown in Figure 2.1.

2.2 Characterization of Vegetable Oils

Vegetable oil molecules are triglycerides with unbranched chains of different lengths and different degree of saturation. Diesel is a complex mixture of thousands of individual compounds, most with carbon numbers between 10 and 22 and mostly of saturates. The natural organic compound in the animal and vegetable fats are made up of various combinations of fatty acids (in sets of three) connected to a glycerol molecule, making them triglycerides. Each molecule of a fatty acid consists of a carboxyl group (oxygen, carbon, and hydrogen) attached to a chain of carbon atoms with their associated hydrogen atoms. The chain of carbon atoms may be connected with single bonds of hydrogen between them, making a saturated fat; or it may be connected with double bonds, making an unsaturated fat. The number of carbon and hydrogen atoms in the chain are what determines the qualities of that particular fatty acid (Agarwal 1999).

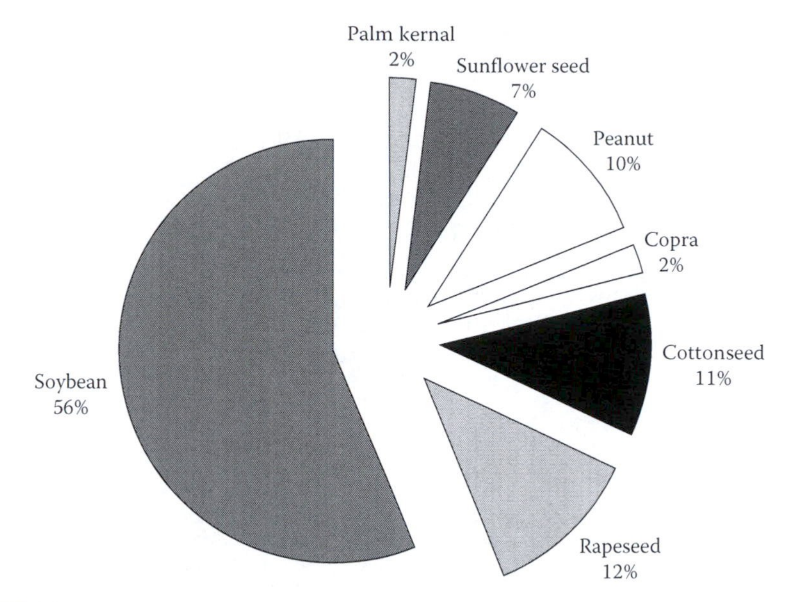

FIGURE 2.1
World oil seed production in 2007.

The general molecular formula of any vegetable oil is given by

$$\begin{array}{l} CH_2 - OOC - R_1 \\ | \\ CH - OOC - R_2 \, . \\ | \\ CH_2 - OOC - R_3 \end{array}$$

Structurally, one molecule of glycerol reacts with three molecules of fatty acids to yield three molecules of water and a molecule of triglyceride. The fatty acids vary in their carbon chain length and in the number of unsaturated bonds. When the three fatty acids are identical, the product is simple triglyceride; when they are dissimilar, the product is a mixed triglyceride. The vegetable oils are of mixed triglycerides. The $C_{18:1}$ is denoted as 18 carbon atoms with one double bond and $C_{18:2}$ is denoted as 18 carbon atoms with two double bonds and $C_{18:3}$ is denoted as C18 carbon atoms with three double bonds. The chemical structure of fatty acids is given in Table 2.1 (Barnwal and Sharma 2005).

Fuel properties can be grouped conveniently into physical, chemical, and thermal properties. The important properties of vegetable oils in groups are

1. Physical properties (viscosity, cloud point, pour point, flash point, etc.)

2. Chemical properties (chemical structure, acid value, saponification value, sulfur content, copper corrosion, oxidation resistance, and thermal degradation, etc.)
3. Thermal properties (distillation temperature, thermal conductivity, carbon residue, and calorific value, etc.)

The test methods to evaluate the fuel properties are given in Table 2.2 (ASTM 2009).

The characteristics of vegetable oils fall within a fairly narrow band and are close to those of diesel. Petroleum-based diesel fuel contains only carbon and hydrogen atoms, which are arranged in straight chain or branched chain structures, as well as aromatic configurations. The straight chain structures

TABLE 2.1

Chemical Structure of Common Fatty Acids

Fatty Acid	Chemical Name	Structure	Formula
Lauric	Dodecanoic	12:0	$C_{12}H_{24}O_2$
Myristic	Tetradecanoic	14:0	$C_{14}H_{28}O_2$
Palmitic	Hexadecanoic	16:0	$C_{16}H_{32}O_2$
Stearic	Octadecanoic	18:0	$C_{18}H_{36}O_2$
Arachidic	Eicosanoic	20:0	$C_{20}H_{40}O_2$
Bcheric	Docosanoic	22:0	$C_{22}H_{44}O_2$
Lignoceric	Tetracosanoic	24:0	$C_{24}H_{48}O_2$
Oleic	Cis-9-octadecenoic	18:1	$C_{18}H_{34}O_2$
Linoleic	Cis-9, cis-12-octadecatrienoic	18:2	$C_{18}H_{32}O_2$
Linolenic	Cis-9, cis-15-octadecatrienoic	18:3	$C_{18}H_{30}O_2$
Erucle	Cis-13-docosenoic	22:1	$C_{22}H_{42}O_2$

Source: From Barnwal, B. K. and Sharma, M. P. *Renewable and Sustainable Energy Reviews,* 9, 363–378, 2005. Reproduced with permission from Elsevier Publications.

TABLE 2.2

Physical and Chemical Properties Test Methods

Property	ASTM Test Method	Unit
Density	D4502	g/ml
Higher heating value	D2015	MJ/kg
Cloud point	D2500	K
Pour point	D97	K
Flash point	D93	K
Cetane number	D613	—
Kinematic viscosity @ 40° C	D445	mm^2/s
Sulfur	D5453	wt%

are preferred for better ignition quality as they have a high cetane number. Diesel fuel contains both saturated and unsaturated hydrocarbons (UHC). The saturated HC are present in large amounts as compared to unsaturated hydrocarbon in diesel. The higher saturation reduces the oxidation problem. Also, the aromatics are good oxidation resistant. On the other hand, resistance to oxidation of vegetable oils is remarkably affected by their fatty acid composition. The quantity of free fatty acids predominantly affects the flash point of vegetable oils and hence its ignition characteristics. The type of fatty acids contained in vegetable oil depends on the plant species and on the growth conditions of the plant. The degree of unsaturation (number of double bonds) determines the oxidation stability. The oxidation stability of fuel decreases with increase in unsaturation. Although vegetable oils are of very low volatility in nature, it may quickly produce volatile combustible compounds upon heating. The olefinic nature of the plant oils can also give rise to thermal and oxidative polymerization reactions. These polymers can no longer volatilize, but may deposit on available surfaces and, upon further heating, char to form coke-like substances. Fatty acid composition of various vegetable oils is depicted in Table 2.3 (Marchetti, Miguel, and Errazu 2007).

Vegetable oils have about 10% less heating value than diesel due to the presence of oxygen content in the molecules. The cetane number of diesel/biodiesel defines its ignition quality. The higher cetane number of fuel is an indication of better ignition properties. The cetane number affects the engine performance parameters like combustion, stability, drivability, white smoke, noise, and emissions of CO and HC. Vegetable oils have a cetane number in the range of 25–45 CN. The molecular weight of vegetable oils is about 3 to 4 times higher than that of diesel. The viscosity of vegetable oil (25–60 cSt) is several times higher than that of diesel (3–6 cSt). The high viscosity, large size vegetable oil molecules are of low volatility in nature that leads poor atomization and incomplete combustion of fuel. Saponification

TABLE 2.3

Fatty Acid Composition of Vegetable Oils

Vegetable Oil	Fatty Acid Composition % by Weight								
	16:1	18:0	20:0	22:0	24:0	18:1	22:1	18:2	18:3
Corn	11.67	1.85	0.24	0.00	0.00	25.16	0.00	60.60	0.48
Cotton seed	28.33	0.89	0.00	0.00	0.00	13.27	0.00	57.51	0.00
Crambe	20.70	0.70	2.09	0.80	1.12	18.86	58.51	9.00	6.85
Peanut	11.38	2.39	1.32	2.52	1.23	48.28	0.00	31.95	0.93
Rapeseed	3.49	0.85	0.00	0.00	0.00	64.4	0.00	22.30	8.23
Soybean	11.75	3.15	0.00	0.00	0.00	23.26	0.00	55.53	6.31
Sunflower	6.08	3.26	0.00	0.00	0.00	16.93	0.00	73.73	0.00

Source: From Marchetti, J. M., Miguel, V. U., and Errazu, A., *Renewable and Sustainable Energy Reviews*, 11, 1300–1311, 2007. Reproduced with permission from Elsevier Publications.

value of an oil increases with the decrease of its molecular weight. The C and H percentages in oil increase with a decrease in molecular weight. The decrease in saponification value results in the increase in heat content of an oil. The increase in iodine value results in a decrease in heat content of oil. The heat content of oil depends on iodine and saponification value. The physiochemical properties of various vegetable oils are depicted in Table 2.4 (Deimbras 2003).

2.3 Methods to Use Vegetable Oils in Engines

2.3.1 Pyrolysis

Pyrolysis is a promising method for the production of environmentally friendly liquid fuels. It is the chemical reaction caused by the application of thermal energy in the absence of air. Vegetable oils and animal fats can be pyrolyzed. Pyrolysis process takes place at higher temperatures of about 250–400°C and at higher heating rates. Heating of vegetable oils breaks the bigger molecules into smaller molecules and a wide range of HC are formed. The pyrolyzed products can be divided into gaseous, liquid fractions consisting of paraffins, olefins and naphthenes, and solid residue. The pyrolyzed vegetable oils contain acceptable amounts of sulfur, water, and sediment and give acceptable copper corrosion values but unacceptable ash, carbon residue, and pour point. Pyrolyzed vegetable oil contains compounds in the boiling range of gasoline. The properties of bio-oil depends upon the nature of the feedstock, temperature of pyrolysis process, thermal degradation degree and catalytic cracking, the water content of the pyrolysis oil, the amount of light ends that have collected, and the pyrolysis process used. The high oxygen content of pyrolysis oil results in a very low energy density in comparison to conventional fuel oils. The comparison of properties of pyrolyzed oil with biodiesel and diesel is given in Table 2.5.

2.3.2 Microemulsification

The formation of microemulsions (cosolvency) is one of the potential solutions for solving the problem of vegetable oil viscosity. Microemulsions are defined as transparent, thermodynamically stable colloidal dispersions. The droplet diameters in microemulsions range from 100 to 1000 Å. The microemulsion can be made of vegetable oils with an ester and dispersant (cosolvent), or of vegetable oils, an alcohol and a surfactant and a cetane improver, with or without diesel fuels. Water (from aqueous ethanol) may also be present in order to use lower-proof ethanol, thus increasing water tolerance of the microemulsions. Microemulsion can improve the spray characteristics

TABLE 2.4

Physiochemical Properties of Vegetable Oils

Vegetable Oil	Kinematic Viscosity (cSt)	Carbon Residue (m, %)	Cetane Number	Higher Heating Value (MJ/kg)	Ash Content (wt, %)	Sulfur Content (wt, %)	Iodine Value	Saponification Value
Cottonseed	33.7	0.25	33.7	39.4	0.02	0.01	113.20	207.71
Poppyseed	42.4	0.25	36.7	39.6	0.02	0.01	116.83	196.82
Rapeseed	37.3	0.31	37.5	39.7	0.006	0.01	108.05	197.07
Safflower	31.6	0.26	42.0	39.5	0.007	0.01	108.05	190.3
Sunflower	34.4	0.28	36.7	39.6	0.01	0.01	91.76	210.34
Sesame seed	36.0	0.25	40.4	39.4	0.002	0.01	91.76	210.34
Linseed	28.0	0.24	27.6	39.3	0.02	0.02	120.96	205.68
Wheat grain	32.6	0.23	35.2	39.3	0.02	0.02	120.96	205.68
Corn narrow	35.1	0.22	37.5	39.6	0.01	0.01	119.41	194.14
Castor	29.7	0.21	42.3	37.4	0.01	0.01	88.72	202.71
Soybean	33.1	0.24	38.1	39.6	0.006	0.01	69.82	220.78
Bay laurel leaf	23.2	0.20	33.6	39.3	0.03	0.02	105.15	220.62
Peanut	40.0	0.22	34.6	39.5	0.02	0.01	119.55	199.80
Hazelnut kernel	24.0	0.21	52.9	39.8	0.01	0.02	98.62	197.63
Walnut kernel	36.8	0.24	33.6	39.6	0.02	0.02	135.24	190.82
Almond kernel	34.2	0.22	34.5	39.8	0.01	0.01	102.35	197.56
Olive kernel	29.4	0.23	49.3	39.7	0.008	0.02	100.16	196.83

Source: From Demirbas, A., *Energy Conversion and Management*, 44, 2093–2109, 2003. Reproduced with permission from Elsevier publications.

TABLE 2.5

Properties of Pyrolysis Oil in Comparison With Diesel and Biodiesel

Property	Test Method	ASTM D975 Diesel	ASTM D6571 (B100)	Pyrolysis Oil
Flash point (°C)	D93	52 °C	130	—
Wear and sediment, max, vol%	D2709	0.05	0.05	0.01–0.04
Kinematic viscosity, mm^2/s	D445	1.3–4.1	1.9–6.0	25–1000
Sulphated ash max, wt%	D874	—	0.02	—
Ash, max, wt%	D482	0.01	—	0.05–0.01
Sulfur, max, wt%	D5453	0.05	0.05	—
Sulfur, max, wt%	D2622/129	—	—	0.001–0.02
Cetane number	D613	40	47	—
Aromatics, max, vol%	D1319	—	35	—
Carbon residue, max, mass%	D4530	—	0.05	0.001–0.0
Carbon residue, max, mass%	D524	0.35	—	—

Source: From Demirbas, A., *Fuel Processing Technology,* 88, 591–97, 2007. Reproduced with permission from Elsevier Publications.

due to explosive vaporization of low-boiling constituents in the micelle. Microemulsion of methanol with vegetable oil can perform nearly the same as diesel. In microemulsion formation, the stability of the emulsion is determined by the energy input into it and the type and amount of emulsifier needed (Ramadhas, Jayaraj, and Muraleedharan 2004). Czerwinski (1994) prepared an emulsion of 53% sunflower oil, 13.3% ethanol, and 33.4% butonal. This emulsion had a viscosity of 6.3 cSt at 40°C and cetane number of 25. Lower viscosities and better spray patterns has been reported with an increase in percentage of butanol.

2.3.3 Dilution

Vegetable oils can be directly mixed with diesel fuel and may be used for running diesel engine. Rudolf Diesel used peanut oil to run his engine. Several studies have been conducted to use straight vegetable oil as fuel for diesel engines. Wang et al. (2006) reported that modern diesel engines that have fuel injection systems are sensitive to viscosity change. High viscosity of vegetable oil may lead to its poor atomization, incomplete combustion, choking of the injectors, ring carbonization, and accumulation of the fuel in the lubricating oils. A way to avoid those problems and to improve the performance is to reduce the viscosity of vegetable oil. Dilution of vegetable oils can be accomplished with such materials as diesel fuel, solvent, or ethanol. There has been an increased advantage when blending vegetable oil with diesel as fuel with minimum processing and engine modification.

Blending of diesel with vegetable oil reduces its viscosity and hence improves its combustion characteristics. Straight vegetable oil (neat vegetable oil) can be used as fuel for IC engines with some minor modifications in the fuel system. Straight (raw) vegetable oil fueled engine can be used to run the generator sets to produce electricity in villages where vegetable oils are available locally. Results of short-term tests conducted by various researchers were found to be successful. However, some problems were experienced on mileage accumulation.

1. High viscosity, low cetane number, and high flash point cause cold starting problems. This can be reduced by preheating the vegetable oils before injection or adding suitable additives to improve cold startability.
2. The high flash point of vegetable oils attributes to lower volatility.
3. Both cloud and pour points are significantly higher than that of diesel fuel. These high values may cause problems during cold weather.
4. Vegetable oils are very low in cetane number (25–35 CN) and hence knocking occurs. However, the use of higher compression ratio in engines reduces the knocking tendency.
5. Vegetable oils are of low oxidation stability and hence form injector plugging and gum formation. Filtering of vegetable oils before injection would reduce the injector plugging.
6. Poor atomization of vegetable oils cause incomplete combustion and crankcase dilution due to blow-by cause excessive engine wear, coking of injectors, ring sticking, lube oil dilution, and an increase in combustion chamber deposits. These can be controlled by operating the engine with vegetable oils at full load only and thereby increases the oil change interval.

2.3.4 Transesterification

Transesterification is a chemical process of transforming large, branched triglyceride molecules of bio-oils and fats into smaller, straight chain molecules, almost similar in size to the molecules of the species present in diesel fuel. This process has been widely used to reduce the viscosity of triglycerides. The transesterification reaction is represented by the general equation

$$\text{R-COOR}' + \text{R}''\text{–OH} \rightarrow \text{RCOOR}'' + \text{R}'\text{OH}.$$

Triglycerides are readily transesterified in the presence of alkaline catalyst at atmospheric pressure and at a temperature of approximately 60–70°C with an excess of methanol. The mixture at the end of a reaction is allowed to settle. The lower glycerol layer is drawn off while the upper methyl ester layer is washed to remove entrained glycerol and is then processed further. The excess methanol is recovered by distillation and sent to a rectifying column for purification and recycled. The transesterification works well when the feedstock oil is of high quality. However, quite often low quality oils are used as raw materials for biodiesel preparation. In cases where the free fatty acid content of the oil is above 1%, difficulties arise due to the formation of soap, which promotes emulsification during the water washing stage and at a FFA content above 2%, the process becomes unworkable. The most important variables that influence transesterification reaction time and conversion are

1. Oil temperature
2. Reaction temperature
3. Ratio of alcohol to oil
4. Type of catalyst and concentration
5. Intensity of mixing
6. Purity of reactants

2.4 Engine Tests

2.4.1 Performance and Emissions Studies

Suitability of vegetable oils for fuel purposes has been investigated by many researchers all over the world during the past few decades. A typical engine test conducted by Ramadhas et al. (2005a) is described as follows. Rubber seed oil, typical nonedible type oil was considered for evaluating engine short-term performance and emission characteristics. Engine tests have been conducted on diesel engines using various blends of rubber seed oil diesel. The variation of brake thermal efficiency of the engine with various blends is depicted in Figure 2.2. The experimental results reveal that with increasing brake power, the brake thermal efficiency of the blends is increased. There was considerable increase in efficiency of the engine with blends as compared to the neat of rubber seed oil. The brake thermal efficiency of 80:20 (rubber seed oil:diesel) blend, closely matches with that of diesel oil. The higher percentage of rubber seed oil in the blends had given lower thermal efficiency as compared to diesel. The reason for the drop in thermal efficiency was attributed to higher viscosity and poor combustion characteristics of rubber seed oil.

Nwafor, Rice, and Ogbonna (2000) conducted tests using rapeseed oil on single cylinder diesel engine and reported that rapeseed oil with standard timing produced the highest mechanical efficiency (Figure 2.3). Vegetable oils with longer ignition delay and longer combustion duration do not require any advanced injection at low engine speed. They reported that diesel fuel

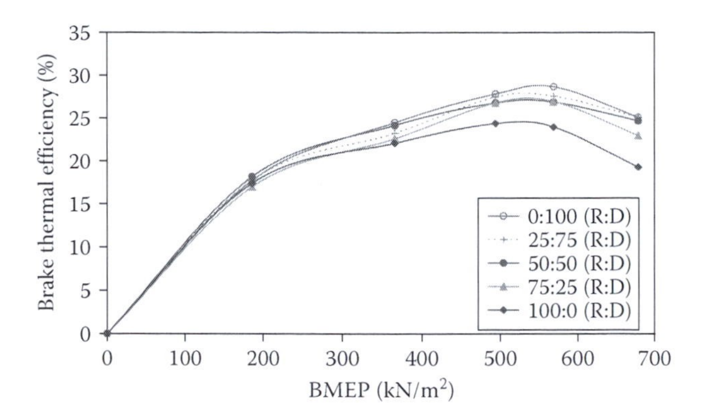

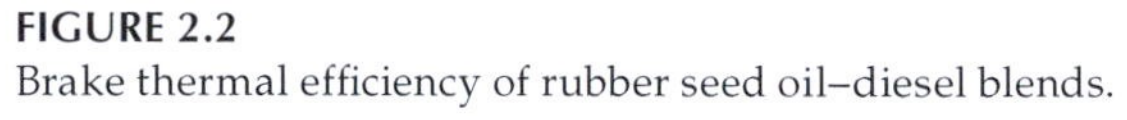

FIGURE 2.2
Brake thermal efficiency of rubber seed oil–diesel blends.

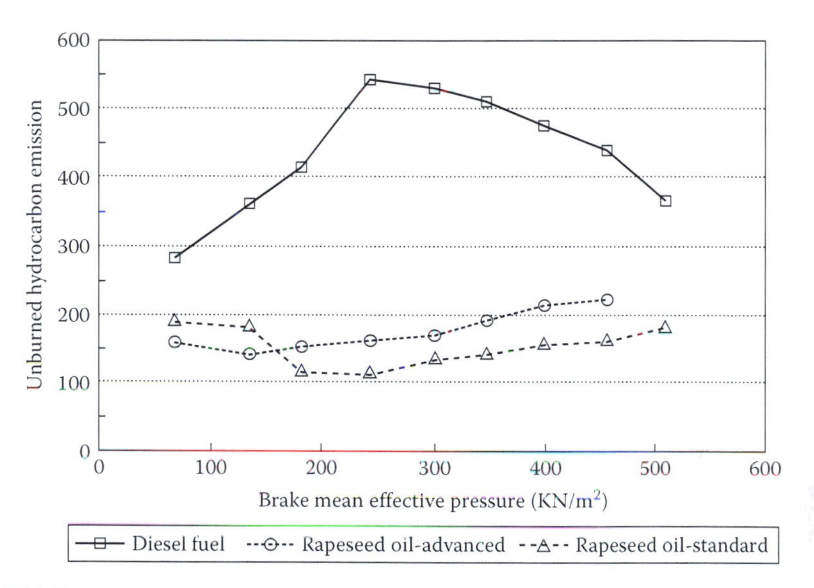

FIGURE 2.3
Unburned hydrocarbon emissions of rapeseed oil. (From Nwafor, O. M. I., Rice, G., and Ogbonna, A. I., *Renewable Energy*, 21, 433–44, 2000. Reproduced with permission from Elsevier Publications.)

operation had showed the highest concentrations of HC in the exhaust (Figure 2.4). The advanced injection unit had given a marginal increase in HC emissions over the standard injection timing unit. The HC emissions may be increased by several factors such as quenched flame, lean combustion, cold starting, and poor mixture preparation. Moreover, high viscosity of vegetable oil might have provided a good sealant between the piston rings

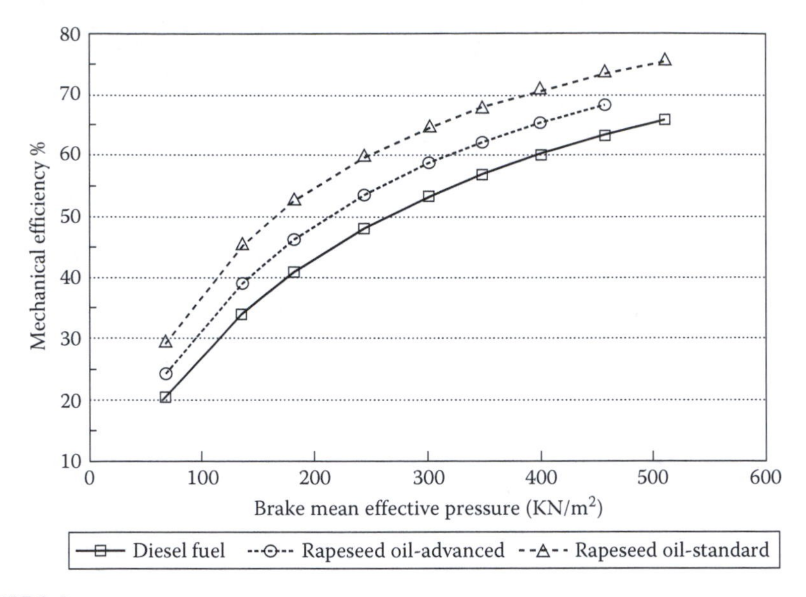

FIGURE 2.4
Mechanical efficiency of rapeseed oil. (From Nwafor, O. M. I., Rice, G., and Ogbonna, A. I., *Renewable Energy*, 21, 433–44, 2000. Reproduced with permission from Elsevier Publications.)

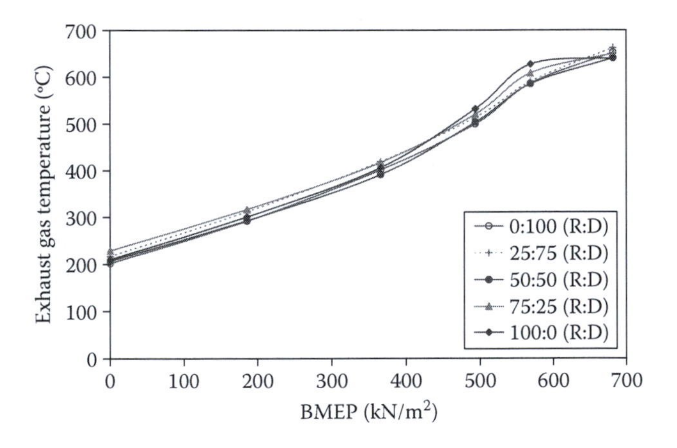

FIGURE 2.5
Exhaust gas temperature of rubber seed oil-diesel blends.

and cylinder wall, and hence improved the engine compression for efficient combustion.

Ramadhas, Jayaraj, and Muraleedharan (2005a) reported that the exhaust gas temperature also increases with increase in rubber seed oil concentration in the blend (Figure 2.5). Rubber seed oil, which contains some amount of oxygen molecules, is also taking part in the combustion. Hence, while using neat rubber seed oil, higher exhaust temperature was attained; this indicates more energy loss in this case. The exhaust gas temperature increases with increase in load for all tested fuels. The NO_x emission is directly related to the engine combustion chamber temperatures, which in turn was indicated by the prevailing exhaust gas temperature. Vegetable oil engines have the potential to emit more NO_x as compared to that of diesel-fueled engines.

2.4.2 Endurance Tests

Endurance tests are conducted to evaluate the effect of fuel on engine components like injectors, piston ring wear, engine combustion chamber deposits, and engine lubricant wear metal analysis. Vegetable oil can be produced locally and can be effectively used in rural areas for water pumping and power generation. For that purpose, diesel engines are suitable for electric generator application used in agricultural water pumping is selected. The engine ran continuously at full load conditions using 50:50 blend (rubber seed oil:diesel) and compared with that of diesel. The engine cylinder head is dismantled for visual inspection and quantitatively analyzed for the carbon particles deposited.

The combustion chamber region of the cylinder head showed uniform, flat carbon buildup for both fueled engines. No differences in the conditions of the intake and exhaust valve seats were detected between the blend and diesel. A thin layer of carbon particles is distributed over the circumference of the valve seats. The valve seats, as well as the valve faces, showed the peening caused by hard particles released from combustion chamber deposits. Carbon buildup on the top surface of the piston is not uniform.

Figures 2.6 and 2.7 show the carbon deposits on the cylinder head of the diesel-fueled and blend-fueled engine conditions, respectively. There was more carbon deposits on the cylinder head of the blend-fueled engine than that of the diesel-fueled engine. A quick carbon buildup of carbon deposits on the blend-fueled engine injector nozzles is observed. Incomplete combustion had promoted the formation of additional solid residues, which were deposited on the combustion chamber walls and cylinder head (Ramadhas, Jayaraj, and Muraleedharan 2005b).

Figures 2.8 and 2.9 shows carbon deposits on the piston surface of the diesel-fueled engine and blend-fueled engine, respectively. More deposits were formed with vegetable oil blend-fueled engine than that of diesel. More carbon deposits formed due to the incomplete combustion of the rubber seed oil diesel blend. The higher viscosity of fuel that leads to poor atomization characteristics causes incomplete combustion of fuel. Moreover, the gumming nature of vegetable oil formed carbon particles on the wall of the combustion chamber (Ramadhas, Jayaraj, and Muraleedharan 2005b).

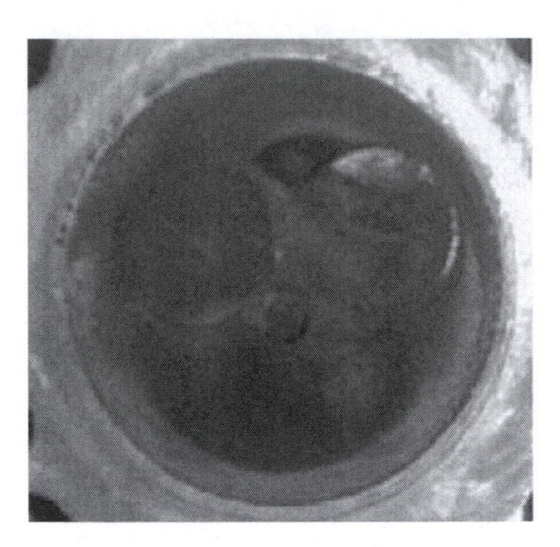

FIGURE 2.6
Cylinder head of diesel-fueled engine.

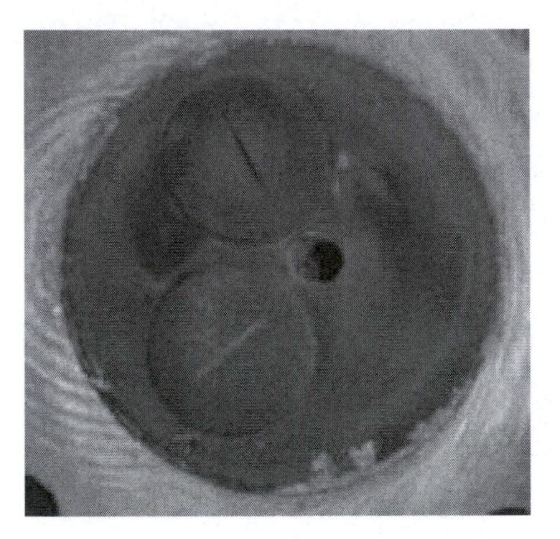

FIGURE 2.7
Cylinder head of vegetable oil fueled engine.

Bruwer et al. (1981) reported that 8% power loss occurred after 1000 hours of operation with neat sunflower oil. The power loss was corrected by replacing the fuel injectors and injector pump. After 1300 hours of operation, the carbon deposits in the engine were reported to be equivalent to a diesel-fueled engine except for the injector tips, which exhibited excessive carbon buildup. Tadashi et al. (1984) evaluated the feasibility of using rapeseed oil and palm oil for diesel fuel in a naturally aspirated direct injection diesel engine. It has been reported that vegetable oil fueled engines gave an acceptable engine performance and exhaust emission levels for short-term operation. However,

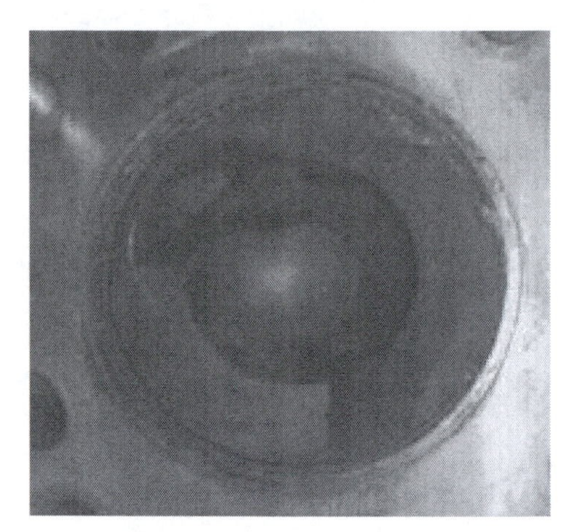

FIGURE 2.8
Piston top of diesel-fueled engine.

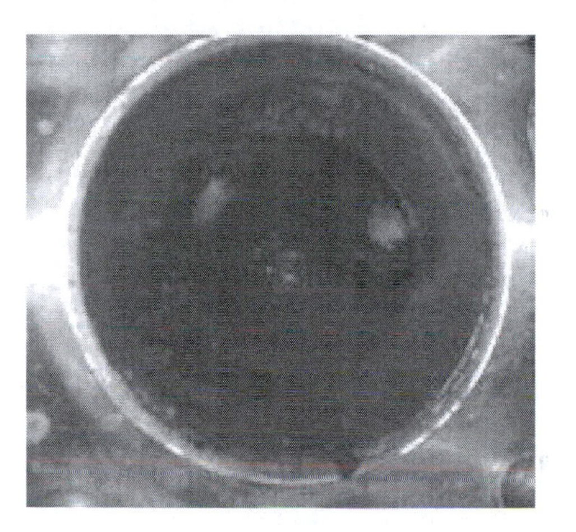

FIGURE 2.9
Piston top of vegetable oil fueled engine.

these vegetable oils caused carbon deposit buildups and sticking of piston rings with extended operation.

German et al. (1985) conducted field trial tests using six John Deere and Case tractors by averaged 1300 hours of operation. Carbon deposits in the combustion chamber were greater for the tractors fueled with sunflower oil/diesel (50:50) than for those fueled with a sunflower oil/diesel fuel blend (25:75). All the test engines that ran with vegetable oil/blends had more carbon buildup than diesel fuel.

Malaysia uses palm oil as the major source of renewable energy for diesel engine purposes. Sapaun, Masjuki, and Azlan (1996) reported that palm oil as a diesel fuel substitute exhibited encouraging results. Power outputs were nearly the same for palm oil, blends of palm oil and diesel fuel, and 100% diesel fuel. Short-term tests using palm oil fuels showed no signs of adverse combustion chamber wear, increase in carbon deposits, or lubricating oil contamination.

The Southwest Research Institute (Reid, Hansen, and Goering 1989) evaluated the chemical and physical properties of 14 vegetable oils. The injection studies pointed out that the oils behave very differently from petroleum-based fuels. These changes in behavior were attributed to the high viscosity of vegetable oils. However, preheating vegetable oils before injection reduced the carbon deposits. Moreover, it has been reported that oils with similar viscosities offered different levels of carbon deposits.

2.4.3 Lube Oil Analysis

Raadnui et al. (2003) conducted wear analysis of 100% conventional petroleum diesel fuel, a 50% refined palm oil (RPO) and 50% diesel fuel, and a 100% RPO in fleet trial. The amounts of wear metal noted in the drained engine lubricating oil samples was not significantly higher when compared to an engine fueled with petroleum diesel fuel. The wear metal in lube oil on mile accumulation is depicted in Figures 2.10 through 2.13.

These studies established aggravate deposit formation and injector coking formation with vegetable oil fueled engine. These vegetable oils should be used after proper filteration, degumming, and dewaxing. For long-term use and for heavy engine applications, blends of diesel and vegetable oils

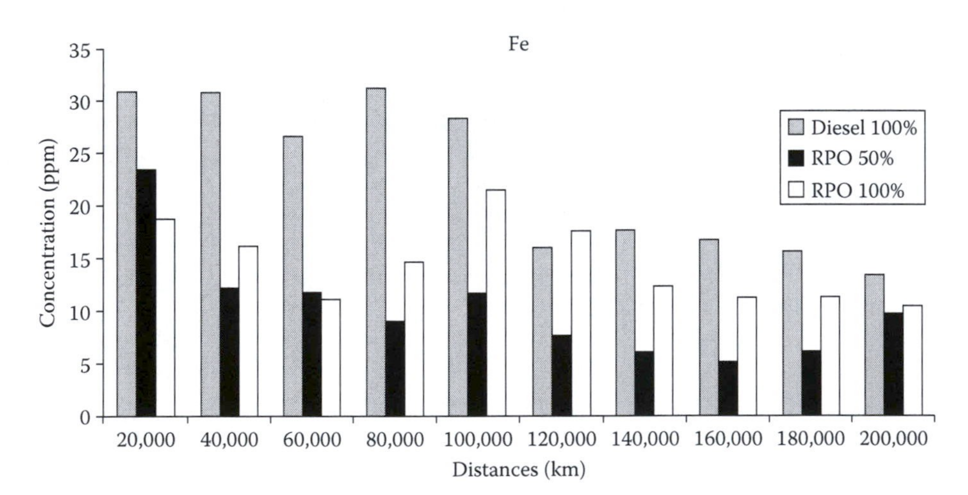

FIGURE 2.10
Fe concentration on mileage accumulation. (From Raadnui, S. and Meenak, A., *Wear*, 254, 1281–88, 2003. Reproduced with permission from Elsevier Publications.)

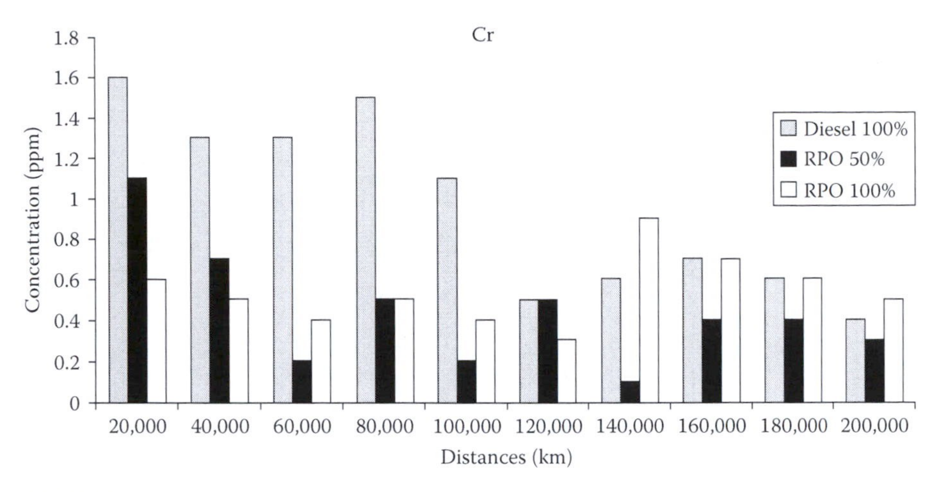

FIGURE 2.11
Cr concentration on mileage accumulation. (From Raadnui, S. and Meenak, A., *Wear,* 254, 1281–88, 2003. Reproduced with permission from Elsevier Publications.)

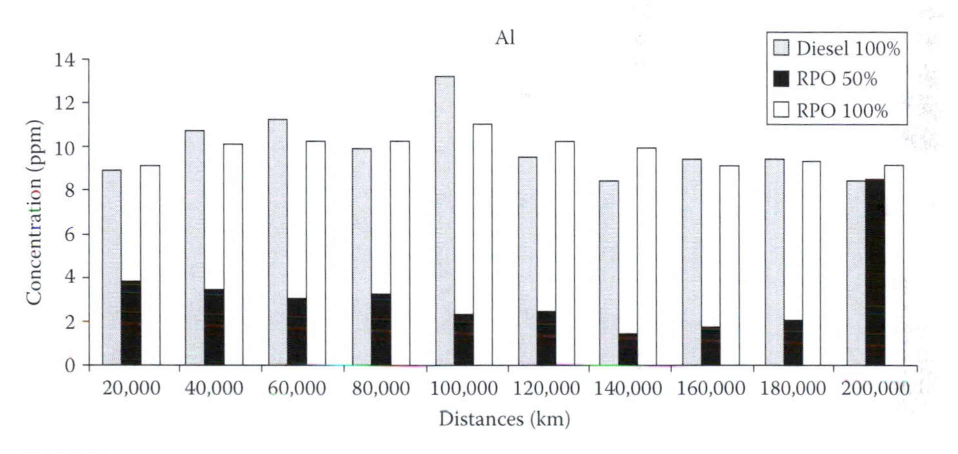

FIGURE 2.12
Al concentration on mileage accumulation. (From Raadnui, S. and Meenak, A., *Wear,* 254, 1281–88, 2003. Reproduced with permission from Elsevier Publications.)

are recommended. The engine should be started and before stopping run by using diesel alone. After warming up it should be shifted to the vegetable oil blend. These problems can be sorted out by routine and preventive maintenance of engine. Heating of vegetable oils could reduce their viscosity and hence improve the spray pattern of vegetable oil atomization that leads to improvement in engine performance. As the engine requires frequent maintenance, this is best suited for stationary engine application particularly

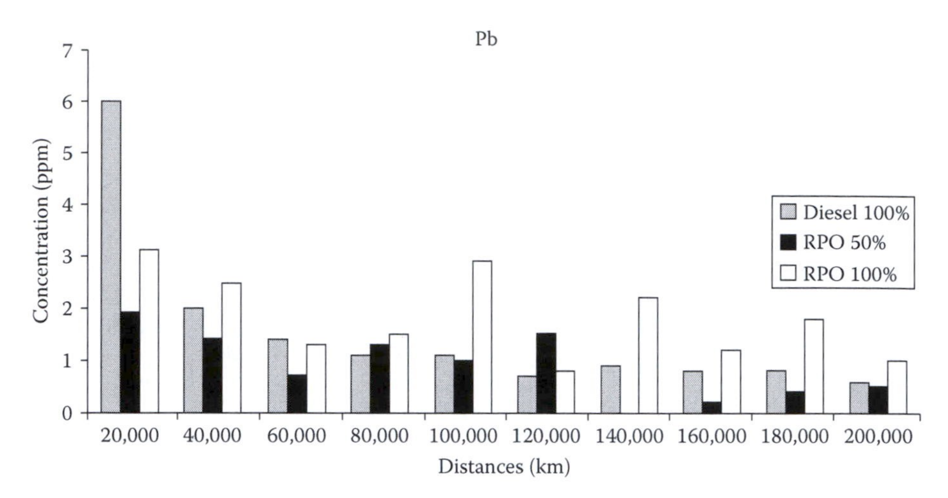

FIGURE 2.13
Pb concentration on mileage accumulation. (From Raadnui, S. and Meenak, A., *Wear*, 254, 1281–88, 2003. Reproduced with permission from Elsevier Publications.)

in power generation and water pumping generator motor set applications rather than automotive engine application.

2.5 Advantages of Vegetable Oils

1. Vegetable oils can be used as substitute fuel for diesel engine application.
2. Use of vegetable oil for fuel purposes reduces the import of costly petroleum and improves the economy of agricultural countries.
3. They are biodegradable and nontoxic.
4. Vegetable oils are of low aromatics and low sulfur content and hence reduce the particulate matter emissions.
5. They have a reasonable cetane number and hence possesses less knocking tendency.
6. Vegetable oils are environmentally friendly fuels.
7. Enhanced lubricity, thereby no major modification is required in the engine.
8. Use of vegetable oils improves the personal safety also (flash point of vegetable oil is above 100°C).
9. These are usable within the existing petroleum diesel infrastructure with minor or no modification in the engine.

2.6 Challenges for Vegetable Oils

The major challenges that are faced by the vegetable oils for fuel purposes are listed below (Peterson, 1999).

1. The price of vegetable oil is dependent on the seed price and it is market based. Moreover, feed stock homogeneity, consistency, and reliability are questionable.
2. Production of vegetable oil derived biofuels are at an optimum cost.
3. Studies are needed to reduce the production cost and identify potential markets in order to balance cost and availability.
4. Studies are needed on oxidation stability and long storage of vegetable oils.
5. Manufacturer warranty and compatibility with IC engine material needs to be studied further.
6. Durability and emission testing with a wide range of feed stocks.
7. Environmental benefits to be offered by vegetable oil over diesel fuel needs to be popularized.
8. Development of additives for improving cold flow properties, material compatibility, and prevention of oxidation in storage, etc.
9. Continuous and long-term availability of the vegetable oils.

References

Agarwal. A. K. 1999. Performance evaluation and tribological studies on biodiesel fueled compression ignition engine. PhD Thesis, IIT Delhi, India.

ASTM International. 2009. *ASTM book of standards 5.0 petroleum products, lubricants, and fossil fuels.* West Conshohocken, PA: ASTM International.

Barnwal, B. K., and M. P. Sharma. 2005. Prospects of biodiesel production from vegetable oils in India. *Renewable and Sustainable Energy Reviews* 9:363–78.

Bruwer, J. J., B. D. Boshoff, F. J. C. Hugo, L. M. DuPlessis, J. Fuls, C. Hawkins, A. N. VanderWalt, and A. Engelbert. 1981. The utilization of sunflower seed oil as renewable fuel diesel engines. *Biomass Energy/Crop Production* 2:4–81.

Czerwinski, J. 1994. Performance of D.I. diesel engine with addition of ethanol and rapeseed oil. *SAE* 940545.

Demirbas, A. 2003. Biodiesel fuels from vegetable oils via catalytic and non-catalytic supercritical alcohol transesterifications and other methods: A survey. *Energy Conversion and Management* 44:2093–109.

German, T. J., K. R. Kaufman, G. L. Pratt, and J. Derry. 1985. Field evaluation of sunflower oil diesel fuel blends in diesel engines. *ASAE* 853078.

Marchetti, J. M., V. U. Miguel, and A. F. Errazu. 2007. Possible methods for biodiesel production. *Renewable and Sustainable Energy Reviews* 11:1300–11.

Nwafor, O. M. I., G. Rice, and A. I. Ogbonna. 2000. Effect of advanced injection timing on the performance of rapeseed oil in diesel engines. *Renewable Energy* 21:433–44.

Peterson, E. 1999. *Conference summary—Commercialization of bio-diesel: Producing a quality fuel.* Boise, Idaho: University of Idaho.

Raadnui, S., and A. Meenak. 2003. Effects of refined palm oil (RPO) fuel on wear of diesel engine components. *Wear* 254:1281–8.

Ramadhas, A. S., S. Jayaraj, and C. Muraleedharan. 2004. Use of vegetable oils as I.C. engine fuels—A review. *Renewable Energy* 29:727–42.

Ramadhas, A. S., S. Jayaraj, and C. Muraleedharan. 2005a. Theoretical modelling and experimentation on different alternative fuels for compression ignition engines. PhDThesis, National Institute of Technology, Calicut, India.

Ramadhas, A. S., S. Jayaraj, and C. Muraleedharan. 2005b. Characterization and effect of using rubber seed oil as fuel in the compression ignition engines. *Renewable Energy* 30:795–803.

Reid, J. F., A. C. Hansen, and C. E. Goering. 1989. Quantifying diesel injector coking with computer vision. *ASAE* 32:1503–6.

Sapaun, S. M., H. H. Masjuki, and A. Azlan. 1996. The use of palm oil as diesel fuel substitute. *Journal of Power Energy* 210:47–53.

Tadashi, Y. 1984. Low carbon build up, low smoke and efficient diesel operation with vegetable oil by conversion to monoesters and blending with diesel oil or alcohols. *SAE* 841161.

Wang, Y. D., T. A. Shemmeri, P. Eames, J. McMullan, N. Hewitt, and Y. Huang. 2006. An experimental investigation of the performance and gaseous exhaust emission of a diesel engine using blends of a vegetable oil. *Applied Thermal Engineering* 26:1684–91.

3

Biodiesel

Arumugam Sakunthalai Ramadhas, Simon Jayaraj, and Chandrasekaran Muraleedharan

CONTENTS

3.1 Introduction

Petroleum products consumption is increasing day-by-day as the number of vehicles on road increases. Combustion of hydrocarbon fuel increases its concentration in the environmental pollution also. There is a need to solve these twin problems—fuel supply and environmental pollution. Nonrenewable fuel emits more hydrocarbons, oxides of nitrogen, sulfur, and carbon monoxides as compared to renewable biofuels. Various alternative fuels are considered as substitute fuels for petroleum products and efforts were made to analyze the suitability of the fuel and its demonstration. Renewable fuels have received more attention as they reduce the environmental pollution (by completing carbon cycle) and reduce the import of petroleum. Hence, researchers and the scientific community worldwide have focused on development of biodiesel and the optimization of the processes to meet the standards and specifications needed for the fuel to be used commercially.

Significant research on the production and application of biomass energy for fuel purposes is being carried out all around the world. Alcohols, vegetable oils, and their derivatives are the promising biomass sources for use in engines. Use of biofuels, particularly biodiesel derived from vegetable oil, is gaining more importance currently as it does not require any engine hardware modification without affecting the engine performance, no sulfur emissions, and a reduction in greenhouse gas emissions. Biodiesel can be derived from vegetable oils, animal fats, and algae materials. Dr. Rudolf Diesel, inventor of the diesel engine created the interest among the public to use vegetable oil as diesel engine fuel, particularly peanut oil. Low cost petroleum products are used for diesel engines because of its abundant availability. However, during the energy crisis period (1970s), vegetable oils and alcohols were widely used as engine fuel. Due to the ever-rising crude oil prices and environmental concerns, there has been a renewed focus on vegetable oils and their derivatives for use as engine fuel.

Vegetable oils can be modified into various, more useful forms that are suitable for diesel engine application. These include microemulsion, pyrolysis, and transesterification. Pyrolysis of the vegetable oil resulted in products with low viscosity, high cetane number, accepted amounts of sulfur, water and sediments, and accepted copper corrosion values. But these are unacceptable in terms of their ash contents, carbon residues, and pour points. Microemulsion of vegetable oil lowered the viscosity of the vegetable oil but resulted in irregular injector needle sticking, heavy carbon deposits, and incomplete combustion.

Among these various conversion methodologies, the transesterification process has become commercial success. Transesterification is a chemical process of transforming large, branched triglyceride molecules of the vegetable oils and fats into smaller, straight chain molecules, almost similar in size to the molecules of the species present in diesel fuel. Many countries started biodiesel production industries and biodiesel is blended with diesel commercially as per their national policy. Engine performance, emission, endurance and metal components wear analysis, fleet studies, and engine oil effects were conducted by various automotive manufacturers and oil companies. Based on the positive results obtained by them, automotive manufacturers extended the warranty of biodiesel operated vehicles also. European Union has become the leader of biodiesel production in the world.

Biodiesel refers to diesel equivalent, processed fuel derived from biological sources such as vegetable oils. It can also be made from animal fats. It is the ester based oxygenated fuel derived from biological sources. Chemically, biodiesel is defined as mono alkyl esters of long-chain fatty acids of lipids. Methanol is preferred over ethanol in the transesterification process as it is less expensive, and considering its polar and short chain molecular structure. Biodiesel used for blending with diesel should meet the ASTM D 6751 or particular country's national fuel standards. Biodiesel can be used as combustion extender for reducing the engine exhaust emissions.

Key advantages of biodiesel include

1. Biodiesel is an alternative fuel that can be used to operate any conventional, unmodified diesel engine. It can be stored anywhere that diesel fuel is stored.
2. Biodiesel can be used alone or mixed in any ratio with petroleum diesel fuel in diesel engines.
3. The life cycle production and use of biodiesel produces approximately 80% less carbon dioxide emissions, and almost 100% less sulfur dioxide than diesel. Combustion of biodiesel alone provides over 90% reduction in total unburned hydrocarbons and 75–90% reduction in aromatic hydrocarbons. Biodiesel further provides significant reductions in particulates and carbon monoxide compared to petroleum diesel fuel. Biodiesel provides a slight increase or decrease in

nitrogen oxides emissions depending on engine family and testing procedures.

4. Biodiesel has about 11% oxygen by weight and no sulfur. The use of biodiesel can extend the life of diesel engines because it is more lubricating than petroleum diesel fuel, while fuel consumption, auto ignition, power output, and engine torque are relatively unaffected by biodiesel.
5. Biodiesel is safe to handle and transport because it is biodegradable as sugar, 10 times less toxic than table salt, and has a high flashpoint about 110°C compared to petroleum diesel fuel whose flash point is 45–55°C).
6. Biodiesel can be made from domestically produced, renewable oil seed crops such as soybeans, jatropha, cottonseed, rubber seed, and mustard seed.

3.2 Biodiesel Production

Vegetable oil reacts with alcohol (typically methanol or ethanol) in the presence of catalyst produced biodiesel. Biodiesel production process consists of three steps; namely,

1. Conversion of triglycerides (TG) to diglycerides and one ester molecule
2. Followed by the conversion of diglycerides (DG) into monoglycerides (MG) and one ester molecule
3. Monoglycerides into glycerol and one ester molecule

$$\text{Triglycerides(TG)} + \text{R'OH} \overset{\text{catalyst}}{\Leftrightarrow} \text{Diglycerides(DG)} + \text{R'COOR}_1$$

$$\text{Diglycerides(TG)} + \text{R'OH} \overset{\text{catalyst}}{\Leftrightarrow} \text{Monoglycerides(MG)} + \text{R'COOR}_2$$

$$\text{Monoglycerides(MG)} + \text{R'OH} \overset{\text{catalyst}}{\Leftrightarrow} \text{Glycerol} + \text{R'COOR}_3.$$

Glycerol, the by-product of the transesterification process, also has commercial value. Stoichiometrically, three moles of alcohol are required for each mole of triglyceride, but in general, a higher molar ratio is often employed for maximum ester production and moves the reaction toward the forward direction. It also depends upon the type of feedstock, amount of catalyst, temperature, and so on. Commonly used alcohols include methanol, ethanol, and butanol. The esters have reduced the viscosity and increased volatility relative to the triglycerides present in vegetable oils. Biodiesel production methods include acid, alkaline, two-step, supercritical methanol, and ultrasonic methods.

Several catalysts were tried for the purpose of transesterification, for example, magnesium, calcium oxides, and carbonates of basic and acidic macroreticular organic resin, alkaline alumina, phase transfer catalysts, sulfuric acids, p-toluene sulfonic acid, and dehydrating agents as co-catalysts (Agarwal, Bijwe, and Das 2003). The catalysts reported to be effective at room temperature were alkoxides and hydroxides (Canakci and Van Gerpan 1999).

The biodiesel blend is referred as Bxx, where xx indicates the percentage amount of biodiesel that is in the B20 blend, which represents the mixture of 20% biodiesel and 80% diesel. Biodiesel is registered as a fuel and fuel additive with the United States Environmental Protection Agency (EPA) and meets clean diesel standards established by the California Air Resources Board (CARB). Neat biodiesel (B100) has been designated as an alternative fuel by the Department of Energy and Department of Transportation of the United States. Most of the countries started using biodiesel as substitute fuel for diesel and made their own national fuel quality standards for biodiesel. Biodiesel has been in use in countries such as the United States, European Union, Germany, Malaysia, Thailand, France, and Italy.

3.2.1 Alkaline Transesterification

Alkaline-catalyzed transesterification process is the commercially well-developed biodiesel production process. Alkaline catalysts (NaOH, KOH) are used to improve the reaction rate and to increase the yield of the process. To complete the transesterification stoichiometrically, 3:1 molar ratio of alcohol to triglycerides is needed. In practice, the ratio needs to be higher to drive the equilibrium to a maximum ester yield. Since the transesterification reaction is a reversible process, excess alcohol is required to shift the reaction equilibrium to the products side. Alcohols such as methanol, ethanol, or butanol are used in the transesterification. Figure 3.1 shows the transesterification equation of vegetable oils with alcohols.

The alkali-catalyzed transesterification of vegetable oils proceeds faster than the acid-catalyzed reaction. The first step is the reaction of the base with

$$
\begin{array}{ccccccc}
CH_2\text{–}OOC\text{–}R_1 & & & & R_1\text{–}COO\text{–}R' & & CH_2\text{–}OH \\
| & & & \text{Catalyst} & & & | \\
CH\text{–}OOC\text{–}R_2 & + & 3R'OH & \rightleftharpoons & R_2\text{–}COO\text{–}R' & + & CH\text{–}OH \\
| & & & & & & | \\
CH_2\text{–}OOC\text{–}R_3 & & & & R_3\text{–}COO\text{–}R' & & CH_2\text{–}OH \\
\text{Glyceride} & & \text{Alcohol} & & \text{Esters} & & \text{Glycerol}
\end{array}
$$

FIGURE 3.1
Transesterification equation.

the alcohol, producing an alkoxide and a protonated catalyst. The nucleophilic attack of the alkoxide at the carbonyl group of the triglyceride generates a tetrahedral intermediate, from which the alkyl ester and the corresponding anion of the diglyceride are formed. Similar process converts diglycerides into monoglycerides and monoglycerides into glycerol (Demirbas 2009).

The vegetable oil charged in the reactor and heated to about 60–70°C with the moderate stirring. Meanwhile, about 0.5–1.0% (w/w) of anhydrous alkaline catalyst (sodium hydroxide or potassium hydroxide) is dissolved in 10–15% (w/w) of methanol. This sodium hydroxide–alcohol solution is mixed with the oil, and heating and stirring are continued. After 25–45 minutes, the reaction is stopped and the products are allowed to settle into two phases. The upper phase consists of esters and the lower phase consists of glycerol and impurities. The mixture of KOH and methanol settles at the bottom of the funnel because of its higher density compared with biodiesel whereas a small amount of catalyst, excess methanol, and glycerol are in the upper biodiesel layer. Washing is a purification process to remove entrained glycerol, catalyst, soap, and excess methanol in the upper layer. The excess methanol in biodiesel corrodes the fuel injection system and should be separated from the biodiesel. The ester layer is washed with water several times until the washing becomes clear.

3.2.1.1 Purification of Biodiesel

The transesterified vegetable oils (i.e., biodiesel/esters) have reduced viscosity and increased volatility relative to the triglycerides present in vegetable oils. A dark viscous liquid (rich in glycerol) is the by-product of the transesterification process. Traces of the methanol, catalyst, and free fatty acids (FFAs) in the glycerol phase can be processed in one or two stages depending upon the level of purity required. Distillation column recovers the excess alcohol and it can be recycled. The by-product glycerol needs to be recovered because of its value as an industrial chemical such as CP glycerol, USP glycerol, and dynamite glycerol. The glycerol is then removed by gravity separation and the remaining ester is mixed with hot water for separation of catalyst. By using silica gel moisture it can be removed.

The triglycerides should have lower acid value and all material used in the process should be substantially anhydrous. The addition of more base catalyst to compensate the higher acidity of vegetable oils results in formation of soap and the increase in viscosity or formation of gels that interferes in the transesterification reaction as well as with the separation of glycerol. Saponification reaction also takes place simultaneously along with the transesterification process but soap formation is not a major problem if presence of water is less than 1%. Prolonged contact with air will diminish the effectiveness of these catalysts through interaction with moisture and carbon dioxide. When the reaction conditions do not meet the above requirements, ester yields are significantly reduced.

The free fatty acid (FFA) and moisture content are key parameters for determining the suitability of the vegetable oil transesterification process. To carry the alkaline-catalyzed reaction to completion, a FFA value lower than 2% (i.e., acid value of less than four) is needed. The higher acidity of the oil reduces biodiesel production efficiency. Ma, Clements, and Hanna (1998) studied the transesterification of beef tallow catalyzed by NaOH in the presence of FFAs and water. Without adding FFA and water, the apparent yield of beef tallow methyl esters (BTME) was the highest. When 0.6% of FFA was added, the apparent yield of BTME reached the lowest, less than 5%, with any level of water added (Meher, Vidya Sagar, and Naik 2006).

3.2.1.2 Alcohol to Oil Molar Ratio

The stoichiometric transesterification requires 3 mol of the alcohol per mol of the triglyceride to yield 3 mol of the fatty esters and 1 mol of the glycerol. However, the transesterification reaction is an equilibrium reaction in which a large excess of the alcohol is required to drive the reaction close to completion in a forward direction. The molar ratio of 6:1 or higher generally gives the maximum yield (higher than 98% by weight). Lower molar ratios require a longer time to complete the reaction. Excess molar ratios increase the conversion rate but lead to difficulties in the separation of the glycerol. At optimum molar ratio only the process gives higher yield and easier separation of the glycerol. The optimum molar ratios depend upon the type and quality of the vegetable oil used (Ramadhas, Jayaraj, and Muraleedharran 2005). Figure 3.2 shows the ester conversion efficiency with respect to molar ratio.

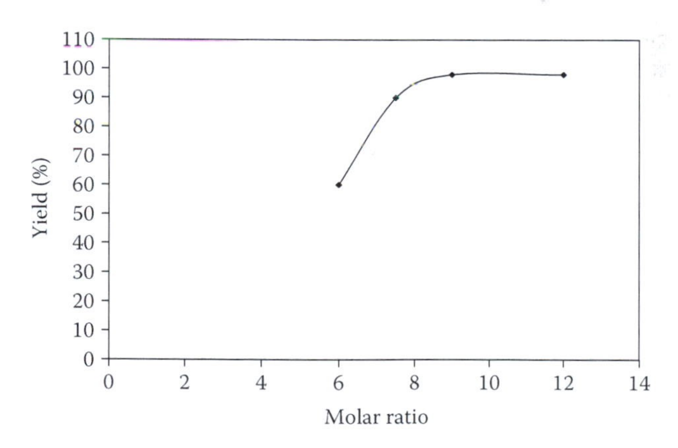

FIGURE 3.2
Conversion efficiency with respect to molar ratio.

3.2.1.3 Catalyst

As a catalyst in the process of alkaline methanolysis, mostly sodium hydroxide or potassium hydroxide have been used, both in concentration from 0.4 to 2% w/w of oil. Refined and crude oils with 1% or less either sodium hydroxide or potassium hydroxide catalyst will be sufficient (Tomasevic and Marinkovic 2003). Figure 3.3 shows the ester conversion efficiency with respect to catalyst amount.

These catalysts increase the reaction rate several times faster than that of acid catalysts. The alkaline catalyst concentration in the range of 0.5–1.0% by weight gives 94–99% conversion efficiency. Further increase in catalyst concentration does not increase the yield, but it adds to the extra cost and increases the complication in the separation process.

3.2.1.4 Reaction Temperature

The rate of the transesterification reaction is strongly influenced by the reaction temperature. Generally, this reaction is carried out close to the boiling point of the methanol (60–70°C) at atmospheric pressure. With further increase in temperature there is more chance in loss of methanol. Pretreatment is not required if the reaction is carried out under high pressure (90 bar) and high temperature (240°C), where simultaneous esterification and transesterification take place with maximum yield obtained (Barnwal and Sharma 2005).

3.2.1.5 Mixing Intensity

The mixing effect is more significant during the slow rate region of the transesterification reaction and when the single phase is established, mixing

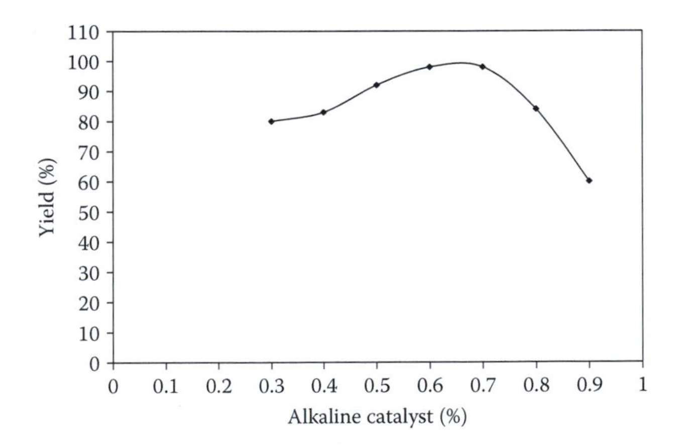

FIGURE 3.3
Conversion efficiency with respect to catalyst amount.

becomes insignificant. The understanding of the mixing effects on the kinetics of the transesterification process is a valuable tool in the process scale-up and design. Generally, after adding the methanol and catalyst to the oil, 5–10 minutes stirring helps in a higher rate of conversion and recovery.

3.2.1.6 Purity of Reactants

The impurities present in the vegetable oil also affect ester conversion levels significantly. The vegetable oil (refined or raw oil) is to be filtered before the transesterification reaction. The oil settled at the bottom of the tank during storage would give lesser yield because of deposition of impurities like wax.

3.2.1.7 Effect of Reaction Time

The conversion rate increases with reaction time. Transesterification of peanut, cottonseed, sunflower, and soybean oils under the condition of methanol to oil ratio of 6:1, 0.5% sodium methoxide catalyst, and 60°C the conversion efficiency of about 80% was obtained after 1 minute for soybean and sunflower oils; and after completion of 60 minutes the conversions were almost the same for all four oils (93–98%; Agarwal 2007).

Ma and Hanna (1999) studied the effect of reaction time on transesterification of beef tallow with methanol. The reaction was very slow during the first minute due to the mixing and dispersion of methanol into beef tallow. From 1 to 5 minutes, the reaction proceeded very fast and the conversion efficiency peaked from 1 to 38%.

3.2.2 Acid Catalyst Transesterification

Alkaline esterification is suitable for the triglycerides having an acid value less than four and all reactants should be substantially anhydrous. If the acid value is greater than four the conversion efficiency would reduce drastically. The alkaline catalyst reacts with the high FFA feedstock to produce the soap and water. Acid catalyzed process can be used for esterification of these FFAs. The nonedible type oil, crude vegetable oils, and used cooking oils typically contain more than 2% FFA, and the animal fats contain from 5 to 30% FFA. Low quality feedstock, such as trap grease, can contain FFA up to 100%. Moisture or water present in the vegetable oils increase the FFA value. FFA content of rice bran rapidly increased within few hours, showing 5% increase in FFA content per day. Heating of the rice bran immediately after the milling inactivates the lipase and prohibits the formation of FFA.

Van Gerpen (2005) advocates that up to 5% FFA, alkaline catalyst can be used for the transesterification reaction; however, additional catalyst must be added to compensate for the catalyst lost to the soap. When FFA value of the vegetable oil is more than 5%, the formation of soap inhibits the separation of methyl esters from the glycerol and contributes to emulsion formation during the water wash. For these cases, acid catalyst such as sulfuric acid is used to esterify the FFAs to methyl esters. Figures 3.4 and 3.5 show the acid esterification reaction equation and its mechanism, respectively. Figure 3.6 shows the ester conversion with respect to molar ratio.

Canakci and Van Gerpan (1999) reported that the standard conditions of the reaction consisted of 60°C reaction temperature, 3% sulfuric acid, 6:1 molar ratio of the methanol to the oil, and reaction duration of 48 hours. Figure 3.6 shows the ester conversion with respect to molar ratio. The ester conversion increased from 87.8 to 95.1% when the reaction time was increased from 48 to 96 hours.

Different alcohols have different boiling points and reaction temperature is chosen nearer to the boiling point. Effect of alcohol type on transesterification of vegetable oils is shown in Table 3.1. The ester conversion is inhibited by the presence of water in the oil. If the water concentration in oil is greater than 0.5%, the ester conversion drops below 90%. The alcohols with higher boiling points increase the ester conversion. Water formed by esterficiation further reduces reactions. It has been reported that more than 0.5% water in the oil would decrease the ester conversion to below 90%.

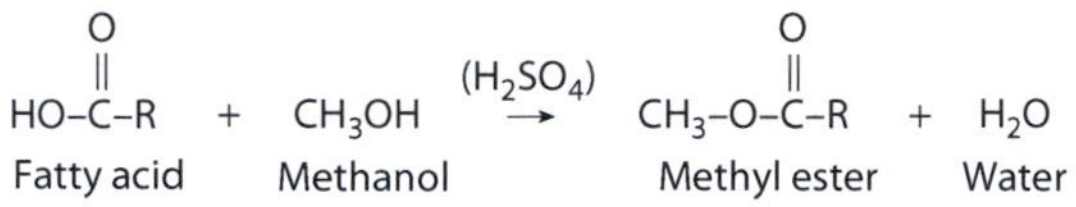

$$\underset{\text{Fatty acid}}{HO\text{–}\overset{\overset{\displaystyle O}{\|}}{C}\text{–}R} + \underset{\text{Methanol}}{CH_3OH} \xrightarrow{(H_2SO_4)} \underset{\text{Methyl ester}}{CH_3\text{–}O\text{–}\overset{\overset{\displaystyle O}{\|}}{C}\text{–}R} + \underset{\text{Water}}{H_2O}$$

FIGURE 3.4
Acid esterification equation.

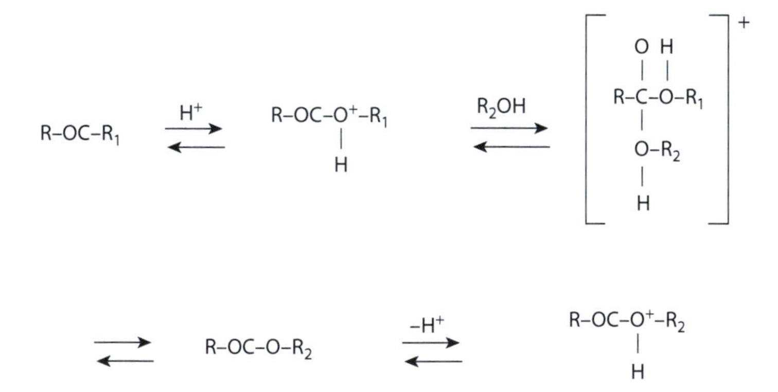

FIGURE 3.5
Mechanism of acid catalyzed transesterification of vegetable oils. (From Demirbas, A., *Energy Conversion and Management*, 50, 14–34, 2009. Reprinted with permission from Elsevier Publications.)

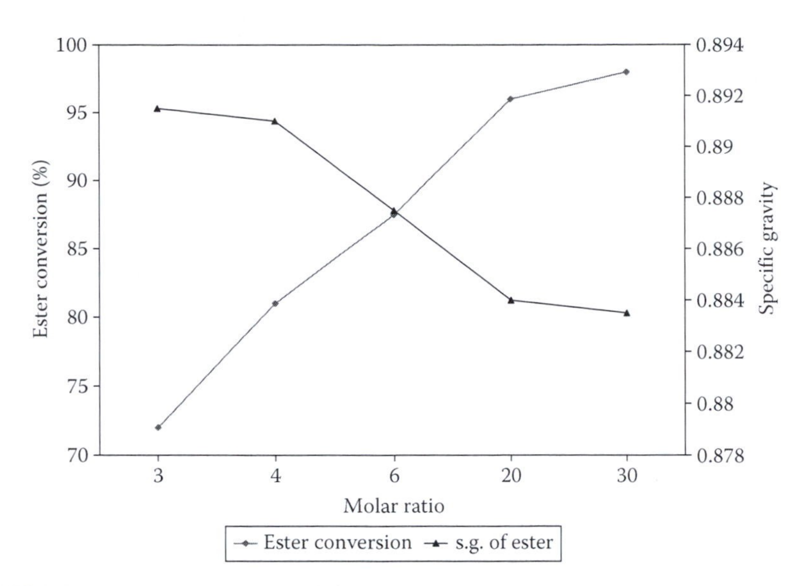

FIGURE 3.6
Ester conversion with respect to molar ratio. (From Canakci, M. and Van Gerpan, J., *ASAE*, 42, 1203–10, 1999. Reprinted with permission from the American Society of Agricultural and Biological Engineers.)

TABLE 3.1
Effect of Alcohol Type on Ester Conversion and Specific Gravity of Ester

Alcohol Type	Boiling Temperature (°C)	Reaction Temperature (°C)	Ester Conversion (%)	Ester Specific Gravity
Methanol	65.0	60	87.8	0.8876
2-propanol	82.4	75	92.9	0.8786
1-butanol	117.0	110	92.1	0.8782
Ethanol	78.5	75	95.8	0.8814

Source: From Canakci, M. and Van Gerpan, J., *ASAE*, 42, 1203–10, 1999. Reprinted with permission from the American Society of Agricultural and Biological Engineers.
Test conditions: molar ratio: 6:1, sulfonic acid amount: 3%, reaction temperature: 48 hours.

3.2.3 Supercritical Alcoholysis

Free fatty acid content and water in the vegetable oil plays a major role in conventional transesterification process. Presence of FFAs and water causes soap formation, consumes catalyst, and reduces the effectiveness of catalyst that leads to reduction in conversion efficiency. Transesterification of the vegetable oil with the help of catalysts reduce the reaction duration but promotes complications in purification of biodiesel from catalysts and saponified products. Purification of biodiesel and separation of glycerol from the catalyst are necessary, but it would increase the cost of overall production process. The supercritical alcohol transesterification process is catalyst-free transesterification process, which completes in a very short time, about a few minutes. Because of the noncatalytic process, the purification products after the transesterification reaction is much simpler and environmentally friendly compared to that of conventional process. Saka and Kusdiana (2001), Warabi, Kusdiana, and Saka (2004), Kusdiana and Saka (2004), Demirbas (2003) developed noncatalytic transesterification with supercritical methanol for producing biodiesel from vegetable oils to overcome these problems.

Saka and Kusdiana (2001) developed a biodiesel production process without the aid of catalysts. Rapeseed oil, methanol mixture (molar ratio up to 42) heated at its supercritical temperature (350–500°C) for different time periods (1–4 minutes). The product (biodiesel) is evaporated at 90°C for about 20 minutes to remove the excess methanol and water produced by esterification. Saka and Kusdiana (2001) reported that optimized process parameters for the transesterification of rapeseed oil were: molar ratio of 42:1, pressure 430 bar, reaction temperature 350°C for 4 minutes that yields 95% conversion efficiency. Figure 3.7 describes the yield of the process with respect to the reaction time.

Kusdiana and Saka (2004) analyzed the effects of water content in vegetable oils on the yield of methyl esters in the transesterification of triglycerides and methyl esterification of fatty acids under supercritical methanol method. Water content in vegetable oils had negative effect on alkaline or acid catalyzed esterification process; that is, it consumes the catalyst and hence reduces the catalyst efficiency as well as conversion efficiency of the process. Catalyst-free supercritical methanolysis process has no influence on water content in vegetable oil. Water addition even up to 50% has not reduced the yield of methyl esters. Hydrolysis reaction is much faster than the transesterification and hence triglycerides are transformed into fatty acids in the presence of water. These FFAs are methyl esterified by supercritical methanolysis. With the addition of water in the supercritical methanol process, separation of the

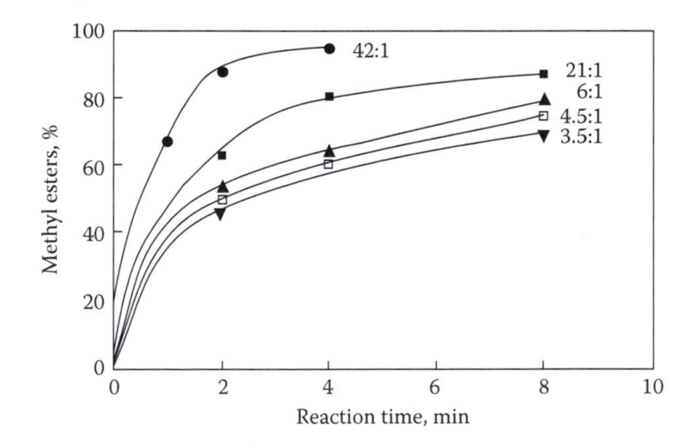

FIGURE 3.7
Yield of the process with respect to reaction time. (From Saka, S. and Kusdiana, D., *Fuel*, 80, 225–31, 2001. Reprinted with permission from Elsevier Publications.)

methyl esters and glycerol from the reaction mixture becomes much easier. The glycerol is more soluble in water than methanol, which moves to the lower portion whereas biodiesel moves to the upper layer. The effect of water content on the yield of methyl esters in the supercritical methanol process is shown in Figure 3.8 (Sharma et al. 2008).

Lee, Hubble, and Saka (2006) developed a supercritical methanolysis process for transesterification of vegetable oils in a continuous process. In the first stage, triglycerides rapidly hydrolyze to FFAs under 100 bar pressure and at 270°C temperature. When the pressure is reduced, the mixture promptly separates into two phases and the water phase can be separated to recover glycerol. Methanol becomes supercritical at a pressure of 100 bar and a temperature of 270°C and the supercritical conditions favor rapid formation of methyl esters from the fatty acids. Figure 3.9 shows two-stage continuous biodiesel production process with supercritical methanol (Lee, Hubble, and Saka 2006).

Wrabi, Kusdiana, and Saka (2004) analyzed the reactivity of the triglyceride and the fatty acids of the rapeseed oil. They reported that with an increase in reaction duration, the yield of the alkyl esters was increased. For the same reaction duration treatment, the alcohols with shorter alkyl chains gave better conversion than that of longer alkyl chains. Increase in the reaction temperature, especially at supercritical temperatures, increases the yield of the process. Advantages of supercritical methanol process include

1. Simpler purification process
2. Lower reaction time
3. Less energy intensive and hence, the supercritical methanol method would be more efficient than the conventional processes

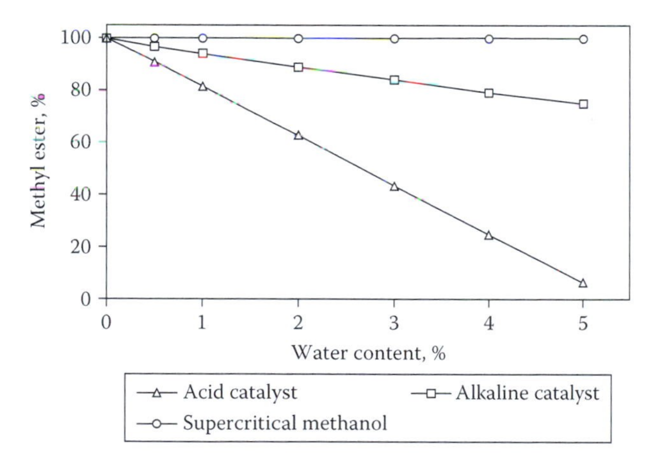

FIGURE 3.8
Effect of water content on yield of the process. (From Sharma, Y. C., Singh, B., and Upadhyay, S. N., *Fuel*, 87, 2355–73, 2008. Reprinted with permission from Elsevier Publications.)

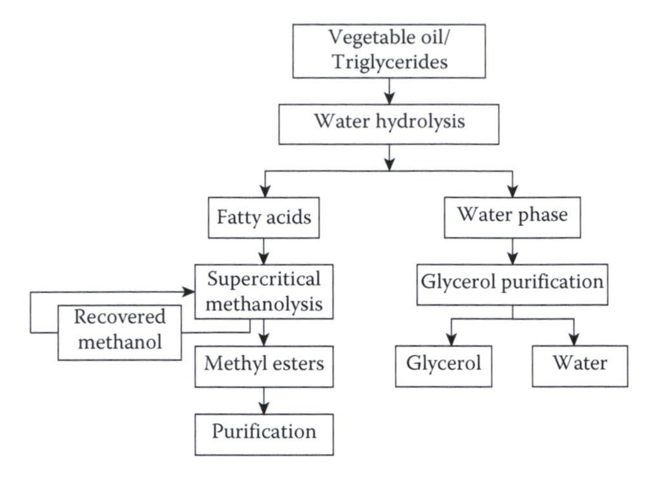

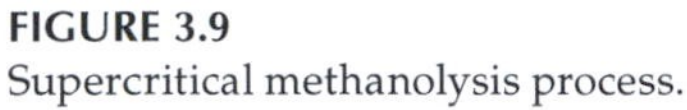

FIGURE 3.9
Supercritical methanolysis process.

3.2.4 Lipase Catalyzed Transesterification

Utilization of lipase as a catalyst for biodiesel fuel production has a very good potential compared with chemical methods using alkaline or acid catalyst as no complex operations are required for removal of catalyst and separation/ purification of glycerol. The enzymatic conversion of the triglycerides has been suggested as a realistic alternative to the conventional physiochemical methods.

Vegetable oils contain FFA, waxes, unsaponified matter, pigments, and lecithin. It is necessary to preheat the oil for the removal of lecithin, pigments by degumming and bleaching for efficient conversions in lipase based transesterification. Lipids have the capability to convert the FFA into triglycerides. The yield of crude vegetable oils is low due to the presence of higher amounts of phospholipids. The interference of interaction of lipase molecule with substrates by phospholipids bound on the immobilized preparation reduces the yield of biodiesel production process.

Du et al. (2004) developed an enzymatic transesterification for the production of biodiesel and the process details are: reaction 9.65 g soybean oil, 4% Novozym 435 (based on oil weight), and 1 molar equivalent of methanol, reaction temperature 40°C. There were 50 microliter samples taken from the reaction mixture at specified times and centrifuged to obtain the upper layer. Figure 3.10 shows the conversion efficiency of lipase esterification process for crude oils.

Noureddini, Gao, and Philkana (2005) developed a typical biodiesel production method using lipase catalyst and the conditions are: 10g soybean oil, 3 g methanol (methanol to oil molar ratio of 8.2), 0.5 g water, 3 g immobilized lipase phyllosilicate sol-gel matrix (PS), 40°C, 700 rpm and 1 hour reaction

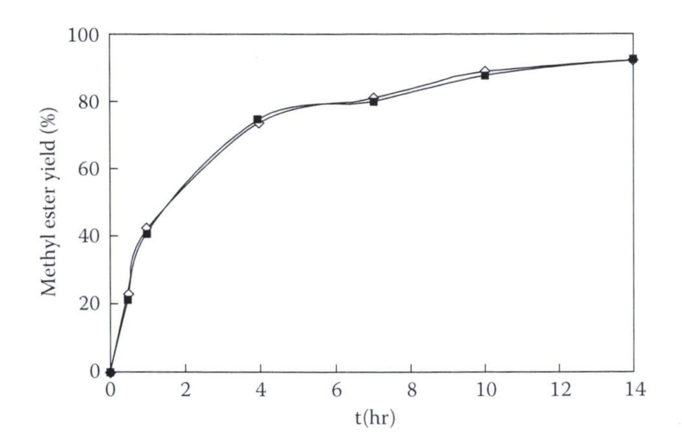

FIGURE 3.10
Yield of the lipase catalyst process with respect to reaction duration. (From Du, W., Xu, Y., Liu, D., and Zeng, J., *Journal of Molecular Catalysis B: Enzymatic*, 30, 125–29, 2004. Reprinted with permission from Elsevier Publications.)

duration. For similar reaction conditions, with ethanol, 0.3 g of water and 5 g of ethanol (ethanol to oil molar ratio of 9.5) were used. It has been reported that using the methyl acetate as acyl acceptor for the biodiesel production from crude soybean oil gave methyl ester yield of 92%, just high of that in the refined soybean oil.

As the lipases can simultaneously esterify and transesterify the FFA and triglycerides in vegetable oils, low quality vegetable oils such as high FFA oils, acid oils, and restaurant greases can be converted into biodiesel. In the lipase process, excess methanol would lead to inactivation of enzyme and glycerol, as a major by-product, could block the immobilized enzyme resulting in low enzymatic activity. These problems could be the limitations for industrial production of biodiesel with enzymes as catalyst.

Advantages of lipased based transesterification include:

1. Possibility of regeneration and reuse of the immobilized residue
2. Use of enzymes in reactors for a longer activation
3. Higher thermal stability due to the native state
4. Easier separation of biodiesel from glycerol

3.2.5 Ultrasonic Transesterification

Biodiesel is commonly produced in batch reactors using heat and mechanical mixing as energy input. Ultrasonic mixing is an effective means to achieve a better mixing in biodiesel processing. Ultrasonic cavitations provide the necessary activation energy for the industrial biodiesel transesterification.

Biodiesel is mainly produced in batch reactors commercially. Ultrasonic biodiesel processing permits for continuous inline processing. Ultrasonication can achieve a biodiesel yield in excess of 99%. Ultrasonic reactors reduce the processing time remarkably. The conventional esterification reaction in a batch is slow and phase separation of the glycerin is time-consuming, often taking more than 4 hours. Ultrasonic process reduces the separation time from 4 hours to less than an hour. Due to the increased chemical activity in the presence of cavitations, the ultrasonic transesterification helps to decrease the amount of catalyst required by up to 50%. Ultrasonic transesterification reduce the amount of excess methanol required and also to improve the purity of glycerin.

Stavarache et al. (2005) used an ultrasonic method for the preparation of the emulsification in the alkaline-catalyzed esterification process. The collapse of cavitation bubbles disrupts the phase boundary and causes the emulsification, by ultrasonic jets that impinge one liquid to another. With increasing the chain length, the miscibility between the oil and alcohol increases, thus, decreasing the reaction time (10–20 minutes) but also making the separation of the esters difficult. The normal chain alcohols react quite rapidly under the ultrasonic irradiation. This behavior is due to increased mass transfer in the presence of ultrasound.

The velocity of an ultrasonic wave through a material depends on its physical properties and hence, the ultrasonic velocity decreases with the increasing density. The droplets of more dense oil move upward and form a cream layer, while the alcohol moves downward, facilitating the mixing and increasing the contact surface between alcohol and oil.

The reaction time at 40 kHz is shorter than that of a reaction at 28 kHz, however the yield was lower. At 40 kHz, the soap is formed in higher amount and acts as phase transfer catalyst leading to formation of the esters more rapidly than at 28 kHz. But during the washing, the soap hinders the separation and some ester is trapped into the soap micelles and thus the yield in the isolated product is decreased.

Table 3.2 describes the comparison of the yield of methyl esters with the mechanical stirring and ultrasonic irradiation (Stavarache et al. 2005). At higher frequencies, the collapse of cavitation bubbles is not very strong and impingement of one liquid to the other is poor. Thus, the mixing between the two immiscible layers (alcohol and oil) is very poor and the emulsification does not occur. The transesterification takes place mainly at the boundary between the two layers.

Low frequency ultrasound is efficient, time saving, and economical offering advantages over the conventional procedure. The ultrasonic aided biodiesel production method can be a valuable tool for the transesterification of fatty acids.

Ji et al. (2006) developed an alkali-catalyzed biodiesel production method with power ultrasonic (19.7 kHz) that allows a short reaction time and high yield because of emulsification and cavitations of the liquid–liquid immiscible

TABLE 3.2
Yield of Isolated Methyl Esters

Method	0.5% (w/w) KOH		1.0% (w/w) KOH		1.5% (w/w) KOH	
	Time (min)	Yield (%)	Time (min)	Yield (%)	Time (min)	Yield (%)
Mechanical stirring	60	86	40	85	20	83
Ultrasonic irradiation 28 kHZ	40	95	40	93	40	93
Ultrasonic irradiation 40 kHZ	20	96	40	92	20	90

Source: From Stavarache, C., Vinatoru, M., Nishimura, R., and Maeda, Y., *Ultrasonics Sonochemistry*, 12, 367–72, 2005. Reprinted with permission from Elsevier Publications.

system. Stavarache et al. (2007) developed a bench scale continuous process for the production of biodiesel from neat vegetable oils under high power, low frequency ultrasonic irradiation. The highest conversion was achieved when short residence time was employed. The transesterification under ultrasonic irradiation is mainly influenced by the residence time in the reactor and alcohol–oil molar ratio. The advantages of ultrasonic process transesterification include

1. Effective means to increase the reaction speed and conversion rate in the biodiesel processing
2. Helps to decrease the amount of catalyst required by up to 50% due to the increased chemical activity in the presence of cavitations
3. Amount of excess methanol required is reduced
4. Results increase in the purity of the glycerin

3.2.6 Comparison of Biodiesel Production Methods

Biodiesel produced by various methods depends upon the feedstock quality and its fatty acid composition. It has been reported that 459 biodiesel plants in the world have the capacity of 76.793 billion liters (www.worldbioplants.com). Each process has its own advantages in comparison with others. It is essential to select the suitable method based on our requirement and feedstock. Table 3.3 shows a comparison of various biodiesel production methods.

3.3 Biodiesel Process Equipments

The basic equipment used in biodiesel production plants are the biodiesel reactor, pumps, settling tanks, distillation columns, and storage tanks.

TABLE 3.3
Comparison of Biodiesel Production Methods

Variable	Alkaline	Acid	Two-Step	Ultrasonic	Lipase	Supercritical
Reaction temperature (°C)	40–70	55–80	40–70	30–40	30–403	240–385
Yield	Normal	Normal	Good	Higher	Higher	Good
Glycerol recovery	Difficult	Difficult	Difficult	Difficult	Easy	—
Purification of ester	Washing	Washing	Washing	Washing	—	—
Production cost	Cheap	Cheap	Cheap	Medium	Expensive	Medium

Biodiesel production process is classified into two: batch and continuous process. In batch production process, measured quantities of reagents are charged and reaction taken place at prescribed conditions. For the base catalyzed transesterification process, stainless steel material is used for the fabrication of the reactor.

In continuous process, reactants are fed into one side of the reactor. The chemical composition changes as the reactants moves in plug flow through the reactor. Gear pumps are used for transferring the reactants and products from one place to another. The separation of biodiesel and glycerol is achieved by using a settling tank. Though settling is cheaper, centrifuge can be used to increase the rate of separation relative to a settling tank. Separation is accomplished by exposing the mixture to a centrifugal force. The denser phase is separated to the outer surface of the centrifuge. Distillation column separates the fluids with similar boiling points.

3.4 Biodiesel Properties

Biodiesel fuel quality depends upon composition of feedstock, production process, storage, and handling. Biodiesel quality is evaluated through the determination of chemical composition and physical properties of the fuel. Contaminants and other minor components, due to incomplete reaction, are the major issues in the quality of biodiesel (i.e., glycerol, mono, di, triglycerides, alcohol, catalysts, and FFA present in the biodiesel). Moreover, biodiesel composition could be changed during storage and handling. Biodiesel can absorb water or undergoes oxidation during storage. Therefore, significance of these parameters and their analytical or engine test methods are addressed in standards. Each country has its own fuel quality testing methods and standards to specify the properties of the fuel. Here the standard methods and limits are described with reference to ASTM/EN/IS standards.

3.4.1 Flash Point

Flash point is defined as the lowest temperature corrected to a barometric pressure of 101.3 kPa (760 mm Hg), at which application of an ignition source causes the vapors of a specimen to ignite under specified conditions of a test. Flash point of the fuel is evaluated as per ASTM D93 test method.

Flash point of biodiesel is higher than that of diesel (>130°C), which makes biodiesel safer than diesel in handling and the storage point of view. A minimum flash point for biodiesel is specified in restricting the alcohol content. Flash point of biodiesel will reduce drastically if the alcohol used in production of biodiesel was not completely removed. Moreover, it reduces the combustion quality of fuel. Excess methanol in the fuel may also affect engine seals and elastomers and corrode metal components.

3.4.2 Water Sediment

Water and sediment is a test that determines the volume of free water and sediment in middle distillate fuels having viscosities at 40°C in the range 1.0–4.1 mm^2/s and densities in the range of 700–900 kg/m^3. ASTM D 2709 test method evaluates the water content in the biodiesel. If any water is present in biodiesel, water can react with the biodiesel making FFAs and can support microbial growth in storage tanks.

The water and sediment content did not change in most cases with increased triglycerides content. While biodiesel is generally considered to be insoluble in water, it actually takes up considerably more water than diesel fuel. However, water is deliberately added during the washing process to remove contaminants from the biodiesel. This washing process should be followed by a drying process to ensure the final product will meet ASTM D 2709. Sediments may plug fuel filters and may contribute to the formation of deposits on fuel injectors and other engine damage. Sediment levels in biodiesel may increase over time as the fuel degrades during extended storage.

3.4.3 Kinematic Viscosity

Kinematic viscosity is the resistance to flow of a fluid under gravity. It is the time taken for a fixed volume of fuel to flow under gravity through the capillary tube viscometer immersed in a thermostatically controlled bath at 40°C. It is the product of measured time flow and calibration constant of viscometer. The kinematic viscosity is a basic design parameter for the fuel injectors used in diesel engines. Fuel viscosity has influence on fuel droplet size and spray characteristics. Viscosity is inversely proportional to temperature. Viscosity of biodiesel increases with chain length and degree of saturation.

Biodiesel specification ensures that at viscosity upper limit, fuel will flow readily during cold starting. Higher viscosity of fuel leads to poor atomization, incomplete combustion, and increases carbon deposits. Moreover, higher viscosity fuel needs higher pumping power also. Fuels with lower viscosity leaks past the plunger through the clearance between plunger and barrel during fuel compression.

3.4.4 Density

Density is mass of the substances occupying unit volume at 15°C. Density of fuel is evaluated as per ASTM D4052. Hydrometers are used to evaluate the density of liquids. Biodiesel is slightly higher than conventional diesel. Diesel engine injectors normally operate on a volume metering system. If the fuel is of higher density, a large mass of fuel is injected and hence more power and emissions.

3.4.5 Cetane Number

Cetane number of fuel is a measure of the ignition performance of diesel fuel obtained by comparing it to the reference fuels in a standardized engine test; that is, cooperative fuel research (CFR) cetane engine, as per ASTM D613 test method. The higher cetane number in fuel is better in its ignition properties. The cetane number affects engine parameters like combustion, stability, drivability, smoke, noise, and emissions. Cetane number of biodiesel is higher than that of diesel. Adequate cetane number is required for good engine performance. High cetane number ensures good cold start properties and minimizes the formation of white smoke.

The cetane number of biodiesel depends on the feedstock used for the biodiesel production. Cetane number of biodiesel decrease with an increase in unsaturation (oleic and linolenic) and increases with an increase in chain length. Esters of saturated fatty acids such as palmitic and stearic acids have higher cetane numbers.

3.4.6 Distillation Characteristics

Biodiesel does not contain any highly volatile components, hence it evaporates at higher temperatures. Biodiesel components have very close boiling points. Boiling point of biodiesel generally ranges from 330–360°C.

3.4.7 Cold Filter Plugging Point

Cold filter plugging point (CFPP) is the temperatures at which a fuel will cause a fuel filter to plug due to fuel components that have begun to crystallize or gel. CFPP of biodiesel is determined as per EN 116 test method. CFPP reflects the cold weather performance of the fuel. At lower temperatures fuel will not flow properly and affects the fuel pumps and fuel injectors. CFPP is less conservative than the cloud point, and is considered by some to be the true indication of low temperature operability.

3.4.8 Pour Point and Cloud Point

Cloud point is the temperature at which a cloud or haze wax crystals appears at the bottom of the test jar when the oil is cooled under prescribed conditions. Pour point is the lowest temperature at which the oil is observed to flow when cooled and examined under prescribed conditions. Biodiesel generally has higher cloud point and pour point than diesel. In saturated fatty acids, the carbon is bound to two hydrogen atoms by double bonds. More double bonds in the biodiesel indicate lower cloud point.

3.4.9 Stability

Biodiesel ages more quickly than diesel due to chemical structure of methyl esters present in biodiesel. Saturated methyl esters in biodiesel increase its cloud point and cetane number and improve its stability. The unsaturated fatty acids reduce cloud point and cetane number and reduce its stability also.

Oxidation stability of biodiesel is evaluated by EN 14214 test method. Rancimat apparatus is used for evaluating the oxidation stability of biodiesel. Methyl esters are heated at 110°C under a constant stream of air at the rate of 10 l/h. The vapors released during the oxidation process, together with the air, are passed into the jar containing 60 ml of water and electrode. When the conductivity begins to increase rapidly, it indicates induction period. This accelerated increase is caused by dissociation of volatile carboxylic acids produced during the oxidation process and absorbed in the water. Rancimat induction period of 6 hours is defined for biodiesel samples as per biodiesel specification.

Most of the biodiesel will not meet this specification unless suitable antioxidants are added at suitable treat rate. Stability of biodiesel is influenced by factors such as presence of air, heat, traces of metal, peroxides, light, and number of double bonds. Oxidation stability of biodiesel will decrease with the increase of polyunsaturated methyl esters. Oxidation stability of biodiesel depends upon the feedstock used in the production of biodiesel. Vegetable oils such as sunflower or soybean (rich in linoleic and linolenic acids) have poor oxidation stability.

Iodine number (ASTM D1520 or EN 14111 test method) indicates the tendency of fuel to be unstable. Iodine number refers to the amount required to convert unsaturated oil into saturated oil. It does not refer to the amount of iodine in the oil but the presence of unsaturated fatty acids in the fuel. It indicates the tendency of fuel to be unstable. Unsaturated fatty acids are prone to oxidation.

$C_{18:3}$ is three times more unstable than $C_{18:0}$. Poor oxidation stability can cause fuel thickening, formation of gums, filter clogging, and injector fouling. Moreover, the high iodine number biodiesel increases the danger of polymerization in the engine oil. Sudden increase in oil viscosity is attributed to oxidation and polymerization of unsaturated fuel entering into oil through dilution. Moreover, fuel having the higher iodine number tends to polymerize and forms deposits on injector nozzles, piston rings, and ring grooves. The higher fuel temperature reduces the oxidation stability. The variation in the induction period of fuels with respect to fuel temperature is shown in Figure 3.11.

The oxidation stability of palm oil is good due to the presence of a higher concentration of saturated fatty acids (43.4 saturates and 56.6 unsaturates), whereas jatropha oil has more unsaturates (21.1 saturates and 78.9% unsaturates). Figure 3.12 shows the induction period of neat jatropha biodiesel and blends of jatropha–palm biodiesel. They have reported that the induction period of neat biodiesel is substantially low and hence antioxidants are required to improve the oxidation stability.

3.4.10 Free Glycerol and Total Glycerol

Degree of completion of esterification process is indicated by the amount of free and total glycerol in the biodiesel. These are evaluated as per ASTM D6584 and EN 14105 test methods.

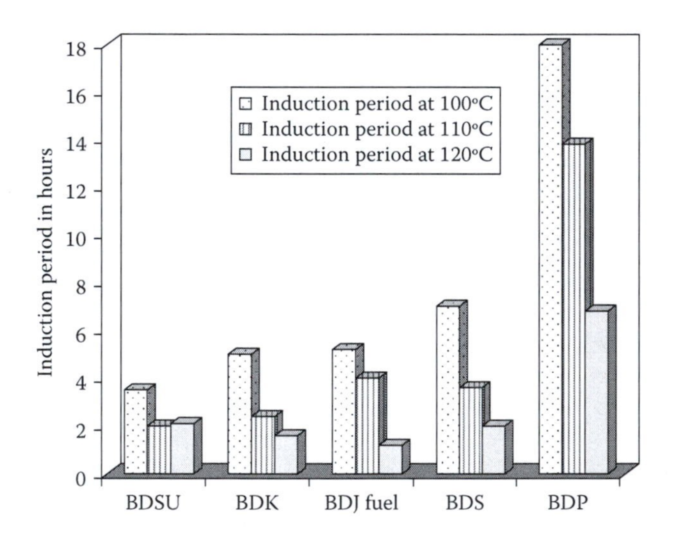

FIGURE 3.11
Variation in induction period of fuels with respect to fuel temperature. (From Sarin, R., Sharma, M., Sinharay, S., and Malhotra, R. K., *Fuel*, 86, 1365–71, 2007. Reprinted with permission from Elsevier Publications.)

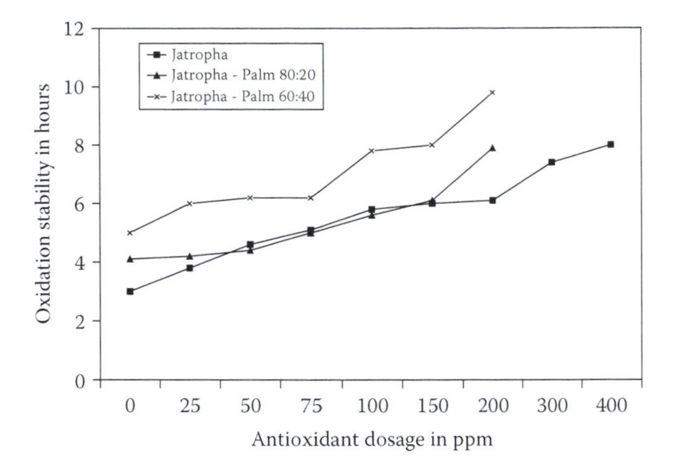

FIGURE 3.12
Effect of antioxidant dosage on fuel induction period. (From Sarin, R., Sharma, M., Sinharay, S., and Malhotra, R. K., *Fuel*, 86, 1365–71, 2007. Reprinted with permission from Elsevier Publications.)

Free glycerol is the glycerides that is left in the biodiesel as suspended droplets are dissolved. Free glycerol results from incomplete separation of the ester and glycerol products after the transesterification reaction. Glycerol should be removed during the water washing process. Water washed biodiesel is generally very low in free glycerol particularly if hot water is used for washing. This may be due to incomplete conversion, or purification/washing that could not completely separate the glycerol from the biodiesel. Free glycerol can be a source of carbon deposits in an engine because of incomplete combustion. Glycerides have a much higher boiling point than biodiesel or conventional diesel fuel and can lead to carbon deposits in the engine and durability problems. High content of free glycerol causes problems in storage and fuel system due to separation of glycerol.

Total glycerol is bound glycerol (glycerol portion of mono, di, and triglycerides) added to free glycerol. Total glycerol is of higher boiling point and it causes engine combustion chamber deposits and durability problems.

3.4.11 Ester Content

The amount of esters in the biodiesel is determined as per EN 14103 test method. The biodiesel contains methyl esters in the range of C_{14}–C_{24}. This is measured by gas chromatography equipment. From this ester content value, it can be ensured that whether the transesterification is complete or not and it is suitable for engine application.

Hu et al. (2005) reported that, in general, unrefined biodiesel contains less than 97% of esters and small amounts of mono, di, and triglycerides, while

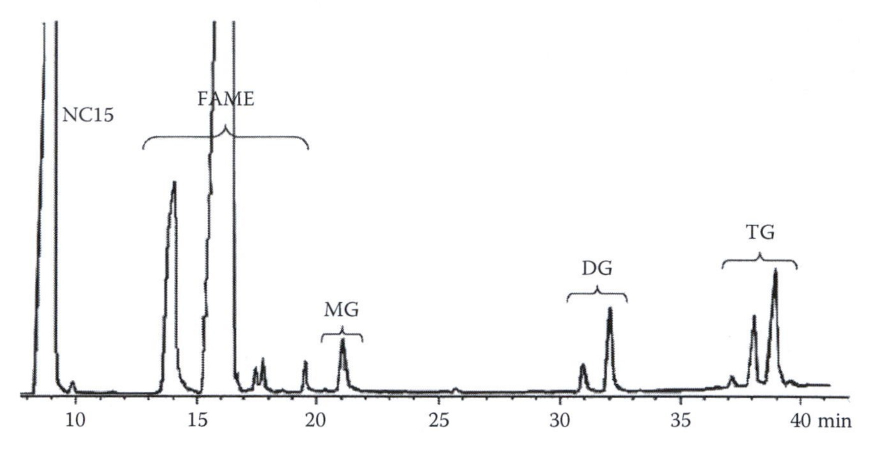

FIGURE 3.13
Chromatogram of biodiesel. (From Hu, J., Du, Z., Li, C., and Min, E., *Fuel*, 84, 1601–06, 2005. Reprinted with permission from Elsevier Publications.)

methyl esters are only detected in refined biodiesel. In order to analyze the purity of biodiesels, acid value was measured. Figure 3.13 shows the chromatogram of biodiesel and Table 3.4 describes the amount of FAME and glycerides in biodiesel.

3.4.12 Sulfur

Sulphur in the biodiesel is measured by titration as per ASTM D5453 test method. Higher sulphur in the biodiesel increases wear between piston rings and cylinder liners, increases combustion chamber deposits, increases particulate matter emissions, and reduces the performance of after treatment devices. Emission norms advocate the reduction in sulfur content in the fuel in order to reduce the exhaust emissions.

3.4.13 Carbon Residue

Conradson carbon residue (CCR) is the measure of tendency of a fuel to produce deposits on injector tips of nozzle and combustion chamber. This is measured as per ASTM D 4530 test method. The test is performed via heating a weighed sample of a fuel to 500°C under a nitrogen atmosphere for a specified duration. At these conditions any volatiles that are formed are purged by nitrogen and the residue that remains is called carbon residue. The maximum limit for carbon residue in biodiesel is 0.05% by mass. The common source of carbon residues in biodiesel is due to an excessive level of total glycerin. This gives an indication of the amount of glycerides, FFAs, soaps, higher unsaturated fatty acids, inorganic impurities, additives used for, and catalyst residues remaining within the sample.

TABLE 3.4

Composition and Acid Values of Biodiesel

Biodiesel	Sunflower Biodiesel	Corn Biodiesel	Canola Biodiesel	Soybean Biodiesel	Refined Sunflower Biodiesel	Refined Canola Biodiesel	Refined Soybean Biodiesel
Methyl esters (w/w/%)	96.46	94.88	96.19	96.70	98.48	98.00	—
Monoglycerides (w/w%)	0.34	0.27	0.35	0.49	ND	ND	—
Diglycerides (w/w%)	0.44	1.64	0.78	ND	ND	ND	
Acid value (mgKOH/g)	0.14	0.21	0.20	0.18	—	0.30	0.38

Source: From Hu, J., Du, Z., Li, C., and Min, E., *Fuel*, 84, 1601–06, 2005. Reprinted with permission from Elsevier Publications.

Note: *ND*: Not detected.

3.4.14 Copper Strip Corrosion

Copper strip corrosion of biodiesel is evaluated as per ASTM D130 test method. The fuel lines of an injection system may be made of copper. Biodiesel may contain alcohols and acids if not purified well. Hence it is necessary to evaluate the corrosiveness of fuel. A polished copper strip is immersed in a sample of biodiesel and heated for a specified length of time at a specified temperature. The copper strip is removed, washed, and then compared to the ASTM certified corrosion standards. The presence of acids or sulfur containing compounds can tarnish the copper strip thus indicating the possibility of corrosion.

3.4.15 Sulfated Ash

Sulfated ash is the percentage of ash by mass in the mass of a sample. Sulfated ash in biodiesel is evaluated as per ASTM D 874. Ash that remains after the sample has been carbonized and residue subsequently treated with sulfuric acid and heated at 775°C, cooled, and weighed. This ensures the removal of catalyst (Na, K, etc.) used in biodiesel. Soluble metallic soap, catalysts, and abrasive solids are the possible sources of sulfated ash. Soluble metallic soaps have little effect on wear but may contribute to filter plugging and engine deposits. Abrasive solids and catalysts contribute to injector, fuel pump, piston and ring wear, and formation of engine deposits.

3.4.16 Acid Value

Acid number is specified to ensure proper aging properties of fuel and reflects the presence of FFAs used in the production of biodiesel. The acid number is an important parameter when feedstocks with high FFAs are used to produce biodiesel. The acid number is directly related to the FFA content and oxidation stability.

Acid number is the quantity of base (mg of KOH) required for neutralizing all acidic constituents present in 1 gram of the sample. It is measured as per EN 14104 test method. A sample dissolved in organic solvent (2.0 gm in 100 ml) is titrated against potassium hydroxide solution of known concentration and phenolphthalein as an indicator.

The acid number can provide an indication of the level of lubricant degradation while the fuel is in service. During the fuel injection process, more fuel returns from an injector than injected into the combustion chamber of the engine. The temperature of the return line fuel may reach more than 60°C and thus accelerates the degradation of biodiesel. High acid value leads to corrosion of engine parts, forms deposits in the fuel system, and affects the life of fuel pump.

3.4.17 Phosphorous

The level of phosphorous in biodiesel is measured as per ASTM D 4951 test method using plasma atomic emissions spectrometry. Phosphorous content in fuel depends upon feedstock. Using emission spectrograph emission standards of sample compared with reference. Phosphorous can damage catalytic converters used in emission control systems and causes injector fouling. Currently, in order to meet the latest emission norms automotive manufacturers are using catalytic converters and hence it is necessary to keep low levels of phosphorous in fuel.

3.4.18 Methanol/Ethanol Content

Alcohol is the major ingredient in the biodiesel production process. In general, the processor will recover excess methanol by a vacuum stripping process. Any methanol remaining after this process should be removed by a water washing process. Therefore, the residual alcohol level should be very low. Even 1% of methanol in biodiesel can lower the flash point from 170 to 40°C. Thereby, including flash point in the specification, the standards limit the alcohol to a very low level, less than 0.1%.

Alcohol content in the biodiesel is quantified as per EN 14110 test method. Sample is heated at 80°C to allow desorption of contained methanol into the gas phase. A defined part of the gas phase is injected into GC where methanol is detected with a flame ionization detector. Methanol amount is determined based on the peak observed and with reference to standards. Alcohol content in the biodiesel accelerates deterioration of rubber seals and gaskets, damage to fuel pumps and injector, and reduces the flash point.

3.4.19 Lubricity

Excessive wear due to lower lubricity shorten the life of fuel pumps and injectors. The lubricity of fuel depends on the crude source, refining process to reduce the sulfur content, and the type of additives used. The ball on cylinder lubricity evaluator (BOCLE) and high frequency reciprocating rig (HFRR) are used to evaluate the lubricity of fuel.

The HFRR method recommends that a limit of 460 micron wear scar diameter (WSD) for diesel. Lower wear scar diameter is an indication of better lubricity property of fuel. Even with 2% biodiesel mixed with diesel WSD comes down to 325 micron. Oil companies reduce the sulfur content to meet the Euro IV/V norms, and loss occurring due to reduction in sulfur level can be compensated by addition of biodiesel in diesel. Even 2% of biodiesel in diesel is sufficient to address the lubricity-related problems.

Hu et al. (2005) conducted studies to evaluate the lubricity properties of biodiesel (refined and unrefined) blended with diesel. Table 3.5 shows the

TABLE 3.5

Wear Scar Diameter of Biodiesel

Biodiesel	Sunflower Biodiesel	Corn Biodiesel	Canola Biodiesel	Soybean Biodiesel	Refined Sun flower	Refined Corn Flower	Refined Canola	Refined Soybean Biodiesel
WSD (μm)	429	366	351	375	528	567	543	540

Source: From Hu, J., Du, Z., Li, C., and Min, E., *Fuel*, 84, 1601–06, 2005. Reprinted with permission from Elsevier Publications.

WSD of diesel blended with 2% w/w biodiesel. When unrefined biodiesel was used as lubricity enhancers, the WSD value of diesel decreased significantly, from 735 to 351–429 microns, and was lower than the acceptable limit of 460 microns of EN590. When refined biodiesels were used, the WSD values decreased to 528–567 microns, but were still over the acceptable limit of 460 microns. It can obviously be seen that all the unrefined biodiesels showed improved lubricity properties over the refined biodiesels. It is expected that MG, DG, and TG are the most probable compounds that influence the lubricity properties of them.

3.4.20 Sodium and Potassium

The presence of Na and K is determined as per EN 14108 test method. Biodiesel (0.001–2 g) sample is diluted with solvent and is allowed to burn in the flame atomic absorption spectrometry. It detects Na and K at 589 nm and 766.5 nm wavelength, respectively. Sodium or potassium is used as catalyst for the base catalyzed transesterification process. These catalyzes are to be removed completely in the purification process, otherwise these may cause engine deposits and high abrasive wear levels. Abrasive solids can contribute to injector, fuel pump, piston, and ring wear, and also to engine deposits. Soluble metallic soaps have little effect on wear but may contribute to filter plugging and engine deposits. High levels of Na and K compounds settled on exhaust particulate collectors can create increased back pressure and increase the maintenance.

3.5 Biodiesel Quality Standards

Biodiesel processing and its quality are closely related. The processes used to convert the feedstock into biodiesel determine whether the product will meet the desired biodiesel production or not. The biodiesel quality can be influenced by several factors:

- Quality of feedstock
- Fatty acid composition of vegetable oils/animal fats
- Production process and ingredients used
- Postproduction parameters

If biodiesel met the desired limits, then only the biodiesel can be used in engines while maintaining the engine manufacturer's durability and reliability. Biodiesel should meet its specification when it is blended with diesel even in small percentages. Table 3.6 shows the biodiesel standards used in

TABLE 3.6

Biodiesel Standards

Country	Standards
Europe	EN 14214
United States	ASTM D 6751
Germany	DIN E 51601
India	IS 15607
Brazil	ANP 42
Japan	JASO M360

TABLE 3.7

Biodiesel Specifications ASTM D 6751

Property	Test Method	ASTM D6751-07b
Flash point (°C)(closed cup), min	D 93	130
Water and sediment (%v), max	D 2709	0.05
Sulfated ash (mass %), max.	D 874	0.02
Kinematic viscosity at 40°C (cSt)	D 445	1.9–6.0
Total sulfur (mass %), max	D 5453	0.05
Carbon residue (mass %), max	D 4530	0.05
Cetane number, min	D 613	47
Acid no (mg KOH/g), max	D 664	0.5
Copper strip corrosion, max	D 130	No.3
Free glycerin (mass %)	D 6584	0.02
Total glycerin (mass %)	D 6584	0.24
Phosphorus content (% mass), max	D 4951	0.001
Distillation temperature (90% recovered (°C)), max	D 1160	360
Calcium and magnesium combined, max	EN 14538	5
Methanol content, % max	EN 14110	0.2
Oxidation stability, hours	EN 14112	3

Source: Annual book of ASTM standards. *ASTM international, United States*. Volume 5.04, 2008.

different countries and Tables 3.7 through 3.9 depict biodiesel specifications of America, Europe, and India, respectively.

3.6 Engine Tests

The properties of biodiesel are almost similar to those of diesel and hence it can be blended with diesel in any percentages and used to run the diesel engines. Various engine tests were conducted to study the engine

TABLE 3.8
Biodiesel Specification EN 14214

Property	Test Method	EN 14214
Flash point (°C)(closed cup), min	ISO 3679	120
Water content, max, mg/kg	12937	500
Sulfated ash (mass %), max.	EN ISO 3987	0.02
Kinematic viscosity at 40°C (cSt)	EN 3104	3.5–5.0
Density, kg/m^3	EN 3675	860–890
Cetane number, min	EN ISO 5165	51
Acid number (mg KOH/g), max	EN ISO 14104	0.5
Iodine number	EN ISO 14111	120
Copper strip corrosion, max	EN ISO 2160	1
Free glycerin (mass %)	EN ISO 14105	0.02
Total glycerin (mass %)	EN ISO 14105	0.25
Phosphorus content (% mass), max	EN ISO 14107	10
Total contamination, max, mg/kg	EN ISO 12662	24
Methanol content, % max	EN ISO 14110	0.2
Oxidation stability, hours	EN ISO 14112	6
Ester content, min, %	EN 14103	96.5

Source: EN 14214. European standard. European committee for standardization, 2002.

performance, emissions, and durability by academic institutes, automotive industries, and oil companies. These studies reported that biodiesel blends or biodiesel-fueled engines could reduce the exhaust emissions remarkably.

3.6.1 Engine Performance and Emission Studies

Engine performance parameters such as thermal efficiency, specific fuel consumption were studied by many researchers and engineers. The observations of these studies were found to be similar. A typical study conducted on a naturally aspirated diesel engine using rubber seed oil methyl esters follows. Brake thermal efficiency of the engine observed to be increased with an increase in applied load. This is due to the reduction in heat loss and increase in power developed with increase in load. The maximum brake thermal efficiency obtained is about 28% for B10, which is higher than that of diesel (25%). At a lower percentage concentration of biodiesel, brake thermal efficiency of the engine is improved (Figure 3.14). This is due to the additional lubricity provided by the biodiesel. The molecules of biodiesel (i.e., methyl esters of the oil) contain some amount of oxygen, which takes part in the combustion process. However, at the higher percentage concentration of biodiesel, the brake thermal efficiency decreases as a function in concentration of blend. This lower brake thermal efficiency was obtained for B100,

TABLE 3.9

Biodiesel Specification IS 15607: 2005

S.No.	Characteristics	Unit	Requirement	Test Method ISO, ASTM, EN/ IS 1448
1	Density at 15°C	kg/m^3	860–900	ISO 3675/P 32
2	Kinematic viscosity at 40°C	cSt	2.5–6.0	ISO 3104/P25
3	Flash point (closed cup), min	°C	120	P21
4	Sulfur, max	mg/kg	50	D 5443/P83
5	Carbon residue (Ramsbottom),max	%m	0.05	D 4530
6	Sulfated ash, max	%m	0.02	ISO 6245/P4
7	Water content, max	mg/kg	500	D 2709/P40
8	Total contamination, max	mg/kg	24	EN 12662
9	Copper corrosion 3 hours @50°C, max	—	1	ISO 2160/P15
10	Cetane number, min	—	51	ISO 5156/P9
11	Acid value, max	mg KOH/g	0.50	P1
12	Methanol, max	%m	0.20	EN 14110
13	Ethanol, max	%m	0.20	—
14	Ester content, min	%m	96.5	EN 14103
15	Free glycerol, max	%m	0.02	D 6584
16	Total glycerol, max	%m	0.25	D 6584
17	Phosphorous, max	mg/kg	10.0	D 4951
18	Sodium and potassium, max	mg/kg	To report	EN 14108
19	Calcium and magnesium, max	mg/kg	To report	—
20	Iodine value	—	To report	EN 14104
21	Oxidation stability at 110°C, min	Hr	6	EN 14112

Source: IS 15607. *Biodiesel (B100) blends stock for diesel fuel specification.* Bureau of Indian Standards, New Delhi, India, 2005.

which could be due to the reduction in calorific value and increase in fuel consumption as compared to B10 or diesel.

Specific fuel consumption of an engine decreases with increase in the load. This is due to the higher percentage increase in brake power with load as compared to an increase in fuel consumption. Using a lower percentage of biodiesel in blends, the specific fuel consumption of the engine is lower than that of diesel for all loads. The variation of specific fuel consumption with respect to brake mean effective pressure of different fuels is shown in Figure 3.15. In case of B50–B100, the specific fuel consumption is found to be higher than that of diesel. At maximum load conditions, specific fuel consumption of

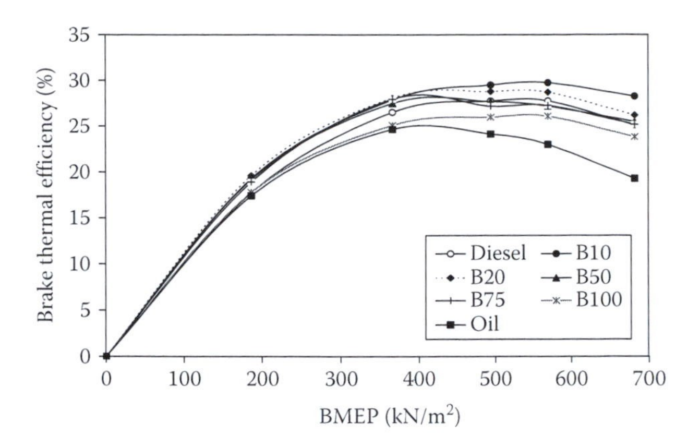

FIGURE 3.14
Brake thermal efficiency of biodiesel–diesel blends.

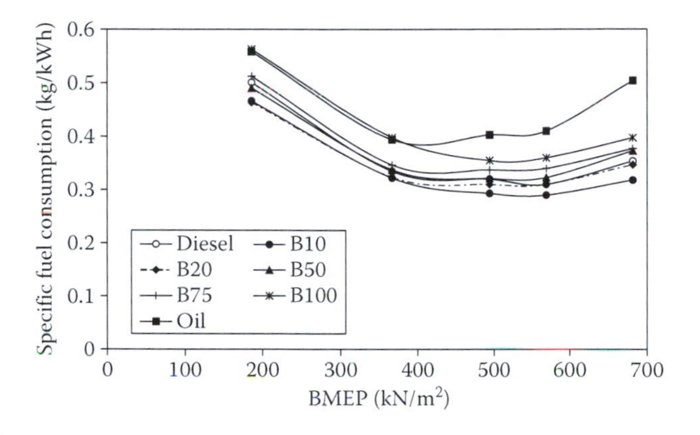

FIGURE 3.15
Specific fuel consumption of biodiesel–diesel engine.

neat biodiesel is about 12% higher than that of diesel. The calorific value of rubber seed oil biodiesel is approximately 14% lower than diesel. Calorific value of biodiesel blends decreases with an increase in percentage concentration of biodiesel. Hence, specific fuel consumption of the higher percentage of biodiesel in blends increases as compared to that of diesel.

Di et al. (2009) conducted engine tests to evaluate the emission reduction potential of biodiesel (Figure 3.16). They used various blends of biodiesel produced from waste cooking oil to ultra low sulfur diesel in a four cylinder diesel engine. Biodiesel blends 19.4%, 39.4%, 59.4%, and 79.4% are denoted as blend 1, 2, 3, and 4, respectively. Hydrocarbon emissions decrease with increase of engine load due to an increase in higher temperature associated with higher engine load. The hydrocarbon emissions decrease with increase in biodiesel percentage in the blend. Lower volatility of biodiesel reduces

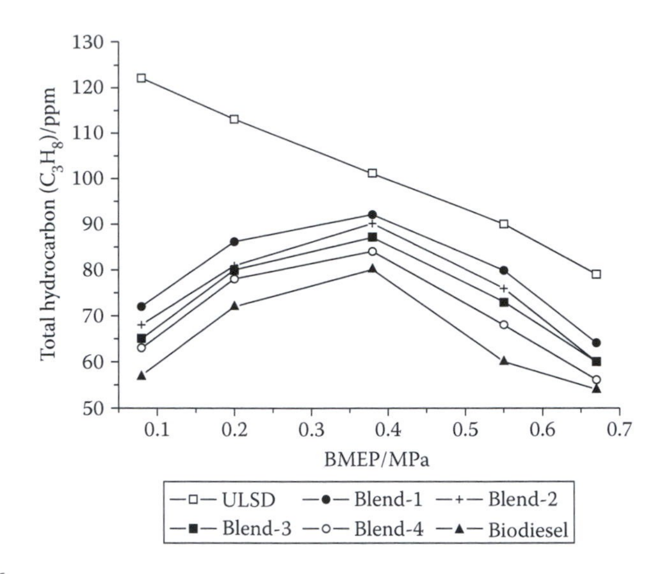

FIGURE 3.16
Hydrocarbon emissions of biodiesel blends. (From Di, Y., Cheung, C. S., and Huang, Z., *Science of the Total Environment*, 407, 835–46, 2009. Reprinted with permission from Elsevier Publications.)

the HC emissions at low load conditions. Lapuerta, Armas, and Jose (2008) reported that the U.S. EPA obtained the following equation for the total hydrocarbon emissions

$$\frac{\text{THC}}{\text{THC}_{\text{diesel}}} = e^{-0.011199\% B}.$$

The CO concentration in the exhaust emission is negligibly small when a homogeneous mixture is burned at stoichiometric air-fuel ratio or on the lean side stoichiometric. With the addition of biodiesel, CO emissions also decrease. Biodiesel itself contains oxygen that promotes complete combustion of fuel and hence reduces the CO emissions (Figure 3.17). Lapuerta, Armas, and Jose (2008) reported that the U.S. EPA obtained the following equation for the CO emissions

$$\frac{\text{CO}}{\text{CO}_{\text{diesel}}} = e^{-0.006561\% B}.$$

Nitrogen oxides emission increases with an increase in load for all load conditions. Biodiesel blends increase NO_x emissions at all load conditions (Figure 3.18). Biodiesel that contains oxygen is taking part in combustion

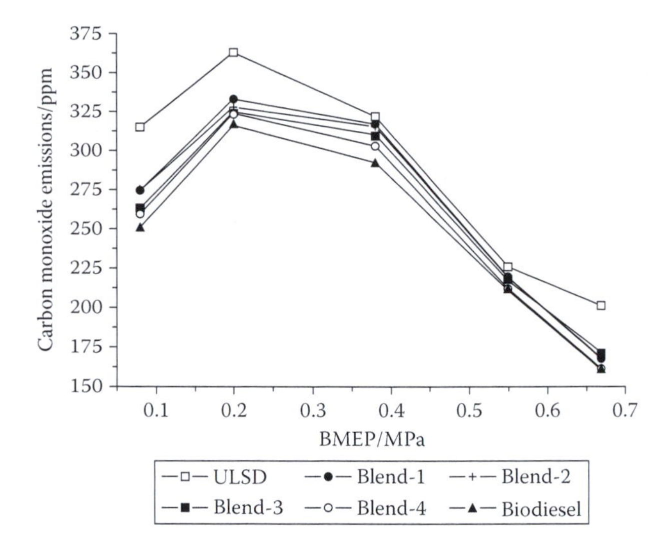

FIGURE 3.17
Carbon dioxide emissions of biodiesel blends. (From Di, Y., Cheung, C. S., and Huang, Z., *Science of the Total Environment,* 407, 835–46, 2009. Reprinted with permission from Elsevier Publications.)

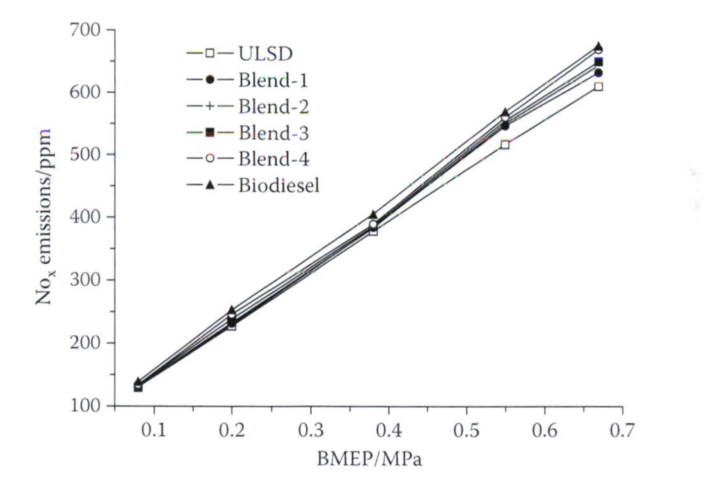

FIGURE 3.18
NO_x emissions of biodiesel blends. (From Di, Y., Cheung, C. S., and Huang, Z., *Science of the Total Environment,* 407, 835–46, 2009. Reprinted with permission from Elsevier Publications.)

and hence increases the temperature of the combustion chamber. Lapuerta, Armas, and Jose (2008) reported that the U.S. EPA obtained the following equation for NO_x emissions,

$$\frac{\mathrm{NO}}{\mathrm{NOx}_{\mathrm{diesel}}} = e^{-0.0009794\%B}.$$

Biodiesel blends reduce both smoke and particle mass concentration of exhaust emissions. Lapuerta, Armas, and Jose (2008) reported that the U.S. EPA obtained the following equation for PM emissions.

$$\frac{\mathrm{PM}}{\mathrm{PM}_{\mathrm{diesel}}} = e^{-0.006384\%B}.$$

Reduction in PM emissions is associated with reduction in soot and sulfate formation. Biodiesel has about 11–14% oxygen content and no sulfur content. Reduction in sulfur in fuel leads to reduction in sulfate formation. This leads to reduction in soot formation and reduces PM emissions (Figure 3.19). Aromatics, polycyclic aromatic hydrocarbons (PAH), are soot precursors and sources of particulate-phase PAH. Biodiesel contains no aromatics and hence this leads to the reduction in PM emission.

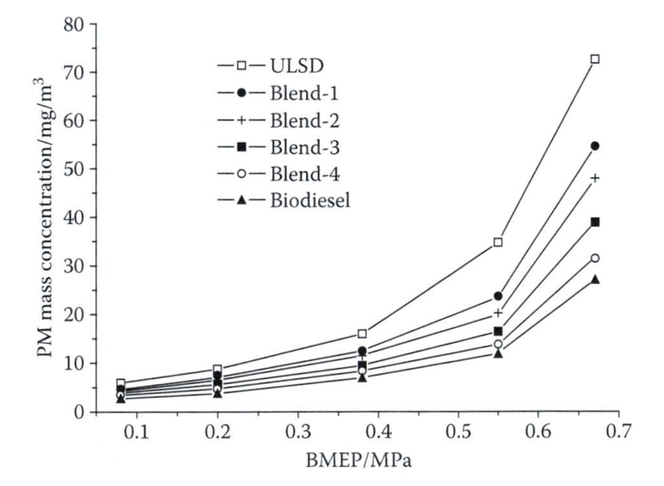

FIGURE 3.19
PM emissions of biodiesel blends. (From Di, Y., Cheung, C. S., and Huang, Z., *Science of the Total Environment*, 407, 835–46, 2009. Reprinted with permission from Elsevier Publications.)

The GMD increases with an increase in load. At high loads, more fuel is consumed in the diffusion mode and hence more particles are formed. With an increase in the number of particles, coagulation rate increases and hence larger particles are formed leading to an increase of GMD. Biodiesel blends favors reduction in soot nuclei and hence particulate number concentration. However, the coagulation and agglomeration of the small particles to form larger particles will also be slowed down, leading to an increase of the smaller-sized particles.

3.6.2 Engine Durability Studies

Engine-related problems can be studied by conducting an engine test for the long duration or endurance tests. In general, biodiesel blends up to 5% do not create engine-related problems. Using higher percentage of biodiesel in engines, creates problems in injectors because the injectors are designed for diesel and materials used also are compatible with diesel only. Poor quality of biodiesel may contain organic acids, water, free glycerol, total glycerol, and any other contaminants that may polymerize and attack engine components. Fuel injectors are prone to be subjected to injector nozzle clogging problems. Though most of fuel injector manufacturers/auto manufactures extend the warranty of parts with the use of biodiesel (B5); it is necessary to store the biodiesel from extreme temperatures in order to avoid oxidation of fuel. The common problems observed on engine components with biodiesel usage are shown in Table 3.10.

TABLE 3.10

Effect of Biodiesel on Engine Components

Fuel Characteristics	Effect
Free methanol	Lowers flash point and corrodes fuel lines and parts made of aluminum and zinc
Water	Increases the electrical conductivity of fuel and corrosion of parts Encourages bacteria growth
Free glycerin	Soaks filters and injector coking Corrodes metal parts
Free fatty acids	Forms organic compounds and salts of organic acids that cause filter plugging and corrosion of fuel injectors
Higher viscosity	Excessive heat generation in rotary distributor pumps and cause higher stress Chances of pump seizures and poor nozzle spray atomization
Catalysts	Na, K compounds cause injector nozzle blockage
Polymerization products	Forms deposits and causes filter plugging
Solid impurities	Creates lubricity problems and hence reduces the service life

The IC engine has many moving components that are normally subjected to wear process including piston, piston ring, bearing, crank shaft, and engine valves. The metal particles upon wear go to the engine lubrication system and will be in suspended form. Agarwal, Bijwe, and Das (2003) carried out detailed investigations to assess the wear of engine parts using biodiesel fuel. They investigated the effect of long-term engine operation with engine wear for a blend of 20% LOME and diesel for wear analysis and lubricating oil analysis. During this test, each engine was run for 512 hours at rated speed and different load conditions.

Fuel dilution is a direct consequence of clearance between piston rings and cylinder liner. If the wear between piston rings and the cylinder increases, more will clear and hence higher fuel dilution. Biodiesel helps in protecting the piston rings from wearing out more effectively. Relatively higher viscosity of biodiesel helps in plugging the clearance between piston rings and cylinder liner effectively, thus reducing blow-by losses and fuel dilution of lubricating oil. Biodiesel has thus proved to be more effective in protecting the moving parts of the engine.

Agarwal (2007) analyzed the concentration of various metal present in the lubricating oil sample by atomic absorption spectroscopy (AAS) for quantitative and qualitative analysis. They reported that AAS studies on lube oils had indicated that biodiesel fuel led to lesser wear of engine moving parts in terms of a lesser amount of metallic debris (such as Fe, Cu, Zn, Mg, Cr, Pb, and Co) present in lube oil samples. These wear metals originated from contact between two moving metal parts. It has been reported that wear was reduced with the use of biodiesel.

3.7 Challenges for Biodiesel

1. Biodiesel-fueled vehicles increase the NO_x emissions.
2. Studies to be conducted when biodiesel used with the lower sulfur fuel will produce any emission benefits in the latest generation of engines.
3. While vehicles that run on B20 and displace some amount of petroleum, they are still dependent on diesel fuel to operate.
4. Biodiesel is significantly more expensive than diesel fuel and hence government incentives are required to promote it.
5. Because there is less energy in a gallon of biodiesel than in a gallon of petroleum diesel, the driving range of vehicles operating on biodiesel blends is less.
6. Requirement of warranty on biodiesel fueled vehicles from manufacturers.

References

Agarwal, A. K. 2007. Biofuels (alcohols and biodiesel) applications as fuels for internal combustion engines. *Progress in Energy and Combustion Science* 33:233–71.

Agarwal, A. K., J. Bijwe, and L. M. Das. 2003. Effect of biodiesel utilization of wear of vital parts in compression ignition engines. *Journal of Power Engineering* 125:604–11.

Annual book of ASTM standards. 2008. *Petroleum products and Lubricants Vol. 5.04*, ASTM International, West Conshohocken, PA.

Barnwal, B. K., and M. P. Sharma. 2005. Prospects of biodiesel production from vegetable oils in India. *Renewable and Sustainable Energy Reviews* 9:363–78.

Canakci, M., and J. Van Gerpan. 1999. Biodiesel production via acid catalysis. *ASAE* 42:1203–10.

Demirbas, A. 2003. Biodiesel fuels from vegetable oils via catalytic and non-catalytic supercritical alcohol transesterification and other methods: A survey. *Energy Conversion and Management* 44:2093–2109.

Demirbas, A. 2009. Progress and recent trends in biodiesel fuels. *Energy Conversion and Management* 50:14–34.

Di, Y., C. S. Cheung, and Z. Huang. 2009. Experimental investigation on regulated and unregulated emissions of a diesel engine fueled with ultra-low sulfur diesel fuel blended with biodiesel from waste cooking oil. *Science of the Total Environment* 407:835–46.

Du, W., Y. Xu, D. Liu, and J. Zeng. 2004. Comparative study on lipase-catalyzed transformation of soybean oil for biodiesel production with different acyl acceptors. *Journal of Molecular Catalysis B: Enzymatic* 30:125–29.

EN 14214. 2002. European standard. European committee for standardization, European standard EN 14214 Biodiesel Specification, Berlin: DIN Deutsches Institut for Normung.

Hu, J., Z. Du, C. Li, and E. Min. 2005. Study on the lubrication properties of biodiesel as fuel lubricity enhancers. *Fuel* 84:1601–6.

IS 15607. 2005. *Biodiesel (B100) blends stock for diesel fuel specification*. Bureau of Indian Standards, New Delhi, India.

Ji., J., J. Wang, Y. Li, Y. Yu, and Z. Xu. 2006. Preparation of biodiesel with the help of ultrasonic and hydrodynamic cavitations. *Ultrasonics Sonochemistry* 44:411–14.

Kusdiana, D., and S. Saka. 2004. Effect of water on biodiesel fuel production by supercritical methanol treatment. *Bioresource Technology* 91:289–95.

Lapuerta, M., O. Armas, and R. F. Jose. 2008. Effect of biodiesel fuels on diesel engine emissions. *Progress in Energy Combustion* 34:198–223.

Lee, S. Y., M. A. Hubble, and S. Saka. 2006. Prospects for biodiesel as by product of wood pulping—A review. *Bioresource* 1:150–71.

Ma, F., L. D. Clements, and M. A. Hanna. 1998. The effects of catalyst, free fatty acids, and water on transesterification of beef tallow. *ASAE* 41:1261–4.

Ma, F., and M. A. Hanna. 1999. Biodiesel production: A review. *Bioresource Technology* 70:1–15.

Meher, L. C., D. Vidya Sagar, and S. N. Naik. 2006. Technical aspects of biodiesel production by transesterification—A review. *Renewable and Sustainable Energy Reviews* 10:248–68.

Noureddini, H., X. Gao, and R. S. Philkana. 2005. Immobilized pseudomonas cepacia lipase for biodiesel fuel production from soybean oil. *Bioresource Technology* 96:769–77.

Ramadhas, A. S., S. Jayaraj, and C. Muraleedharan. 2005. Biodiesel production from high FFA rubber seed oil. *Fuel* 84:335–40.

Saka, S., and D. Kusdiana. 2001. Biodiesel fuel from rapeseed oil as prepared in supercritical methanol. *Fuel* 80:225–31.

Sarin, R., M. Sharma, S. Sinharay, and R. K. Malhotra. 2007. Jatropha–Palm biodiesel blends: An optimum mix for Asia. *Fuel* 86:1365–71.

Sharma, Y. C., B. Singh, and S. N. Upadhyay. 2008. Advancements in development and characterization, of biodiesel: A review. *Fuel* 87:2355–73.

Stavarache, C., M. Vinatoru, R. Nishimura, and Y. Maeda. 2005. Fatty acids methyl esters from vegetable oil by means of ultrasonic energy. *Ultrasonics Sonochemistry* 12 (5): 367–72.

Stavarache, C., M.Vinatoru, Y. Maeda, and H. Bandow. 2007. Ultrasonically driven continuous process for vegetable oil transesterification. *Ultrasonics Sonochemistry* 14:413–17.

Tomasevic, A. V., and S. S. Marinkovic. 2003. Methanolysis of used frying oils. *Fuel Process Technology* 81:1–6.

Van Gerpen, J. 2005. Biodiesel processing and production. *Fuel Processing Technology* 86:109–1107.

Warabi. Y., D. Kusdiana, and S. Saka. 2004. Reactivity of triglycerides and fatty acids of rapeseed oil in supercritical alcohols. *Bioresource Technology* 91:283–87.

4

Methanol

Mustafa Canakci and Oguzhan Ilgen

CONTENTS

4.1 Introduction

Diesel engines are widely preferred to gasoline engines for heavy-duty applications in the agriculture, construction, industrial, and on-highway transport sectors because of their higher thermal efficiency. However, they are also known to be major sources of the emissions such as particulate matter (PM), soot, smoke, and nitrogen oxides (NO_x). On the other hand, emission regulations in many countries have caused the research on improving engine fuel economy and reducing exhaust emissions due to increasing concern of environmental protection and fuel shortage. Using alternative fuel is one of the most attractive methods to improve fuel economy and to reduce environmental pollution. Among the alternative fuels, alcohols (such as methanol and ethanol), vegetable oils, animal fats, biodiesel, and liquefied petroleum gas (LPG) are receiving interest (Chao et al. 2001; Jothi, Nagarajan, and Renganarayanan 2007; Nwafor 2003; Ozsezen, Canakci, and Sayin 2008; Sayin and Uslu 2008; Senthil et al. 2005). These alternative fuels are largely environment-friendly, but they need to be evaluated for their advantages and disadvantages if they were used in engine applications. Despite its low cetane number and poor solubility in diesel fuel, methanol is one of the attractive alternative fuels because it is renewable and oxygenated. Practically, to reduce engine emissions without engine modification, adding oxygenated compounds to fuels seems to be more attractive. Therefore, methanol may provide potential to reduce emissions in diesel engines. This chapter reviews the potential, production methods, and fuel properties of methanol as well as its engine performance, combustion, and emissions.

4.2 Potential of Methanol

Alternative fuels can be derived from noncrude oil resources. Crude oils are petroleum-based fuels. Among the popular alternative fuels are biodiesel, natural gas, propane, methanol, biodiesel, ethanol, and hydrogen (Zhu 1997). The use of alternative fuels in the future is inevitable with rising oil prices and global warming being a dominant environmental issue. The leading goals for both energy security and the clean-air projects have increased interest in the worldwide utilizations of alternative fuels in burners and engines (Liao et al. 2006).

Methanol is the chemically simplest alcohol, containing one carbon atom per molecule. It is a toxic, colorless, tasteless liquid with a very faint odor and commonly known as "wood alcohol." Because it is produced as a liquid, methanol is stored and handled like gasoline. In general, methanol is currently made from natural gas, but it can also be made from a wide range of renewable sources, such as wood, waste paper, coal, and biomass. Methanol is a good candidate for alternative fuel investigations because of its abundances, physical, and chemical properties (Hamid and Ali 2004; Herz 2001).

As an alternative fuel, methanol is receiving more attention and has become a potential application in both the transportation and the power generation sectors (Hong et al. 2005). Methanol has the advantage of being a liquid fuel that can be synthesized from gasification and reforming in a variety of feedstocks like coal and natural gas using well-established thermochemical technology. Methanol can also be produced from a wide range of biomass including municipal and industrial waste. It can thereby serve as a low carbon emitting fuel. Methanol can be an important option for the replacement of gasoline (Bromberg and Cohn 2008). The main uses of methanol as an alternative fuel are: direct use: blend with diesel or gasoline and indirect use: conversion of methanol to dimethyl ether, ingredient in production of biodiesel and hydrogen for use in fuel cell vehicles (Herz 2001).

4.2.1 Methanol: Gasoline Blends

Methanol may be used directly as an engine fuel just as LPG and ethanol. The direct use of methanol as an engine fuel has advantages and disadvantages. The relative very high-latent heat of vaporization of methanol results in a lower burning temperature in the engine cylinders as it does for ethanol. The direct use of methanol as an engine fuel in passenger vehicles would require nontrivial engine modifications and substantial changes in the lubrication system. Due to its high Reid Vapor Pressure (RVP) that is a measure of affected volatility of blended gasoline, use of methanol as gasoline blending is limited even though it has a high octane rating and is an excellent candidate of oxygenated hydrocarbons (Ingersoll 1996; Lee 1997; Lee Speight, and Loyalka 2007). When pure methanol is used, cold start problems can occur because it lacks the highly volatile compounds (butane, iso-butane, propane) that provide ignitable vapor to the engine even under the most frigid conditions. The addition of more volatile components to methanol is usually the preferred solution (Cheng and Kung 1994; Olah, Goeppert, and Prakash 2006).

4.2.2 Methanol: Diesel Blends

Methanol and diesel are not very miscible, which makes it difficult to mix the PM together as a diesel engine fuel. Methanol–diesel blends gave good performance when the amount of methanol in the mixed fuel does not exceed 30% by weight (Shi et al. 1983) and combustion characteristics when the maximum methanol mass fraction was 20% by weight (Huang et al. 2004c). Methanol can be blended with diesel fuel although its corrosive nature creates a need for caution in the design of the engine and fuel system like in ethanol. Methanol has half the volumetric energy density relative to gasoline or diesel. It is usually mixed in percentages ranging from 5 to 85%. The combination rate of methanol in the fuel is 85%, which is called M85. At the same time, the aldehyde compound that comes along with burning methanol forms a strong acid. Therefore, the researchers should pay more attention to this fuel (Pons 2009; Tzeng, Lin, and Opricovic 2005).

4.2.3 Dimethyl Ether

Methanol can be used to make methyl tertiary-butyl ether (MTBE), which is blended with gasoline to enhance octane and create cleaner burning fuel. But MTBE production and use also has disadvantages because it contaminates groundwater (http://www.ceert.org 2009). The closely related derivative of methanol, dimethyl ether (DME) is a highly desirable alternative fuel. It is generally produced by dehydration of methanol. Also it has been known as an ultra-clean fuel, which can be used in diesel engines, households, power generation, and for other purposes. Due to the huge market potential, the research on DME synthesis and its utilization have been attracting more and more interest. DME is also produced mainly from synthesis gas through methanol synthesis and methanol dehydration known as the two-step process (Jia, Tan, and Han 2006). Because of its high cetane number and favorable combustion properties, it is a particularly effective fuel for diesel engines. DME blends well with gasoline or diesel to be used as fuels in internal combustion engines or electricity generators. DME is also a potential substitute for liquefied natural gas (LNG) and LPG for heating homes and in industrial uses (Olah and Prakash 2006). Another methanol derivative is dimethyl carbonate (DMC), which can be obtained by converting methanol with phosgene or by oxidative carbonylation of the methanol. DMC has a high cetane rating, and can be blended into diesel fuel in a concentration up to 10%, reducing fuel viscosity and improving emissions (Olah and Prakash 2006).

4.2.4 Biodiesel Production Ingredient

Another way to use methanol in diesel engines and generators is through biodiesel. Biodiesel can be produced from a large variety of vegetable oils and animal fats that react with methanol in a transesterification process to produce biodiesel. Biodiesel can be blended without major problems with regular diesel oil in any proportion (Olah, Goeppert, and Prakash 2006). Methanol and its derivatives (such as DME, DMC, and biodiesel) have many existing and potential uses. They can be used, for example, as a substitute fuel for diesel engine powered cars with only minor modifications to the existing engines and fuel systems (Olah and Prakash 2006).

4.2.5 Fuel Cell Fuel

Methanol can also be used in fuel cells—for fuel cell vehicles—which are considered to be the best alternative to internal combustion engines in the transportation field. One of the most efficient uses of methanol is in fuel cells, particularly in direct methanol fuel cells (DMFC), in which methanol is directly oxidized with air to carbon dioxide (CO_2) and H_2O while producing electricity (Olah and Prakash 2009). DMFC systems are currently being developed as battery replacements for the portable power market. With the exception of compressed hydrogen and the DMFC, all fuel cell vehicle designs require some kind of steam reforming or partial oxidation to release the hydrogen in the fuel. These processes can create NO_x emissions that, even in very small quantities, can compromise air quality when multiplied by millions of vehicles in use. NO_x is an essential precursor to the formation of ground level ozone, or smog. Because the DMFC breaks methanol into hydrogen and oxygen directly without requiring steam reforming or partial oxidation, there are no NO_x emissions from the vehicle (Nowell 2009).

4.3 Methanol Production

Methanol is one of the top 10 chemicals produced globally (Huber, Lborra, and Corma 2006) because it can be produced from various feedstocks, including fossil fuels (e.g., natural gas, crude oil, coal) and renewable resources (e.g., wood and municipal solid wastes).

4.3.1 Methanol Production Via Synthesis Gas

Synthesis gas can be used as a feedstock for a range of products including methanol. Methanol production via synthesis gas depends on the feedstock of choice in varying proportions. The synthesis gas may come from any number of different routes, including coal gasification, partial oxidation of heavy oils, steam reforming of natural gas (with or without CO_2 injection), steam reforming of LPG feedstocks and naphtha, combined or oxygen-enhanced reforming, and heat-exchange reforming. This list names the principal routes by which methanol synthesis gas may be produced (Cheng and Kung 1994).

Methanol was first produced as a by-product in the manufacture of charcoal through the destructive distillation of wood, with yields of 12–24 litter per ton of wood (National Research Council 1983). Methanol can be produced in a number of different ways, but currently most of the methanol produced is from natural gas. The gas is first compressed and then purified by removing sulfur compounds. The purified natural gas is saturated with heated water. The mixed natural gas and water vapor then goes to the reformer to be partially converted to synthesis gas, a mixture of CO_2, carbon monoxide (CO), and hydrogen (H_2). The synthesis gas undergoes a second step under high temperatures and pressures to combine CO and H_2 to produce methanol. Often times additional CO_2 is added in this step for more methanol end product (http://www.methanol.org 2004; Bradley 2000; Roan et al. 2004). In principle, many carbon-containing materials may be substituted for natural gas as starting materials. These include coal, lignite, and even municipal wastes in addition to wood. Each of these raw materials, however, must first be converted to synthesis gas; for this step, each alternative feedstock requires process modifications that increase capital investment costs over those required for natural gas (National Research Council 1983).

Natural gas is the largest source of synthesis gas. Methane (CH_4) is the chief constituent of natural gas. Methanol is made from CH_4 in a series of three reactions.

- Steam reforming reaction

$$CH_4 + H_2O \leftrightarrow CO + 3H_2 \qquad \Delta H\text{r} = 206 \text{ kJ/mol}$$

  In steam reforming, CH_4 reacts in a highly endothermic reaction with steam over a catalyst, typically based on nickel, at high temperatures (800–1000°C, 20–30 atm) to form CO and H_2.

- Water–gas shift (WGS) reaction

$$CO + H_2O \leftrightarrow CO_2 + H_2 \qquad \Delta H\text{r} = 206 \text{ kJ/mol}$$

  A part of the CO formed reacts consequently with steam in the WGS reaction to yield more H_2 and also CO_2. The gas obtained is thus a mixture of H_2, CO, and CO_2. The compressed synthesis gas enters the converter containing copper zinc and catalyst and the methanol synthesis occurs according to the methanol synthesis reaction.

- Methanol synthesis reaction

$$2H_2 + CO \leftrightarrow CH_3OH \qquad \Delta H\text{r} = -92 \text{ kJ/mol}$$

  If CO is used up in the methanol synthesis reaction the WGS reverses producing more CO.

$$H_2 + CO_2 \leftrightarrow CO + H_2O \qquad \Delta H\text{r} = 41 \text{ kJ/mol}$$

This reaction combines to produce approximately 40% conversion of CO to methanol. The gas mixture is cooled and methanol and water condense out. The remaining gas is returned to the circulator, mixed with incoming compressed synthesis gas and recycled through the methanol converter. Hence, the overall reactions by which methanol are produced from synthesis gas may be summarized into the following equation (http://nzic.org.nz 2009; Olah, Goeppert, and Prakash 2006):

$$CO_2 + CO + 5H_2 \rightarrow 2CH_3OH + H_2O + \text{heat.}$$

Synthesis gas is usually made from CH_4 or natural gas in a process called CO_2 or dry reforming, because it does not involve steam. This reaction is more endothermic ($\Delta H = 206$ kJ/mol) than steam reforming with a reaction enthalpy of $\Delta H = 246$ kJ/mol (Bhattacharyya, Chang, and Schumacher 1998; Olah, Goeppert, and Prakash 2006):

$$CH_4 + CO_2 \leftrightarrow 2CO + 2H_2 \qquad \Delta H_r = 246 \text{ kJ/mol.}$$

Another way of making synthesis gas from natural gas is the direct oxidation of CH_4. This reaction is exothermic and H_2 to CO ratio of the product is more desirable (2:1) for downstream process (Bhattacharyya, Chang, and Schumacher 1998):

$$CH_4 + \tfrac{1}{2} O_2 \leftrightarrow CO + 2H_2 \qquad \Delta H_r = -39 \text{ kJ/mol.}$$

The production of methanol from coal is not that simple. Coal contains many other compounds and impurities that would interfere with the methanol synthesis process. Coal often contains such compounds as nitrogen, sulfur, ash, oxygen, and water. Before coal can be gasified, it must first be dried. Methanol production from coal involves several steps that are coal gasification, acid gas removal, WGS reaction, methanol synthesis, and methanol refining.

The first step in the production of methanol is the gasification of coal to CO and H_2. The overall reaction for coal gasification is

$$C_xH_y + (x/2)\, O_2 \rightarrow x\, CO + (y/2)\, H_2,$$

with x and y depending on the actual coal composition. The main by-product of coal gasification is acid gas, which is composed of hydrogen sulfide (H_2S) and CO_2. The H_2S needs to be removed from the synthesis gas.

Sulfur recovery can be applied via a method known as the Claus process; products that result are either liquid or solid elemental sulfur, or sulfuric acid. A modified Claus process involves a two-stage process. The first stage involves the combustion of one-third the total H_2S to form SO_2 and water; the second is a low temperature catalytic stage that involves the reaction of SO_2

with the remaining H_2S to ultimately yield water and elemental sulfur. The reaction is presented as follows:

$$H_2S + \frac{3}{2}O_2 \leftrightarrow SO_2 + H_2O$$

$$2H_2S + SO_2 \leftrightarrow 2H_2O + \frac{3}{8}S_8$$

$$3H_2S + \frac{3}{2}O_2 \leftrightarrow 3H_2O + \frac{3}{8}S_8.$$

The following step of the WGS reaction usually utilizes a high-temperature shift reaction with Fe_3O_4 as a catalyst. The purpose of the WGS reaction is to adjust the ratio of H_2 to CO to 2:1. In order to synthesize methanol, the synthesis gas needs to have a composition of 2 parts H_2 to 1 part CO. The WGS reaction moves the composition of synthesis gas to this desired ratio. In the last step, CO and H_2 produce methanol by the methanol synthesis reaction. The most widely used catalyst is a mixture of copper, zinc oxide, and alumina, and we can use it to catalyze the production of methanol with better selectivity. The WGS and methanol synthesis reactions are represented previously (Chau 2008).

4.3.2 Methanol from Renewable Resources

Biomass is unique in providing the only renewable source of fixed carbon, which is an essential ingredient in meeting many of our fuel and consumer goods requirements. Biomass is an organic material, such as urban wood wastes, primary mill residues, forest residues, agricultural residues, and dedicated energy crops (e.g., sugar cane and sugar beets) that can be made into fuel. Biomass can be converted to synthesis gas by a process called partial oxidation and later converted to methanol (De Alwis, Mohamad, and Mehrotra 2009).

While most methanol are produced from natural gas and coal, technologies for the production of methanol from renewable feedstocks are already in commercial use, with prices approaching that of conventional feedstock production (http://www.methanol.org 2004). For more than 350 years, scientists have been converting wood into methanol. Historically, however, low efficiency rates from wood alcohol extractions made methanol extraction economically unrealistic. But today's technology allows foresters to convert tree waste into methanol at return rates of up to 50%. Methanol is also a more efficient product to produce, because most of the chemical compounds found in wood are converted. One ton of dry wood could produce up to 186 gallons of methanol, which represents an efficiency rate of nearly 50%. While methanol can also be produced from natural gas, using wood as a methanol source could prove more renewable and economically viable than using natural gas (http://www.vmanswers.com 2009).

The biomass alternative would make methanol a renewable resource by the following simplified reaction:

$$\text{biomass} \rightarrow \text{synthesis gas } (CO, H_2) \rightarrow CH_3OH.$$

Biomass can be converted to synthesis gas by a process called partial oxidation and later converted to methanol. In the first step biomass undergoes gasification to produce synthesis gas. Carbon source is reacted with steam (or steam and oxygen) at very high temperatures to produce CO, and H_2 in the gasification is a process that may be summarized into the following equations (Ebbeson, Stokes, and Stokes 2000; Kaneko et al. 2003; Watkins 2008):

$$CH_xO_y + H_2O \leftrightarrow CO + H_2 + CO_2.$$

Reaction can also be carried out as

$$CH_xO_y + O_2 + H_2O \leftrightarrow CO + H_2 + CO_2,$$

where the O_2 burns some of the biomass to supply the heat for the reaction. The following equation shows that the synthesis gas is then reacted in the presence of a catalyst to yield CH_3OH by methanol synthesis reaction that represented previously (Watkins 2008):

$$2H_2 + CO \leftrightarrow CH_3OH \qquad \Delta Hr = -92 \text{ kJ/mol}.$$

4.4 Methanol Economics

Since its commercial implementation in 1923, methanol synthesis has undergone numerous improvements. These are mainly driven by reduction of investment costs that dominate in the production cost of methanol (Lange 2001). In the early 1980s, methanol gradually emerged as the clear front-runner in studies focusing on fossil fuel replacements and was largely considered as one of the most probable solutions to the energy issue. The methanol production capacities installed throughout the world increased from 18 million tons per year in 1985 to 25 million tons per year in 1990 (De Alwis, Mohamad, and Mehrotra 2009; Vucins 2006). Since the early 1980s, less efficient small facilities are being replaced by larger plants using new efficient low-pressure technologies. The industry has also moved from supplying captive customers, especially for the production of formaldehyde (CH_2O) that typically represents one-half of world demand and serving primarily the home market, to large globally oriented corporations. Demand patterns too have been changing such as in Europe where methanol was once blended into gasoline when its value was around one-half that of gasoline. Offsetting this was the phasing out of leaded gasoline in developed countries that required the use of reformulated gasoline, as in the United States, which promoted the use of MTBE derived from methanol (Chemlink Consultants 2009).

The global market in methanol has seen considerable structural changes since the early 1990s. By the 1990s, methanol awareness seems to have had gradually diminished along with the strong urgency that was created during the by then long-forgotten oil crises. Global demand for methanol, however, continued to increase in accordance to growing industrial production of chemical commodities and the widespread use of MTBE as a gasoline additive. By the mid-1990s, however, market outlook prognoses began to view the long-term prospects for this major methanol outlet as a growing uncertainty. The global methanol demand increased about 8% per year from 1991 to 1995, then 3–4% per year over the next 10-year period following 1995. The methanol production capacities installed throughout the world increased from 25 million tons per year in 1990 to 33.5 million tons per year in 1997. The industry has evolved from a relatively large number of fragmented producers to a group of much larger, sophisticated international distributors and marketers. This structural change has seen the emergence of such market leaders as Methanex and SABIC and consolidation of production in major hub countries such as Chile, Trinidad, Saudi Arabia, and New Zealand.

World methanol demand reached 31 million tons in 2000, having grown at an average rate of nearly 6% in the 1990s. World methanol capacity growth has more than kept pace with world methanol demand over 1990s. In 2000 the world capacity for methanol production was about 40 million tons per year while capacities in the United States and Western Europe were 6.6 and 4.1 million tons, respectively. The capacity additions focused on methanol production in areas with access to low cost or stranded gas. During the period 1995–2000, more than 5 million tons of new methanol capacity was added in Chile, Trinidad, and Saudi Arabia. Significant expansions also occurred in the countries of Norway, New Zealand, and Venezuela during the same period (Hamid and Ali 2004; Weissermel, Arpe, and Lindley 2003). In 2005, the global demand for methanol amounted to about 32 million tons per year while it was 29.4 million tons in 2001. In 2006, worldwide methanol production capacity was over 46 million tons and nearly 35 million tons are consumed per year. A detailed analysis of methanol market is published annually by Chemical Market Associates, Inc. (CMAI). The report provides information on supply, demand, production, history, and forecasts for methanol capacity, trade, and pricing (Lee Speight, and Loyalka 2007; Nouri and Tillman 2005).

The CMAI consultants assess that the methanol industry stumbled in 2008 although year-over-year growth was seen in the last few years. Highly crude oil prices have resulted in a disastrous housing market in the world and a visibly slumping automobile sector have negatively impacted 2008 methanol demand. China's appetite for methanol across much of the derivative slate, plus methanol blending into gasoline and DME will continue to fuel

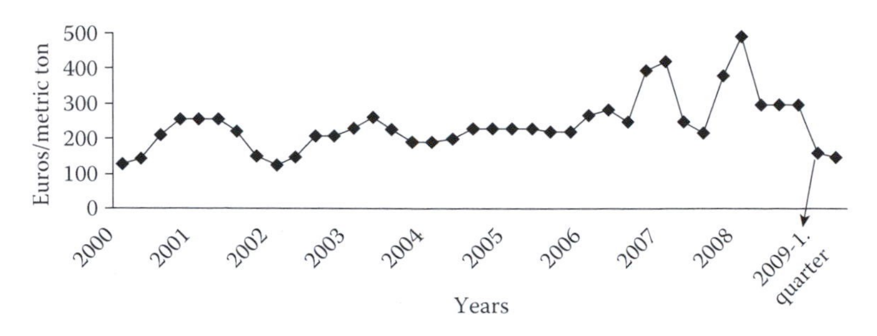

FIGURE 4.1
Price development of methanol. (From INEOS, *Paraform, Methanol Price,* 2009. Online available at http://www.ineosparaform.com/55-Methanol_price.htm.)

methanol growth. Demand in 2009, 2010, and beyond will likely be in double digits due to more gasoline blending, DME demand and "methanol to olefin" and "methanol to propylene" feedstock requirements (CMAI 2009).

Figure 4.1 shows the price development of methanol in Euros/metric ton since early 2000. During the past two years, the methanol price has shown a unique pattern. The price of most chemical feedstocks skyrocketed the same way as the price of crude oil, whereas the price of methanol oscillated between 167 and 869 U.S. $/ton (1 Euro is approximately 1.38 U.S. $, June 2009) in the U.S. spot market. Presently, it can be found in the middle of this range, with no recognizable upward trend. Methanol price is of particular importance in the search for processes and feedstocks as an alternative to classical oil-based chemistry (Schrader et al. 2009).

Methanol is one of the basic chemicals that have a variety of derivatives, and is mainly used as a raw material for CH_2O, acetic acid (CH_3COOH), and so on. The breakdown of uses of methanol is given in Figure 4.2. It is also expected that the methanol demand will grow with having the potential applications such as biodiesel and DME in the future. Methanol is already playing a key role in many applications; from fuel cells to chemicals and some authors, like Nobel Prize laureate G. Olah, see a future methanol-based economy as more than possible (Nouri and Tillman 2005).

4.5 Methanol Safety Aspects

Methanol is a clear, colorless, and volatile liquid with a faint alcohol-like odor, though it is difficult to detect at concentrations below 10 ppm. Methanol is the simplest of the alcohols, having only one carbon atom, and is completely miscible in water. Methanol easily dissolves in other alcohols and chlorinated hydrocarbons, but has limited solubility in diesel fuel, vegetable oils, and aliphatic hydrocarbons. The properties of methanol are compared to the

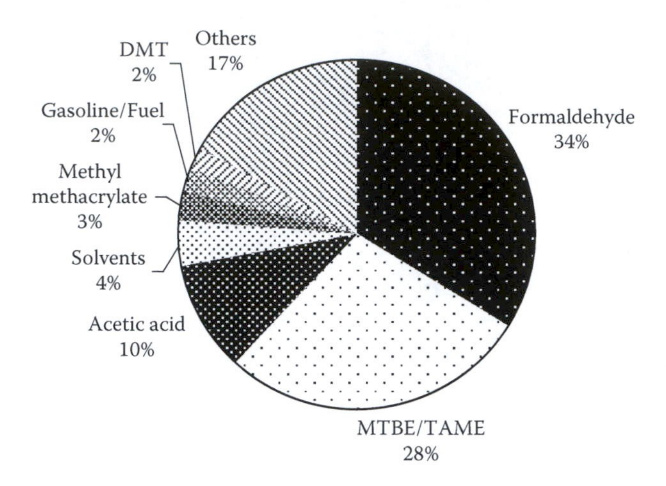

FIGURE 4.2
Breakdowns of uses for methanol. (From ESMAP, *Strategic Gas Plan for Nigeria*. Joint UNDP/World Bank Energy Sector Management Assistance Programme, 2004. Online available at http://wbln0018.worldbank.org/esmap/site.nsf/files/ESM27910paper.pdf/$FILE/ESM27910paper.pdf.)

properties of conventional gasoline in Table 4.1. Several potential safety concerns regarding methanol use as a fuel source have been identified based on its chemical and physical properties required for widespread use of methanol fuel (Pirnie 1999).

Methanol is a flammable liquid and, as such, represents a fire hazard. Pure methanol is much harder to ignite than gasoline and burns about 60% slower. Methanol burns with a barely visible flame in bright sunlight, but in most situations a methanol fire would be visible. M85 produces a flame of adequate visibility under normal lighting conditions. Methanol also burns with little or no smoke and M85 produces only low levels of smoke. This result is a significant safety benefit, since it reduces the risk of smoke inhalation injury and allows for increased visibility around the fire when compared to conventional fuels (Methanol Fact Sheets 1992). Methanol also burns much cooler, releasing its energy at one-fifth the rate of burning gasoline (Nowell 2009; http://www.methanol.org 2009). Hazard assessments use of methanol would result in a much lower frequency of vehicle-related fires and a lower hazard to people and property when a fire does occur when compared to gasoline fires. The basis for these findings are that methanol has fuel properties that make it more difficult to ignite than gasoline and, when ignited, methanol burns in a more controlled manner with less heat and smoke generated than with conventional fuels. Methanol vapor can form a combustible mixture with air in an enclosed fuel tank. Simple modifications to the fuel tanks or the addition of a volatile additive eliminate this possible hazard.

Most evaluations of potential environmental impacts due to releases of methanol have focused primarily on the fate of methanol and circumvented

TABLE 4.1

Properties of Methanol and Gasoline

Characteristics	Methanol	Gasoline
Molecular weight [g/mole]	32.04	~100
Elemental composition by weight		
% Oxygen	50.0%	(mix of C_4 to C_{14}
% Carbon	37.5%	hydrocarbons)
% Hydrogen	12.5%	
Specific gravity (@ 15.5°C)	0.79	0.72–0.78
Boiling point (°C)	64.7	27–225
Water solubility (mg/L)	miscible	100–200
Vapor pressure		
(mm Hg)(@ 25°C)	126	—
(psi) (@ 100°F)	4.63	7–15
Heat of combustion (kJ/kg)	19,930	43,030
Henry's Law constant (atm m^3 g^{-1} $mole^{-1}$)	4.55×10^{-6}	—
Henry's Law constant (@ 25°C)	1.087×10^{-4}	—
Liquid dispersion coefficient @ (m^2/s) 25°C	1.65×10^{-9}	—
Flammability limits:		
Lower (LFL)	6.0	1.4
Volume percent	7	–43
Temperature (°C)		
Upper (UFL)		
Volume percent	36.5	7.6
Temperature (°C)	43	–30 to –12
Flash point (°C)	12	–43
Vapor density (@1 atm; 10°C)	1.4	2–5

Source: From Pirnie, M., *Evaluation of the Fate and Transport of Methanol in the Environment,* Washington, DC: Technical memorandum, prepared for American Methanol Institute, 1999.

any potential ancillary impacts (Pirnie 1999). The main physical and chemical properties of methanol that affect its fate, transport, distribution, and persistence in surface water and groundwater are its miscibility, its affinity for other materials (partition coefficients), and biodegradation (http://www.methanol.org 2008).

Methanol may be released into the environment in significant amounts during its production, storage, transportation, and use. It is also naturally occurring in the environment and is biodegradable in aquatic habitats. As is the case with gasoline and diesel fuels, improper handling and storage of methanol—particularly from leaking underground storage tanks—has the potential to contaminate groundwater (Fishbein 1997; Nowell 2009; OPPT 1994).

An increased number of people could be potentially exposed to environmental methanol as a result of the projected expanded use of methanol in methanol-blended fuels. Exposures would principally arise from exhaust, evaporative emissions, and normal heating of the engine. Simulation models based on 100% of all vehicles powered by methanol-based fuels predict concentrations of methanol in urban streets, expressways, railroad tunnels, or parking garages ranging from a low of 1 mg/m^3 (0.77 ppm) to a high of 60 mg/m^3 (46 ppm). Predicted concentrations during refueling of vehicles range from 30 to 50 mg/m^3 (23–38.5 ppm). For comparison and reference purposes, a current occupational exposure limit for methanol in many countries is 260 mg/m^3 (200 ppm) for an 8 hour working day. There are limited data on human dermal exposure to methanol, but the potential expanded use of methanol in automotive fuels would increase the potential for dermal exposure in a large number of people.

The toxicity of methanol is well known. Ingestion of relatively small amounts will lead to blindness and slightly larger amounts, to death. While individual responses to methanol vary widely, one report claims that ingestion of as little as 300–1000 mg/kg (0.85–2.85 ounces for a 150 pound person) can cause death (Kavet and Nauss 1990). Methanol is easily and rapidly absorbed by all routes of exposure and distributes rapidly throughout the body. Humans absorb 60–85% of the methanol that is inhaled. A small amount is excreted by the lungs and kidneys without being metabolized. The rate of metabolism for methanol in the body is (25 mg/kg hour), which is seven times slower than for ethanol and is independent of concentrations in the blood. Humans metabolize methanol into CH_2O as the first step. The CH_2O is then converted to formic acid (CH_2O_2), which can be toxic at high concentrations, and finally to CO_2 and H_2O. The half-life of methanol elimination in expired air after oral or dermal exposure is 1.5 hours. Due to their limited capability to metabolize CH_2O_2 to CO_2, humans accumulate CH_2O_2 in their bodies from high-dose methanol exposure. If CH_2O_2 generation continues at a rate that exceeds its rate of metabolism, methanol toxicity sets in. Background levels of methanol in the human body will not result in CH_2O_2 accumulation or adverse health effects. Studies have shown that short-term inhalation exposure to 200 ppm methanol results in blood methanol concentrations of less than (10 mg/l) with no observed increase in blood CH_2O_2 concentration (www.methanol.org 2008).

Once methanol exists in the vapor phase in the atmosphere, it reacts with photochemically produced hydroxyl radicals to produce CH_2O. Methanol can also react with NO_x in polluted air to form methyl nitrite. Half-lives of 7–18 days have been reported for the atmospheric reaction of methanol with hydroxyl radicals. Biodegradation is the major route of removal of methanol from soils. Several species of methylobacterium and methylomonas isolated from soils are capable of utilizing methanol as a sole carbon source. In a wide variety of environmental media methanol is readily biodegradable under both aerobic and anaerobic conditions. It dissolves quickly to low concentrations that are eliminated much more rapidly than gasoline by natural bacteria, both when exposed to air and when air is limited such as under the ground. The use of double-walled containment tanks and leak detection monitors greatly reduces the likelihood of methanol spills. Also most methanol are removed from water by biodegradation. The degradation products of CH_4 and CO_2 were detected from aqueous cultures of mixed bacteria isolated from sewage sludge. Aerobic, gram-negative bacteria (65 strains) isolated from seawater, sand, mud, and weeds of marine origin utilized methanol as a sole carbon source. Aquatic hydrolysis, oxidation, and photolysis are not significant fate processes for methanol (Fishbein 1997; Nowell 2009; OPPT 1994).

4.6 Properties of Methanol

In order to understand the combustion and emissions of a diesel engine, it is important to understand the basic fuel properties of the fuel used in the engine. Therefore, some physicochemical fuel properties of methanol and diesel fuel are shown in Table 4.2. As mentioned above, methanol is considered to be one of the favorable fuels for engines and it has been studied in

TABLE 4.2

Properties of Methanol and Diesel Fuel

Property	Methanol	Diesel
Chemical formula	CH_3OH	$C_{14}H_{28}$
Molecular weight (g)	32	196
Boiling temperature (°C)	64.7	190–280
Density (g/cm^3, at 20°C)	0.79	0.84
Flash point (°C)	11	78
Autoignition temperature (°C)	316	464
Lower heating value (MJ/kg)	20.27	42.74
Cetane number	4	56.5
Octane number	110	—
C/H ratio	0.25	0.50
Viscosity (mm^2/s, at 25°C)	0.59	3.35
Carbon content (wt.%)	37.5	86
Hydrogen content (wt.%)	12.5	14
Oxygen content (wt.%)	50	0
Sulfur content (ppm wt)	0	<50
Stoichiometric air–fuel ratio	6.66	14.28
Heat of vaporization (MJ/kg)	1.11	0.27
Flame temperature (°C)	1890	2054

Source: From MERCK, *Product Specification*, Germany, 2006; TUPRAS, *Product Specification*, Turkey, 2005.

spark ignition engine application. However, it is more difficult to fuel diesel engines with methanol because of its very low cetane number, high heat of vaporization, and some other physicochemical properties.

Methanol (CH_3OH) is a simple compound. It does not contain sulfur or any complex organic compounds. However, diesel fuel is a complex mixture of a large number of hydrocarbons (such as C_3–C_{25} hydrocarbons). For this reason, its fuel properties can change depending on the proportion of hydrocarbon types used in the fuel mixture. Methanol contains an oxygen atom so that it is accepted as a partially oxidized hydrocarbon. It has lower energy content than diesel fuel. Therefore, more fuel is needed to obtain the same amount of power with that of a diesel-fueled engine. Its low stoichiometric air–fuel ratio, high oxygen content, and high H/C ratio may be beneficial to improve the combustion and to reduce the soot and smoke.

Methanol has higher latent heat of vaporization than diesel fuel so that it extracts more heat as it vaporizes. Therefore, it can lead to a cooling effect on the cylinder charge. Since methanol has very low viscosity compared to diesel fuel, it can be easily injected, atomized, and mixed with the air introduced into the cylinder. Methanol has poor ignition behavior due to its low cetane number, high latent heat of vaporization, and high ignition temperature. Therefore, it can cause some increase in ignition delay. However, ignition improver, such as diethyl ether, can be added to the blended fuel to compensate for the cetane number (Murayama et al. 1982).

The autoignition temperature of methanol is higher than that of diesel fuel, which makes it safer for transportation and storage. On the other hand, methanol has a much lower flash point than that of diesel fuel; this is a safety disadvantage. At the same time, an additive should be added to methanol to improve its lubrication (Adelman 1979; Kowalecwicz 1993; Ristinen and Kraushaar 2006; Wagner et al. 1979). Due to the low solubility of methanol in diesel fuel, a solvent such as oleic acid and iso-butanol (Huang et al. 2004a, 2004b), 1-dodecanol (Bayraktar 2008), iso-propanol (Murayama et al. 1982) are added to methanol–diesel blends. In some studies, a mixer was used to prevent phase separation (Canakci, Sayin, and Gumus 2008; Sayin et al. 2009). Another disadvantage of methanol is its corrosivity that is more than diesel fuel on copper, brass, aluminum, rubber, and many plastics. This puts some restrictions on the designs and manufacturing of engines to be used with this fuel.

For diesel engines, combustion and emission characteristics are influenced by fuel spray characteristics, nozzle geometry, injection pressure, and so on. Therefore, Yanfeng, Shenghua, and Yu (2007) investigated the effects of opening pressure, ambient density, and nozzle diameter on penetration length and cone angle of methanol sprays. The results obtained by these researchers are shown in Figures 4.3 through 4.6. They found that the methanol spray penetration length increased with the increase of the opening pressure. The methanol spray penetration and tip velocity decreased quickly with the increase of ambient density; on the contrary, they increased

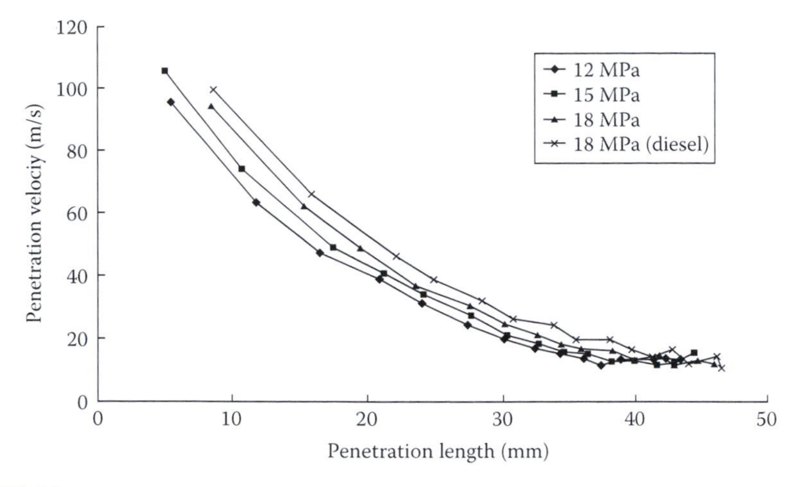

FIGURE 4.3
Effect of opening pressure on spray velocity. (From Yanfeng, G., Shenghua, L., and Yu, L., *Energy and Fuels*, 21, 2991–97, 2007. Reprinted with permission from ACS Publications.)

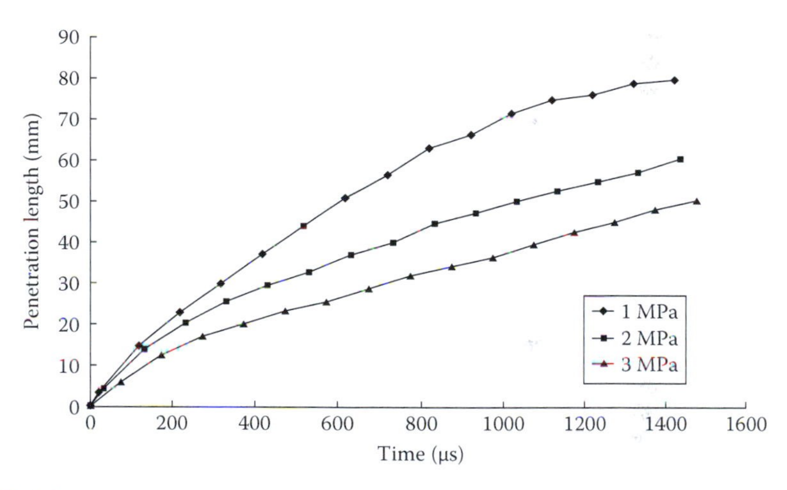

FIGURE 4.4
Effect of ambient density on spray penetration. (From Yanfeng, G., Shenghua, L., and Yu, L., *Energy and Fuels*, 21, 2991–97, 2007. Reprinted with permission from ACS Publications.)

with the increase of nozzle diameter. Between 12 and 18 MPa, the opening pressure had little influence on the spray angle and the angle remained nearly constant during the whole injection process. The spray cone angle increased with the increase of the ambient density or the nozzle diameter. Compared with diesel fuel, they found that the penetration of methanol was shorter and the cone angle of methanol was larger under the same experimental conditions.

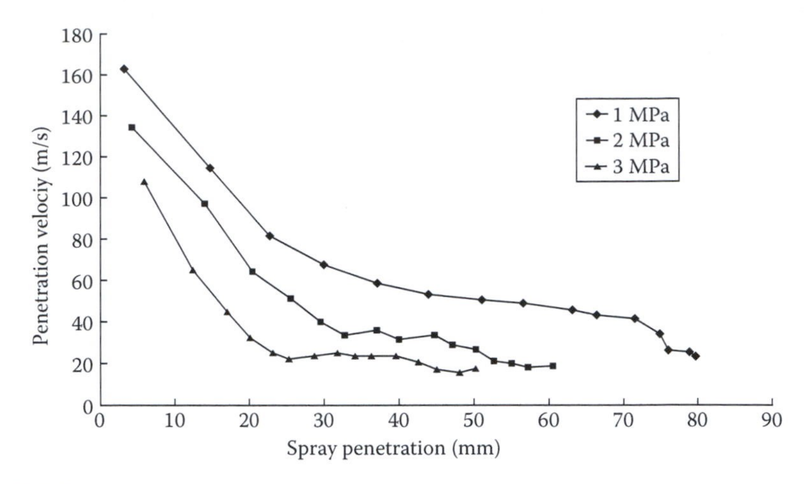

FIGURE 4.5
Effect of ambient density on spray velocity. (From Yanfeng, G., Shenghua, L., and Yu, L., *Energy and Fuels*, 21, 2991–97, 2007. Reprinted with permission from ACS Publications.)

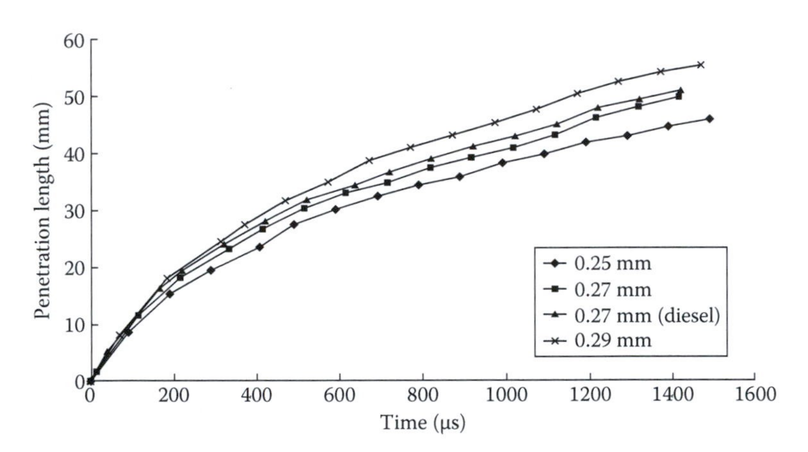

FIGURE 4.6
Effect of nozzle diameter on spray penetration. (From Yanfeng, G., Shenghua, L., and Yu, L., *Energy and Fuels*, 21, 2991–97, 2007. Reprinted with permission from ACS Publications.)

4.7 Engine Tests

The exhaust emissions from diesel engines are very complex mixtures containing many types of pollutants, which can be found in different forms, such as particulate, semivolatile, and gaseous phases. Using methanol can reduce the air pollution caused by these engines. Therefore, many researchers have studied the influence of this fuel on the engine performance and exhaust emissions. At this point, it is important to emphasize that some test results obtained by the researchers can be contradictory to each other since different engine test conditions and engine technologies have been applied in the experiments.

The simplest method of using methanol in a CI engine is to blend it with diesel fuel using an additive to prevent phase separation (Bayraktar 2008; Huang et al. 2004b; Murayama et al. 1982). This application requires no modification on the engine and fuel system. However, other methods such as dual-fueling (Kumar et al. 2005; Song et al. 2008; Wang et al. 2008) and fumigation (Abu-Qudais, Haddad, and Qudaisat 2000; Cheng et al. 2008a; Popa, Negurescu, and Pana 2001; Udayakumar, Sundaram, and Sivakumar 2004; Yao et al. 2007) require additional equipments such as another fuel-injection system and storage tank, which means additional cost. In the fumigation mode, diesel fuel is injected through the original high-pressure fuel injectors into the engine cylinder while the methanol is injected in the air intake for each cylinder through low-pressure fuel injectors.

4.7.1 Engine Performance Tests

Bayraktar (2008) studied the effect of methanol-blended diesel fuel between 2.5 and 15 vol% on the engine performance and found that the methanol could reduce the effective power and brake thermal efficiency (BTE) to some degree and moderately increase brake specific fuel consumption (BSFC).

Sayin et al. (2009) studied the effect of methanol-blended diesel fuel on the BSFC and brake thermal efficiency (BTE) of a single cylinder direct injection (DI) diesel engine. The reference diesel fuel was blended with methanol from 0 to 15% with an increment of 5%. The engine was run at constant speed (2200 rpm) and four different loads (5 Nm, 10 Nm, 15 Nm, and 20 Nm) for three different injection timings (15°, 20° and 25° CA BTDC). The original injection timing of the test engine is 20° CA BTDC.

The BSFC is defined as the ratio of the fuel consumption to the brake power. The effects of methanol–diesel fuel blends and injection timings on the BSFC are shown in Figures 4.7 and 4.8 for different engine load and injection timing, respectively. The results showed that increasing methanol ratio in the fuel blend caused it to increase in the BSFC. This behavior is attributed to lower heating value (LHV) per unit mass of the methanol, which is noticeably lower than that of the diesel fuel as seen in Table 4.2. Therefore, the amount of fuel introduced into the engine cylinder for a desired fuel energy input has to be greater with the methanol. BSFC decreased about two times as the engine load increased from 5 to 20 Nm constant load. This decrease in BSFC could be explained by the fact that, as the engine load increases, the rate of increasing brake power is much more than that of

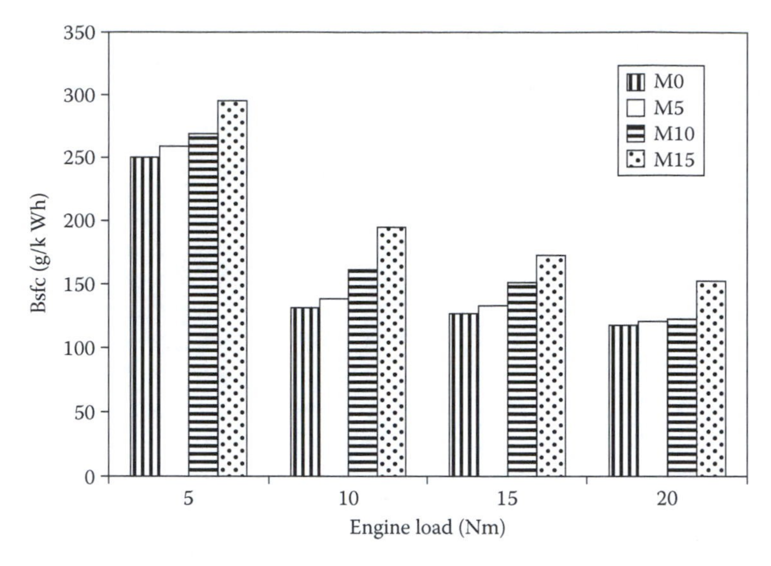

FIGURE 4.7
BSFC results at different loads (ORG injection timing). (From Sayin, C., Ilhan, M., Canakci, M., and Gumus, M., *Renewable Energy*, 34, 1261–69, 2009. Reprinted with permission from Elsevier Publications.)

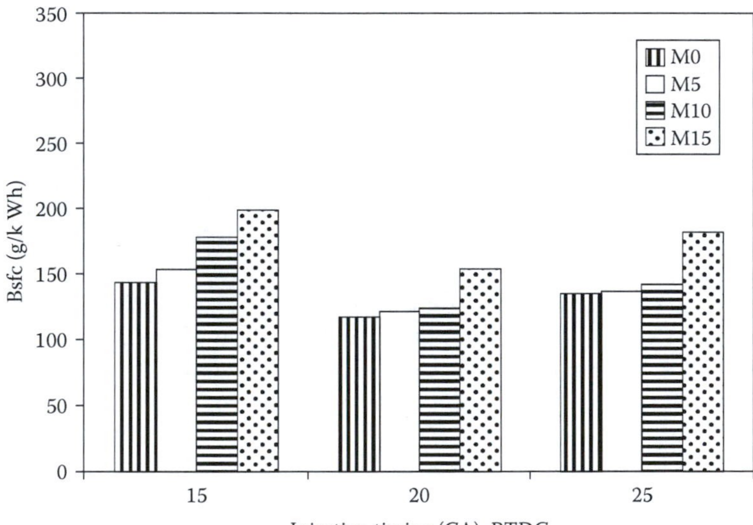

FIGURE 4.8
BSFC results at different injection timings and 20 Nm load. (From Sayin, C., Ilhan, M., Canakci, M., and Gumus, M., *Renewable Energy*, 34, 1261–69, 2009. Reprinted with permission from Elsevier Publications.)

the fuel consumption. Figure 4.8 indicates the variations of BSFC for different methanol-blended diesel fuels under different injection timing at 20 Nm constant loads. When the injection timing was retarded 5° CA BTDC compared to ORG injection timing, BSFC increased by 30% for M15. As will be discussed later, with advancing injection timing, the ignition delay will be longer and speed of the flame will be shorter. This causes reduction of maximum pressure and engine output power. Thus, fuel consumption per output power will increase. On the other hand, retarding injection timing means later combustion, and therefore pressure rises only when the cylinder volume is expanding rapidly and results in a reduced effective pressure to work. In that study, the minimum BSFC was obtained at ORG injection timing for all the fuel blends.

Brake thermal efficiency indicates the ability of the combustion system to accept the experimental fuel, and provides comparable means of assessing how efficient the energy in the fuel was converted to mechanical output. BTE results are presented in Figures 4.9 and 4.10 for different engine loads and injection timings, respectively. The maximum BTE was recorded with M0 for all the engine loads. The M0 fuel at 20 Nm with ORG injection timing produced the highest BTE. The higher BTE of M0 operation can be attributed to its LHV. Figure 4.10 shows the variations of the BTE with different methanol-blended diesel fuels for different injection timings at 20 Nm constant loads. The best results in terms of BTE were obtained

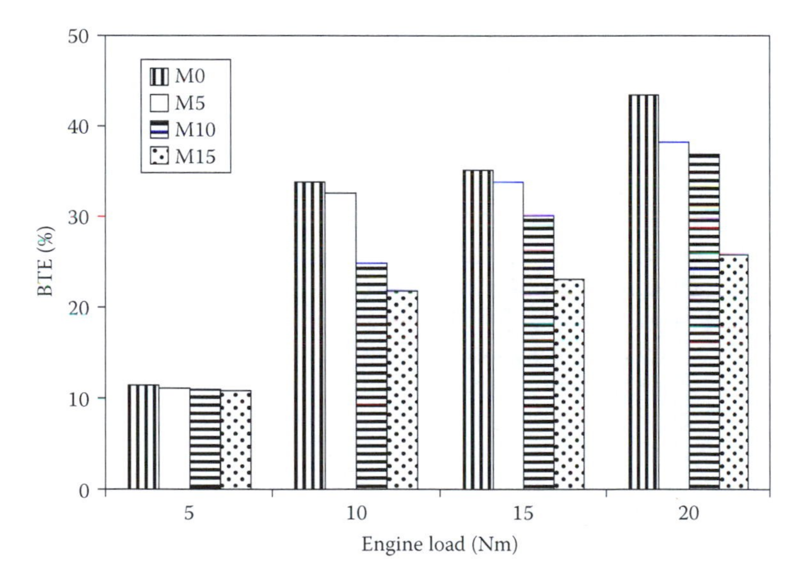

FIGURE 4.9
BTE results at different loads (ORG injection timing). (From Sayin, C., Ilhan, M., Canakci, M., and Gumus, M., *Renewable Energy*, 34, 1261–69, 2009. Reprinted with permission from Elsevier Publications.)

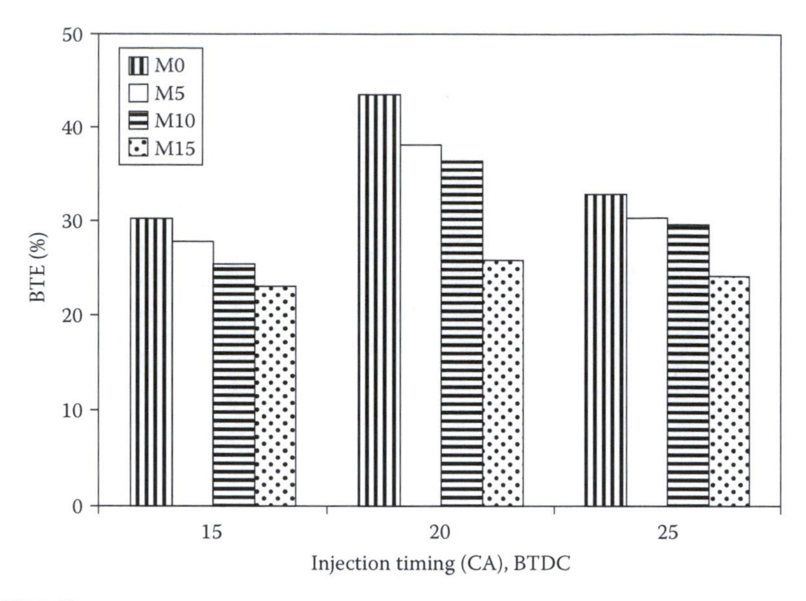

FIGURE 4.10
BTE results at different injection timings and 20 Nm load. (From Sayin, C., Ilhan, M., Canakci, M., and Gumus, M., *Renewable Energy*, 34, 1261–69, 2009. Reprinted with permission from Elsevier Publications.)

at ORG injection timing. This means that retarded or advanced injection timing diminished BTE values.

Canakci, Sayin, and Gumus (2008) experimentally investigated the brake specific energy consumption (BSEC) and combustion efficiency of the same engine at the same test conditions given by Sayin et al. (2009). BSEC is described as the product of BSFC and LHV. The results obtained in that study showed that the BSEC increases with increasing methanol content. It is well known that the LHV of the fuel affects the engine power. The lower heat content of the methanol–diesel fuel blend causes some reductions in the engine power. In addition, the theoretical air/fuel ratio of diesel fuel is about two times higher than that of methanol, as shown in Table 4.2. For these reasons, the effective power should decrease with the increase of the methanol amount in the fuel mixture. Thus, the engine needs to consume more heat to maintain the same amount of power output. BSEC reduced as the engine load increased because of noticeably diminishing BSFC for all fuel blends and injection timing. The BSEC reduced by 15% as the engine load increased from 10 to 20 Nm constant loads for M15 at retarded injection timing. When the injection timing changed from ORG injection timing, BSEC values increased because of the increase in the energy requirement to sustain the same amount of power output at ORG injection timing. The increments for the advanced and the retarded injection timings were 10 and 13% for M0 at 10 Nm, respectively.

The fuel's chemical energy is not fully released inside the engine during the combustion process. Therefore, it is useful to compare the combustion efficiencies of the test fuels used in the engine. Canakci, Sayin, and Gumus (2008) also compared the combustion efficiencies of methanol-blended diesel fuels to M0. The results showed that the combustion efficiency increased with increasing methanol ratio in the fuel blend at all engine loads and injection timings. When methanol is added into diesel fuel, the fuel contains more oxygen, which reduces CO and UHC emissions and increases NO_x emissions. These effects caused an increase in combustion efficiency. Combustion efficiency slightly increased with an increasing engine load from 5 to 20 Nm for all test fuels because of better volumetric efficiency and atomization rate. The increase in combustion efficiency with advancing the injection timing was attributed to increases in the NO_x emissions and decreases in CO and UHC emissions, as mentioned below.

At this point, it is good to mention the change in exhaust temperature when methanol-blended diesel fuel was used. As shown in Figure 4.11, the exhaust temperature increased with an increasing methanol ratio in the fuel mixture (Canakci, Sayin, and Gumus 2008). As mentioned before, methanol has a higher latent heat of vaporization than diesel fuel. This can lead to a cooling effect on the cylinder charge. On the other hand, methanol has poor ignition behavior because of its low cetane number, high latent heat of vaporization, and high ignition temperature. Thus, it can cause a longer ignition delay. It is clear from the figure that the cetane number and oxygen content are more effective than latent heat of vaporization with regard to increasing peak temperature in the cylinder. Therefore, the concentration of NO_x

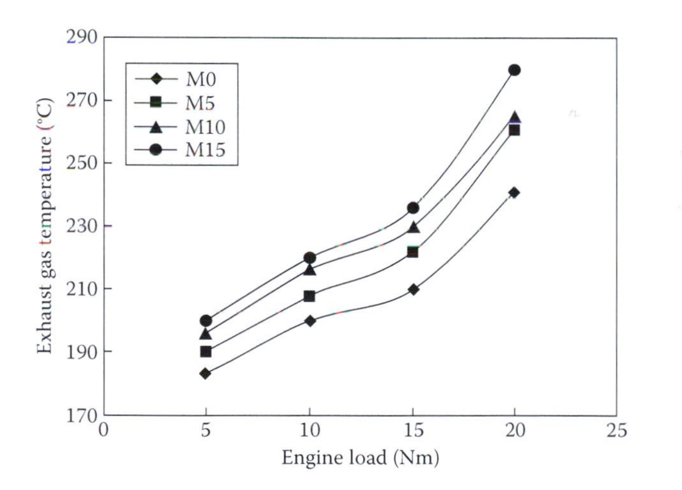

FIGURE 4.11
Exhaust gas temperatures at different engine loads and ORG injection timing. (From Canakci, M., Sayin, C., and Gumus, M. *Energy and Fuels,* 22, 3709–23, 2008. Reprinted with permission from ACS Publications.)

increases as the methanol content was increased in the fuel blend (Nwafor, Rice, and Ogbonna 2000).

4.7.2 Engine Emissions Tests

Engine emissions can be divided into two groups: regulated and unregulated emissions. Regulated emissions are carbon monoxide (CO), nitrogen oxides (NO_x), and unburned fuel or partly oxidized hydrocarbons (HC). The levels of these emissions are specified by legislations. Unregulated emissions include polycyclic aromatic hydrocarbons (PAHs), methane, aldehydes, carbon dioxide (CO_2), other trace organic emissions, and carbon deposits.

Chao et al. (2000) studied the effect of a methanol-containing additive (MCA) on the emissions of carbonyl compounds (CBCs) generated from a heavy-duty diesel engine. When either 10 or 15% MCA was used, the emission factors of the CBCs acrolein and isovaleraldehyde increased by at least 91%. Song et al. (2008) studied dual-fuel operation on a DI, four stroke, single-cylinder, water-cooled diesel engine. In that work, an electronically controlled low-pressure common rail system was employed to deliver methanol to the inlet port, while the engine's original high-pressure diesel injection system was used to deliver a suitable quantity of diesel fuel for ignition. The experimental results showed that smoke was reduced significantly, while a modest reduction in NO_x was observed under the dual-fuel conditions. The equivalent BSFC was improved under high-load operating conditions. Especially, the dual-fuel engine showed a better fuel economy when run at a high rate of methanol addition. However, unburned HC and CO emissions for dual-fuel operation increased when methanol was added. Song et al. (2008) claimed that it is better for the dual-fuel engine to run with a high rate of methanol under high-load operating conditions and with pure diesel under low-load operating conditions. In this case, an improved thermal efficiency as well as an increased alternative ratio could be reached.

Huang et al. (2004a) used various blend rates of methanol–diesel fuels in the engine tests. The results indicated that the increase of methanol content decreased smoke number (SN) and CO and UHC emissions but increased BSFC and NO_x emissions. Cheng et al. (2008a) researched the effects of the fumigation methanol on the engine performance, emissions, and particulates. In that study, the fumigation methanol was injected to top up 10, 20, and 30% of the power output under different engine operating conditions. The experimental results showed that there is a decrease in the BTE when fumigation methanol is applied, except at the highest load of 0.67 MPa. At low loads, the BTE decreased with the increase in fumigation methanol; but at high loads, it increased with the increase in the fumigation methanol. The fumigation methanol resulted in a significant increase in UHC, CO, and NO_x emissions.

Cheng et al. (2008b) investigated the performance and exhaust emissions of a four-cylinder diesel engine operating on biodiesel with methanol in either the blended or fumigation mode. They compared the results to those operating on pure biodiesel and pure diesel fuel. Experiments were performed on a four-cylinder naturally aspirated DI diesel engine operating at a constant speed of 1800 rpm with five different engine loads. The results indicated reductions in CO_2, NO_x, and particulate mass emissions and reductions in the mean particle diameter, in both cases, compared to diesel fuel.

The effects of methanol–diesel blends on CI engine emissions were investigated by Ilhan (2007). In that study, methanol-blended diesel fuels were prepared using 99% pure methanol with the volumetric ratios of 0–15%. The results demonstrated that NO_x and CO_2 emissions increased as CO and UHC emissions decreased with increasing amount of methanol in the fuel mixture. Chao et al. (2001) investigated the effect of MCA on the regulated and unregulated emissions from a diesel engine. The engine was tested on a series of diesel fuels blended with five additive levels (0%, 5%, 8%, 10%, and 15% of MCA by volume). Results showed that MCA addition slightly decreased PM and NO_x emissions but generally increased both UHC and CO emissions.

Yao et al. (2008) studied the effect of diesel–methanol (DMMC) compound combustion on diesel engine combustion and emissions. The amount of methanol injected is controlled by an electronic control unit and depends on engine output. Experiments were carried out at idle and five different engine loads at two levels of engine speeds to compare engine emissions resulted from pure diesel and DMMC use, with and without the oxidation catalytic convertor. The results showed that the diesel engine operating with DMMC could simultaneously decrease the soot and NO_x emissions but increase the UHC and CO emissions compared with the pure diesel.

Udayakumar, Sundaram, and Sivakumar (2004) investigated the effect of fumigation methanol on exhaust emissions by using the injection method into the inlet manifold. They carried out their tests with the inlet air heated to 70°C. They reported that smoke and NO_x emissions were both decreased with methanol injection.

Chu (2008) investigated the influence of M0, M5, and M15 methanol/diesel fuel mixture on a single-cylinder diesel engine without changing the engine parameters. Test results showed that methanol addition caused less engine power but improved fuel economy. Although the NO_x, smoke, and CO emissions were significantly reduced, UHC emissions increased. Kulakoglu (2009) investigated engine performance and exhaust emissions using methanol–diesel blends in a DI diesel engine, and the maximum methanol mass fraction was 15%. The results showed that there is an increase in the fuel consumption and NO_x emissions and a decrease in the thermal efficiency, CO, and UHC emissions when the methanol amount is increased in the mixture.

Popa, Negurescu, and Pana (2001) investigated the effect of the methanol–diesel blend on the exhaust emissions. This paper presented the experimental results obtained by providing two different methods of diesel fuel and methanol engine supplied. The first method consists in the methanol admission through a carburetor combined with the classic diesel fuel injection, and the second one refers to the separate fuels injection. From the exhaust emissions measurements, it was shown that the smoke and NO_x levels significantly reduced for all the engine loads with an increasing amount of methanol in the blend.

For a diesel engine, fuel injection timing is a major parameter that affects the combustion and exhaust emissions. The state of air into which the fuel is injected changes as the injection timing is varied and, thus, ignition delay will vary. If the injection starts earlier, the initial air temperature and pressure are lower; therefore, the ignition delay will increase. If the injection starts later (when the piston is closer to TDC), the temperature and pressure are initially slightly higher and a decrease in ignition delay results. Hence, variation in injection timing has a strong effect on the engine performance and exhaust emissions, because of changing the maximum pressure and temperature in the engine cylinder. Therefore, Canakci, Sayin, and Gumus (2008) investigated exhaust emissions of a single–cylinder diesel engine under different injection timings when methanol-blended diesel fuel was used. The tests were conducted at three different injection timings (15°, 20°, and 25° CA BTDC). All tests were conducted at four different loads (5, 10, 15, and 20 Nm) at constant engine speed of 2200 rpm. The following part summarizes the emission results of that study.

4.7.2.1 Carbon Monoxide (CO) Emissions

CO is a colorless, odorless, poisonous gas, and it must be restricted. CO results from incomplete combustion of fuel and is emitted directly from vehicle tailpipes. Besides the ideal combustion process that combines carbon (C) and oxygen (O_2) to CO_2, incomplete combustion of carbon leads to the formation of CO. The formation of CO takes place when the oxygen present during combustion is insufficient to form CO_2 (Heywood 1984). In general, while the engine is running under fuel-rich mixture conditions, the exhaust will contain a large amount of CO emission, because there is not sufficient oxygen to convert all of the carbon atoms of the fuel into CO_2. Thus, the most important parameters that affect CO emissions are an insufficient amount of air and an insufficient time in the cycle for complete combustion (Ganesan 1994).

Concerning the effect of different fuels on CO emissions, it was uncovered that increasing the methanol ratio in the fuel–blend lessened CO emissions. In comparison to M0, the change in CO emissions was around 19%, 32%, and 39% for M5, M10, and M15, respectively, at 5 Nm load and advanced injection timing, as demonstrated in Figure 4.12a. Methanol is an oxygenated fuel and leads to more complete combustion; hence, CO emissions reduce in the exhaust. CO emission decreased gradually when the engine load increased.

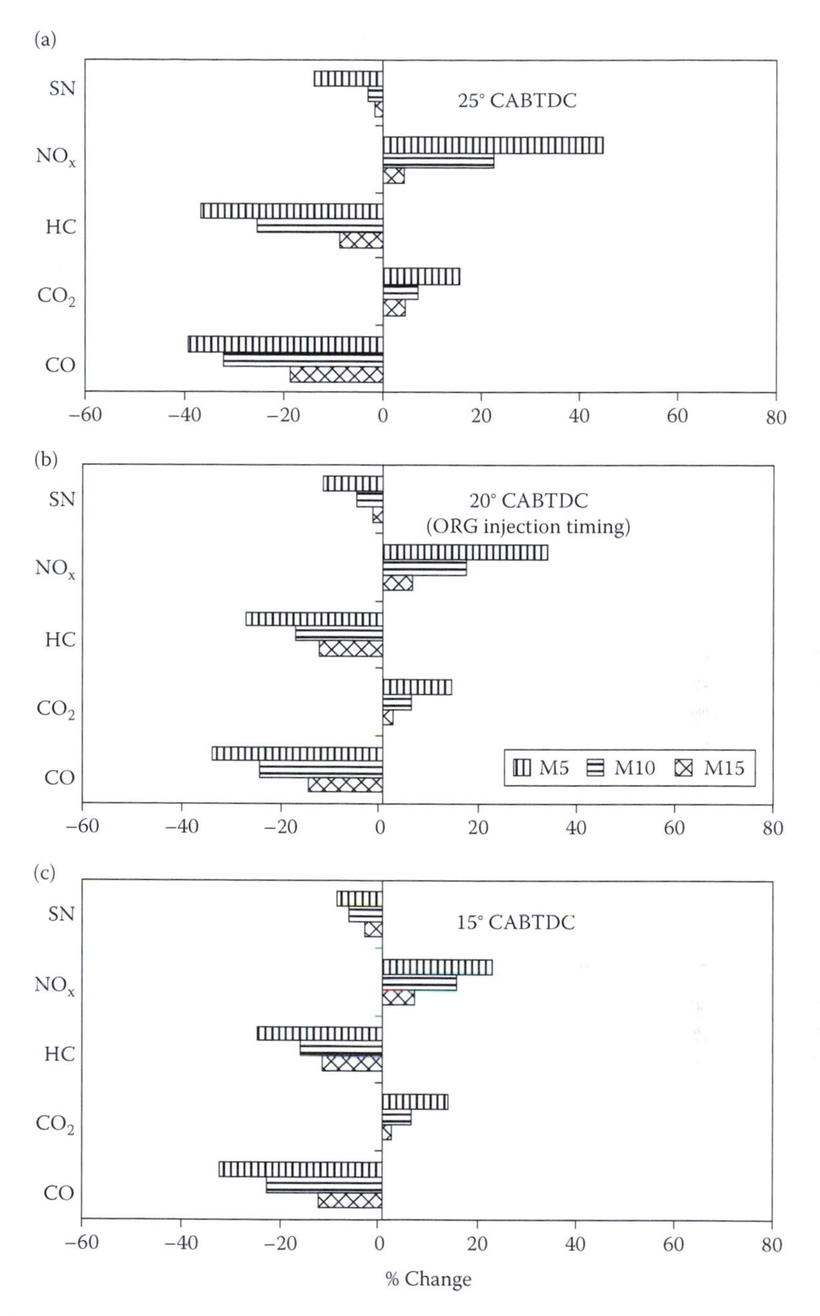

FIGURE 4.12
Changes in the emissions relative to M0 at 5 Nm load. (From Canakci, M., Sayin, C., and Gumus, M. *Energy and Fuels,* 22, 3709–23, 2008. Reprinted with permission from ACS Publications.)

When the engine load increases, the combustion temperature increases and CO emissions start to decrease (Abdel-Rahman 1998).

For example, in comparison to Figures 4.13b and 4.14b, it was seen that the percentage change in CO emissions diminished by around 21% for M5 as the engine load increased from 10 to 15 Nm constant loads at ORG injection timing. Figure 4.13a through c illustrates the percentage change in CO emissions with different methanol blends at different injection timings for 10 Nm load. From these figures, it was concluded that advanced injection timing decreased the CO emission by 8% and retarded injection timing increased the CO emission by 7% compared to ORG injection timing for M10, respectively. The advanced injection timing produced the higher cylinder temperature and increasing oxidation process between carbon and oxygen molecules. These lead to a decrease in the percentage change in CO emissions (Gumus 2008).

4.7.2.2 Unburned Hydrocarbon (UHC) Emissions

The UHC emissions consist of fuel that is incompletely burned. Most of the UHC is caused by an unburned fuel–air mixture, whereas the other source is the engine lubricant and incomplete combustion. The term UHC means organic compounds in the gaseous state; solid HCs are the part of the PM. Typically, HCs are a serious problem at low loads in CI engines. At low loads, the fuel is less apt to impinge on surfaces; but, because of poor fuel distribution, large amounts of excess air and low exhaust temperature, lean fuel–air mixture regions may survive to escape into the exhaust (Canakci 1996; Sayin et al. 2007).

With regard to the effect of different methanol contents on UHC emission, it was found that increasing the methanol ratio in the fuel-blend reduced UHC emissions. For instance, the UHC emissions compared to M0 at ORG injection timing decreased by 14%, 24%, and 40% for M5, M10, and M15, respectively, at 10 Nm load and retarded injection timing, as seen in Figure 4.13c. When methanol was added to the diesel fuel, it provided more oxygen for the combustion process and led to the improving combustion. In addition, methanol molecules are polar and cannot be absorbed easily by the nonpolar lubrication oil, and therefore, methanol can lower the possibility of the production of UHC emissions (Alla et al. 2002). UHC emissions lessened reasonably with increasing load, which was the same trend as with CO. For example, in comparison of Figures 4.13a through 4.15a, it was observed that the change in UHC emissions diminished by 7% for M10 as the engine load increased from 10 to 20 Nm constant loads at advanced injection timing. Figures 4.15a through c shows the change in UHC emissions with different methanol blends at different injection timings for 20 Nm load compared to M0. From these figures, it was found that advanced injection timing caused a reduction in UHC emission by 9% and retarded injection timing boosted the UHC emission by 3% compared to ORG timing for

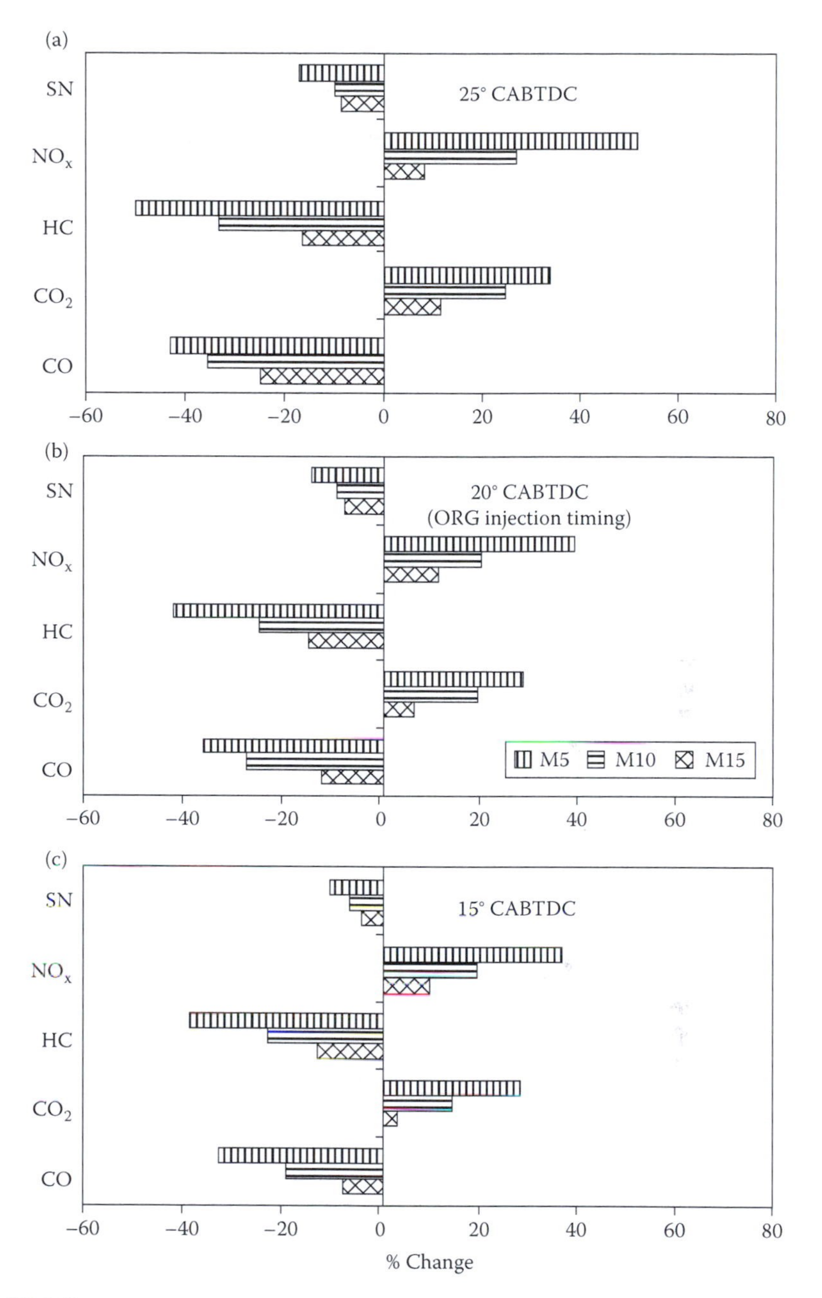

FIGURE 4.13
Changes in the emissions relative to M0 at 10 Nm load. (From Canakci, M., Sayin, C., and Gumus, M., *Energy and Fuels,* 22, 3709–23, 2008. Reprinted with permission from ACS Publications.)

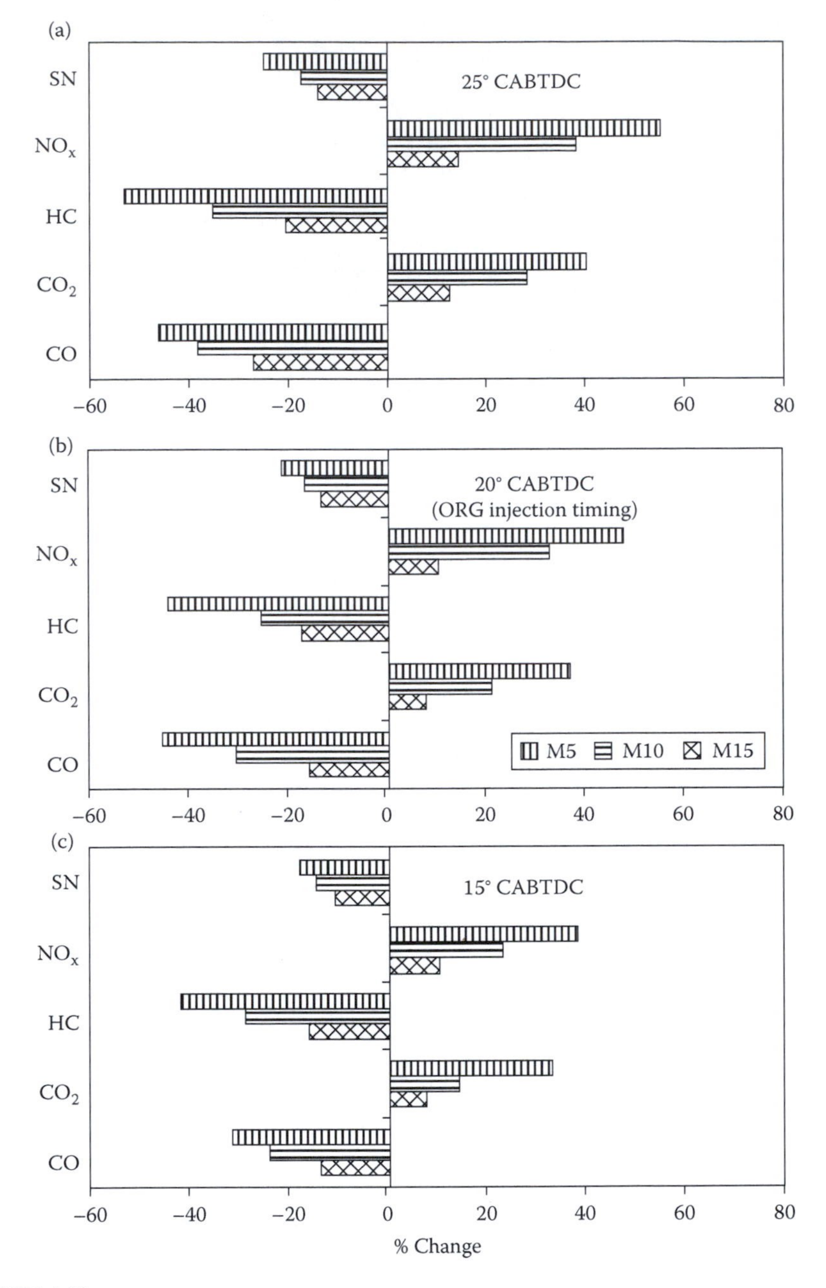

FIGURE 4.14
Changes in the emissions relative to M0 at 15 Nm load. (From Canakci, M., Sayin, C., and Gumus, M., *Energy and Fuels*, 22, 3709–23, 2008. Reprinted with permission from ACS Publications.)

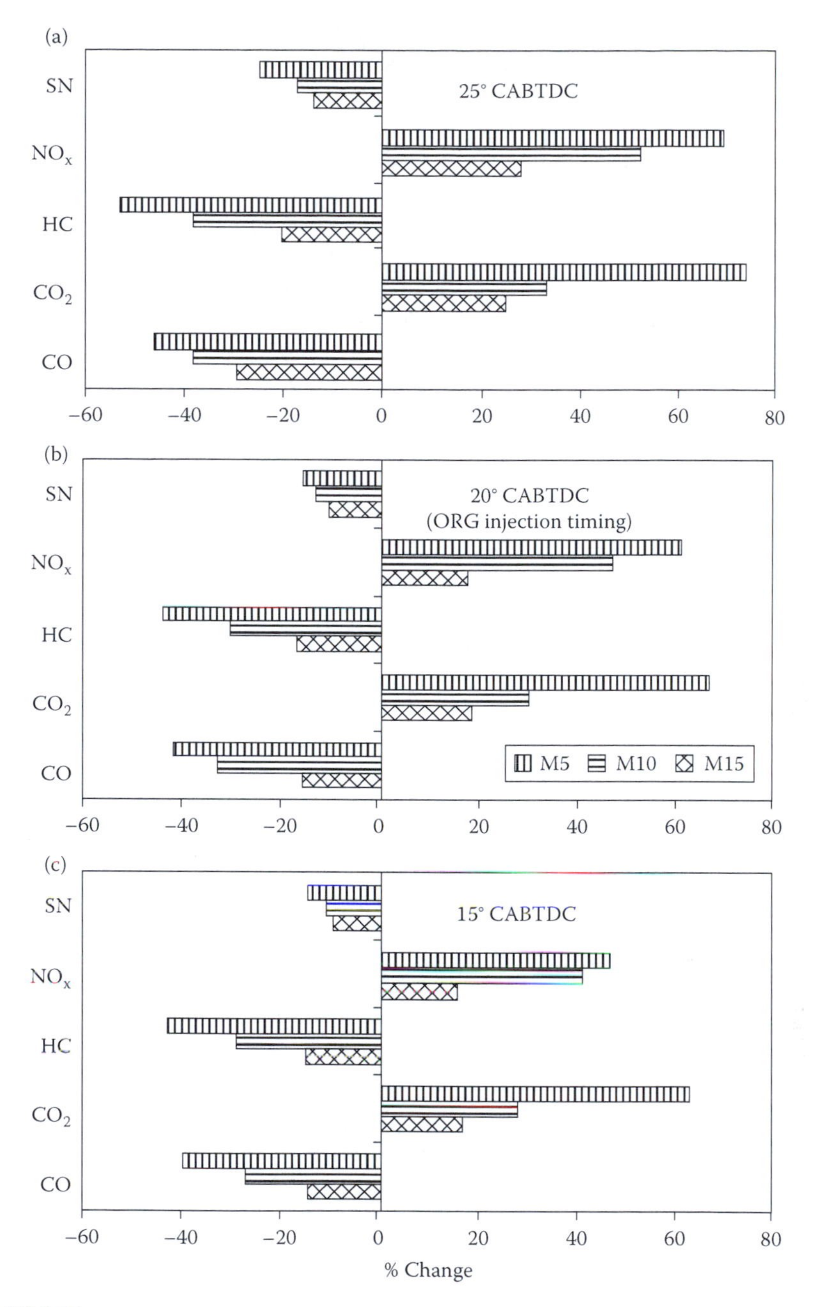

FIGURE 4.15
Changes in the emissions relative to M0 at 20 Nm load. (From Canakci, M., Sayin, C., and Gumus, M., *Energy and Fuels,* 22, 3709–23, 2008. Reprinted with permission from ACS Publications.)

M15, respectively. Advancing the injection timing caused an earlier start of combustion relative to the TDC. Because of this, the cylinder charge, being compressed as the piston moved to the TDC, had relatively higher temperatures and thus lowered the UHC emissions (Ajav, Singh, and Bhattacharya 1998; Pukrakek 1997).

4.7.2.3 Nitrogen Oxides (NO_x) Emissions

The most troublesome emissions from CI engines are NO_x emissions. The oxides of nitrogen in the exhaust emissions contain nitric oxide (NO) and nitrogen dioxides (NO_2). The formation of NO_x is highly dependent on in-cylinder temperatures, the oxygen concentration and residence time for the reaction to take place. In a diesel engine, the fuel distribution is nonuniform. The pollutant formation process is strongly dependent upon the changes in the fuel with time because of mixing. The oxides of nitrogen form in the high-temperature burned region, which is nonuniform, and the formation rates are highest in the regions closest to the stoichiometric region (Agarwal 2007).

Canakci, Sayin, and Gumus (2008) found that increasing the methanol ratio in the blend raised NO_x emissions. For example, as presented in Figure 4.14b, the change in NO_x emissions was compared to M0 and showed that NO_x augmented by 14%, 35%, and 49% for M5, M10, and M15, correspondingly, at 15 Nm load and ORG injection timing. Methanol contains 34% oxygen, and its cetane number is lower than diesel fuel, which boost peak temperature in the cylinder. On the other hand, the LHV of methanol is nearly two times lower than diesel fuel and latent heat of vaporization of methanol is about four times greater than diesel fuel, which decreases peak temperature in the cylinder. However, as shown in Figure 4.11, the exhaust temperature increased with an increasing methanol ratio in the fuel mixture. It is clear from the figure that the cetane number and oxygen content are more effective than LHV and latent heat of vaporization with regard to increasing peak temperature in the cylinder. Therefore, the concentration of NO_x increased as the methanol content was increased in the fuel blend (Nwafor, Rice, and Ogbonna 2000).

Unlike CO and UHC emissions, NO_x emissions increased with an increasing engine load. For example, in comparison to Figures 4.13c and 4.14c, it was detected that NO_x emissions increased by around 6% for M5 as the engine load augments from 10 to 15 Nm constant loads at the retarded injection timing. Figure 4.14a through c demonstrates the percentage change in NO_x emissions with different methanol blends at different injection timings for 15 Nm load. As shown in the figures for M15, retarded injection timing reduced in NO_x emission by 10% and advanced injection timing increased in NO_x emission by 6% compared to ORG injection timing, correspondingly. When the injection timing was retarded, it was observed that NO_x emissions decreased for all fuel mixtures. Retarding the injection timing decreased the peak cylinder pressure because more fuel burned after TDC. Lower peak cylinder pressures resulted in lower peak temperatures. As a consequence, the NO_x concentration diminished (Sayin and Uslu 2008).

4.7.2.4 Smoke Number (SN)

The emitted PM is essentially composed of soot, though some hydrocarbons, generally referred to as a soluble organic fraction (SOF) of the particulate emissions, are also adsorbed on the particle surface or simply emitted as liquid droplets. Smoke opacity formation occurs at the extreme air deficiency. This air or oxygen deficiency is present locally in the very rich core of the fuel sprays in the combustion chamber. It increases as the air–fuel ratio decreases. Soot is produced by oxygen deficient thermal cracking of long-chain molecules (Challen and Baranescu 1999).

Regarding the effect of different methanol contents on SNs, it was observed that increasing methanol ratio in the blend reduced SNs. The change in SNs compared to M0 implied that they diminished by 15%, 19%, and 26% for M5, M10, and M15, respectively, at 20 Nm load and advance injection timing as illustrated in Figure 4.15a. The presence of atomic-bound oxygen in methanol satisfies positive chemical control over soot formation. The tendency to generate soot by the fuel dense region inside a diesel diffusion flame sheath is reduced, so that soot-free spray combustion could be achieved (Can, Celikten, and Usta 2004). The formation of smoke is most strongly dependent upon the engine load. As the load increases, more fuel is injected, and this increases the formation of smoke. The results obtained in the study of Canakci, Sayin, and Gumus (2008) supported this statement. For instance, in comparison to Figures 4.14c and 4.15c, it was seen that the change of SNs increased by 6% for M10 as the engine load increased from 5 to 15 Nm at retarded injection timing. Advancing the injection timing reduced the smoke emissions. The earlier injection led to higher temperatures during the expansion stroke and more time in which oxidation of the soot particles occurred (Challen and Baranescu 1999). Figure 4.13a through c presents the percentage change in SNs at different injection timings for 10 Nm load. As seen from these figures for M15, advanced injection timing lowered in SNs by 3% and retarded injection timing raised in SNs by 3% compared to ORG timing for M15, respectively.

Ozaktas et al. (2000) studied the effect of the compression ratio on the soot formation for methanol–diesel fuel blends on an ASTM-CFR engine and they compared the results with baseline diesel fuel. In that study, the selected compression ratios were 14.8:1, 16:1, 17.4:1, 19:1, and 21:1. They expressed that soot emission decreased with decreasing compression ratio. Soot emissions of the blend fuels were generally lower than those of diesel fuel. The difference was more outstanding for higher compression ratios. Soot emissions of diesel fuel were very high for these compression ratios and the methanol component is relatively very effective on the decrease of soot emission.

On the other hand, it seems that the methanol ratio has little effect on the soot emission.

4.7.2.5 Carbon Dioxide (CO_2) Emission

CO_2 emissions are released into the atmosphere when fuel is completely burned in an engine. As illustrated in the Figure 4.13a, when the methanol amount increased in the fuel mixture, the percentage change in CO and UHC decreased. The percentage change in CO_2 had an opposite behavior when compared to the CO concentrations, and this was due to improving the combustion process as a result of the oxygen content in the methanol. The maximum increase in the change of CO_2 was observed at 24%, 33%, and 74% for M5, M10, and M15, respectively, compared to M0 at 20 Nm engine load and advanced injection timing. CO_2 emissions increased with the advanced injection timing for all fuel mixtures. As shown in Figure 4.14a through c for M10, advanced injection timing increased the change in CO_2 by 6% and retarded injection timing diminished in CO_2 by 8% compared to ORG and retarded injection timing, respectively.

4.7.3 Engine Combustion Studies

Combustion and heat release analysis can yield valuable information about the effect of engine design, fuel injection system, fuel type, and engine operating conditions on the combustion process and engine performance. Wang et al. (2008) investigated the effects of methanol mass fraction and pilot diesel injection timing on the ignition to understand the variation in the ignition delay and detailed combustion characteristics. With the increase of methanol mass fraction, the maximum cylinder pressure, the maximum rate of pressure rise, and the maximum heat release rate increased.

Huang et al. (2004b), studied combustion and heat release characteristics of a CI engine using methanol–diesel fuel blend. According to the experimental results, increasing the methanol mass fraction in the methanol–diesel fuel blends resulted in an increase in the heat release rate at the premixed burning phase and shortened the combustion duration at the diffusive burning phase. Huang et al. (2004c) and Ozaktas et al. (2000) found that there was an increase in ignition delay and increase in premixed heat release with the combustion of methanol-blended diesel fuel.

Ozaktas et al. (2000) studied the effect of the compression ratio on ignition delay for methanol–diesel fuel blends on an ASTM-CFR engine and they compared the results with baseline diesel fuel. They expressed that, with increasing compression ratio, ignition delay decreased almost linearly for all the fuel blends tested. Ignition delays of the blend fuels were longer than those of diesel fuel but had the same tendency as diesel fuel (i.e., they decreased with increasing compression ratio). An increase in delay period was obtained with increasing methanol amount in the fuel mixture. However the difference between the delay times of the blend fuels diminished for the compression ratios 16 and 14.8. They explained this result by the lower temperature during the ignition period as a result of the cooling effect of vaporized methanol. At the low compression ratios where the compression temperature is already low, this cooling effect becomes dominant. Although relatively higher pressures are expected for the fuel blends due to the longer ignition delay periods, they found that there was no significant difference between the peak pressures of fuel types. This was attributed to the lower energy content of methanol used in the fuel blends.

Canakci, Sayin, and Gumus (2008) also investigated combustion and heat release characteristics of a single-cylinder diesel engine under different injection timings when methanol-blended diesel fuel was used. The following part summarizes the combustion and heat release characteristics results obtained in that study. The measured starts of combustion and ignition delay for each fuel blend are shown in Table 4.3.

Figures 4.16 and 4.17 show the cylinder gas pressure, and Figures 4.19 and 4.20 demonstrate the rate of heat release (ROHR) for different fuel blends and ORG injection timing at 20 and 10 Nm loads, respectively. Figures 4.18 and 4.21 illustrate the cylinder gas pressure and ROHR for M0 and M15 at different injection timing and 20 Nm load, respectively.

4.7.3.1 Peak Cylinder Gas Pressure

Figure 4.16 shows the cylinder gas pressure with respect to the crank angle at 20 Nm load and ORG injection timing. As seen in the figure, peak cylinder gas pressure slightly decreased with the increase of the methanol supplement rate. The researchers showed that the peak cylinder pressure occurred at 7.96 MPa (at 3.20° CA ATDC), 7.86 MPa (at 3.28° CA ATDC), 7.78 MPa (at 3.32° CA ATDC), and 7.77 MPa (at 3.44° CA ATDC) for M0, M5, M10, and M15 at 20 Nm load and ORG injection timing, respectively. They claim that lowering the cetane number by methanol addition was responsible for the increase in the ignition delay. The increase in the ignition delay would burn more fuel in the premixed burning phase. Because of this, the rate of pressure rise increased and peak cylinder gas pressure diminished (Heywood 1984).

When Figure 4.16 is compared to Figure 4.17, it is seen that the cylinder gas pressure increased with an increasing engine load. Experimental results showed that the increase in the cylinder pressure was approximately 5.5% for M10 when the engine load increased from 10 to 20 Nm. The fuel consumption per unit time was measured between 0.42 g/s and 0.81 g/s at these

TABLE 4.3

Combustion Characteristics of the Fuels at Different Injection Timing

	M0	M5	M10	M15
20° CA BTDC (ORG Injection Timing), 20 Nm				
Start of injection (°BTDC)	20	20	20	20
Start of combustion (°BTDC)	9.27	8.92	8.34	8.01
Ignition delay (degree)	10.73	11.08	11.16	11.99
15° CA BTDC, 20 Nm				
Start of injection (°BTDC)	15	15	15	15
Start of combustion (°BTDC)	5.98	5.67	5.34	5.01
Ignition delay (degree)	9.02	9.33	9.66	9.99
25° CA BTDC, 20 Nm				
Start of injection (°BTDC)	25	25	25	25
Start of combustion (°BTDC)	12.60	11.61	10.51	9.12
Ignition delay (degree)	12.40	13.39	14.49	15.88
20° CA BTDC (ORG Injection Timing), 10 Nm				
Start of injection (°BTDC)	20	20	20	20
Start of combustion (°BTDC)	5.97	5.54	5.24	4.91
Ignition delay (degree)	14.03	14.46	14.76	15.09
15° CA BTDC, 10 Nm				
Start of injection (°BTDC)	15	15	15	15
Start of combustion (°BTDC)	2.68	2.27	2.04	1.52
Ignition delay (degree)	12.32	12.73	12.96	13.48
25° CA BTDC, 10 Nm				
Start of injection (°BTDC)	25	25	25	25
Start of combustion (°BTDC)	9.27	8.421	6.76	5.37
Ignition delay (degree)	15.73	16.59	18.24	19.63

Source: From Canakci, M., Sayin, C., and Gumus, M., *Energy and Fuels*, 22, 3709–23, 2008. Reprinted with permission from ACS Publications.

test conditions (10 and 20 Nm load at ORG injection timing for M10, respectively). Because of increasing fuel consumption per unit time with increasing engine load, this behavior provided an increase in the maximum cylinder gas pressure. The locations of maximum cylinder gas pressure approached to TDC at 20 Nm engine load because starting the fuel injection occurred earlier than that of 10 Nm.

Figure 4.18 shows a comparison of the changes in the cylinder gas pressures with respect to the crank angle obtained for M0 and M15 at different injection timings and 20 Nm load. The peak cylinder gas pressure was obtained at 8.5 MPa (at 1.68° CA ATDC), 8.0 MPa (at 3.20° CA ATDC), and

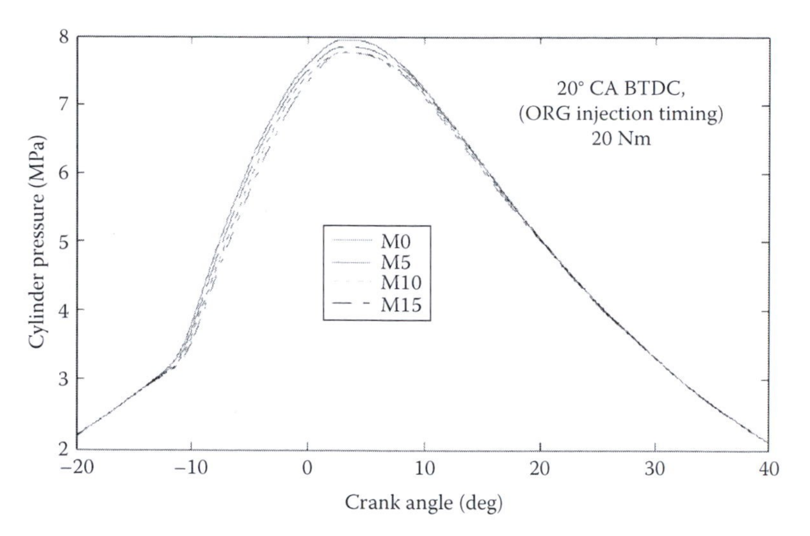

FIGURE 4.16
Cylinder gas pressure versus CA at 20 Nm load and ORG injection timing. (From Canakci, M., Sayin, C., and Gumus, M., *Energy and Fuels*, 22, 3709–23, 2008. Reprinted with permission from ACS Publications.)

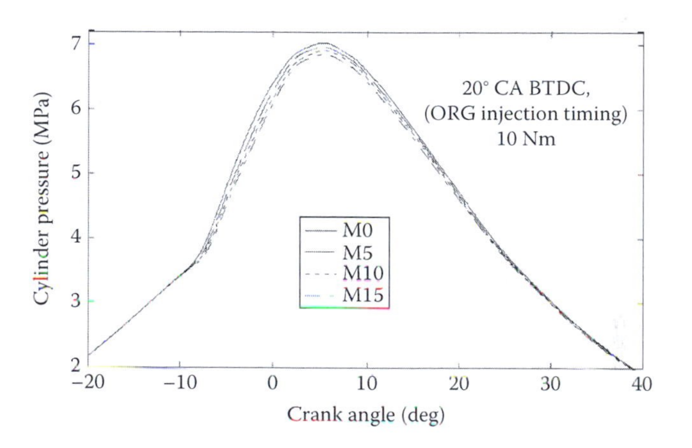

FIGURE 4.17
Cylinder gas pressure versus CA at 10 Nm load and ORG injection timing. (From Canakci, M., Sayin, C., and Gumus, M., *Energy and Fuels*, 22, 3709–23, 2008. Reprinted with permission from ACS Publications.)

7.2 MPa (at 5.08° CA ATDC) for advanced ORG and retarded injection timings, respectively. The trend was such that, as injection started earlier, peak pressures became higher, which applied to all fuel blends. Also, the peak pressures occurred earlier with advancing injection timings (Ozsezen, Canakci, and Sayin 2008; Pukrakek 1997).

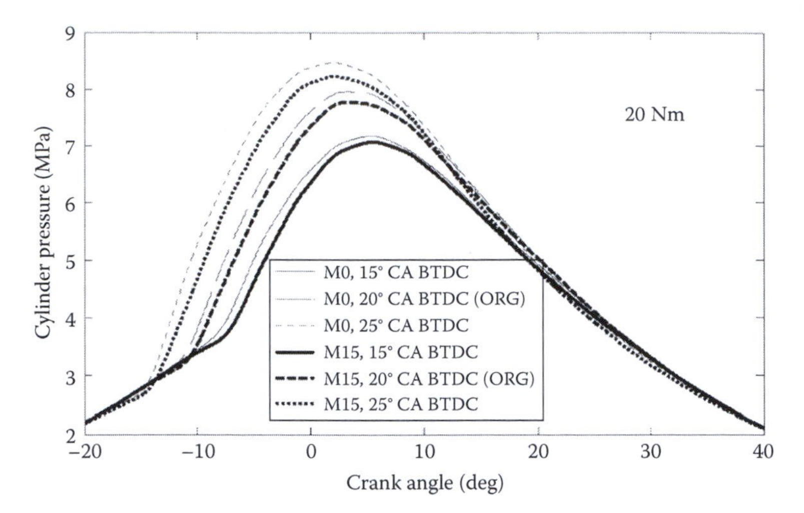

FIGURE 4.18
Cylinder gas pressure versus CA at different injection timing and 20 Nm load. (From Canakci, M., Sayin, C., and Gumus, M., *Energy and Fuels*, 22, 3709–23, 2008. Reprinted with permission from ACS Publications.)

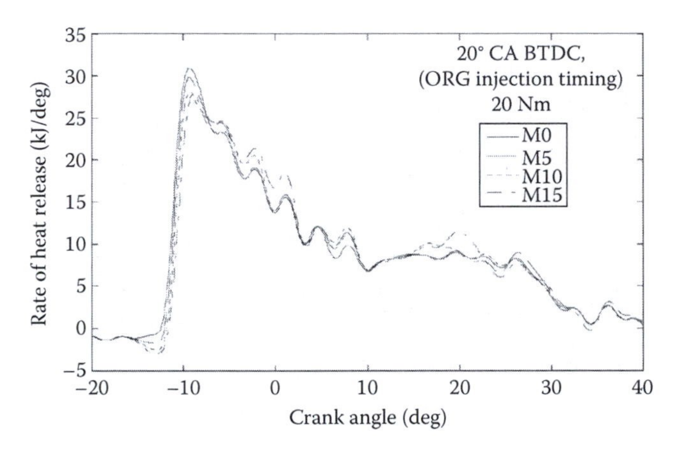

FIGURE 4.19
Rate of heat release versus CA at 20 Nm load and ORG injection timing. (From Canakci, M., Sayin, C., and Gumus, M., *Energy and Fuels*, 22, 3709–23, 2008. Reprinted with permission from ACS Publications.)

4.7.3.2 Rate of Heat Release (ROHR)

As illustrated in the Figure 4.19, ROHR decreased with the increase of the methanol amount in the fuel blend. The maximum ROHR was obtained at 31.1 kJ/deg (at 8.93° CA BTDC), 29.9 kJ/deg (at 8.90° CA BTDC), 27.3 kJ/deg (at 8.86° CA BTDC), and 27.0 kJ/deg (at 8.85° CA BTDC) for M0, M5, M10, and

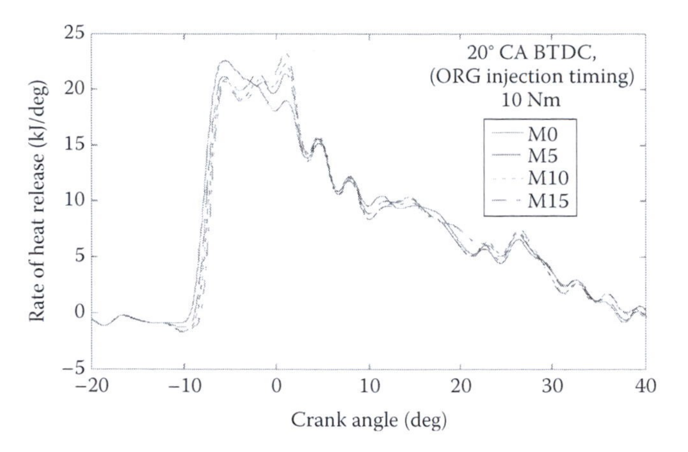

FIGURE 4.20
Rate of heat release versus CA at 10 Nm load and ORG injection timing. (From Canakci, M., Sayin, C., and Gumus, M., *Energy and Fuels*, 22, 3709–23, 2008. Reprinted with permission from ACS Publications.)

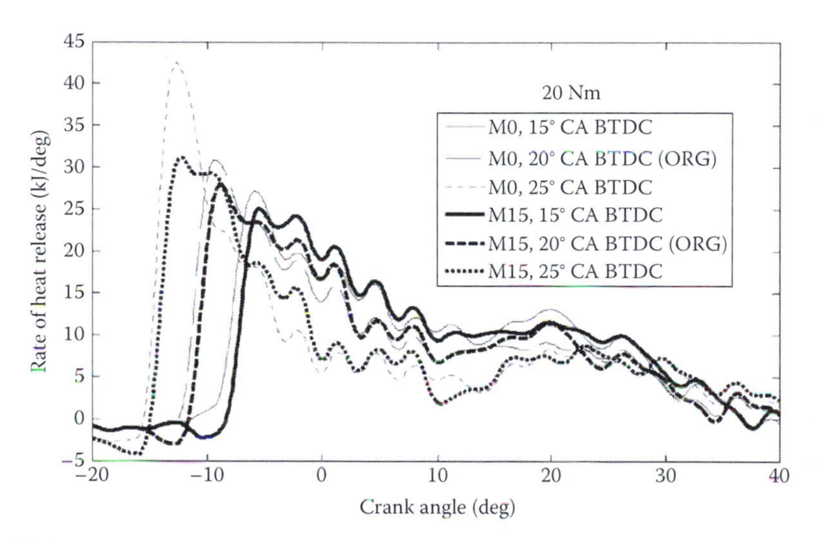

FIGURE 4.21
Rate of heat release versus CA at different injection timing and 20 Nm load. (From Canakci, M., Sayin, C., and Gumus, M., *Energy and Fuels*, 22, 3709–23, 2008. Reprinted with permission from ACS Publications.)

M15 at 20 Nm load and ORG injection timing, respectively. Methanol does not evaporate as easily as diesel fuel hence the ignition delay increases with an increasing methanol ratio, as seen in Table 4.3. The increase in the ignition delay may cause more fuel to be burned in the premixed burned phase and an increase in the ROHR. Conversely, the LHV of methanol is lower than diesel fuel, which reduces ROHR. It is apparent from the figure that the LHV of methanol was more significant than the heat of vaporization with regard to ROHR (Ghojel and Honnery 2005).

Figures 4.19 and 4.20 show the ROHR for different fuel blends and ORG injection timing at 20 and 10 Nm loads, respectively. As seen in these figures, ROHR increased with the rise in the engine load because of the increase in the quantity of fuel injected. Figure 4.21 demonstrates a comparison of the changes in the cylinder gas pressures with respect to the crank angle obtained for M0 and M15 at different injection timings and 20 Nm load. The ignition delay in a diesel engine is defined as the time between the start of fuel injection and the start of combustion. As seen in Table 4.3, the ignition delay was raised with advanced injection timing for M0 and M15 because the fuel was injected earlier into the combustion chamber. This led to greater accumulation of the fuel in the ignition delay period and an increasing premixed heat release. This is the reason of increasing ROHR. Retarding injection timing led to a lower accumulation of fuel and poor combustion. However, both physical and chemical reactions must take place before a significant fraction of chemical energy in the fuel is released. These reactions need a finite time to occur. However, as ignition delay proceeds, the in-cylinder pressures and temperatures decrease and reduce the favorable conditions for ignition. The most favorable timing for ignition lies between these two conditions (Kumar, Ramesh, and Nagalinman 2003).

Fuel injection pressure is a significant operating parameter affecting the performance and combustion characteristics in CI engine. When fuel injection pressure is low, fuel particle diameters will enlarge and ignition delay period during the combustion will increase. CO, UHC, and NO_x emissions can increase since the combustion process goes to a bad condition. When injection pressure is increased, fuel particle diameters will become small. Since formation of mixing of fuel to air becomes better during the ignition period, smoke level, and CO emissions can be less. But, if injection pressure is too high, the ignition delay period becomes shorter. So, possibilities of the homogeneous mixing decrease. Therefore, smoke can be seen at the exhaust of the engine. See Canakci et al. (2009), for further information related the effect of injection pressure on the emission and combustion characteristics of a CI engine using methanol-blended diesel fuel.

4.8 Methanol Benefits and Challenges

Methanol is one of the most promising sources of hydrogen for fuel cell systems and synthetic fuel in the future with the advantages of high energy density, easy availability, and safe storage. The using of methanol is in the point of view of its power and other characteristics similar to those using gasoline or diesel. Methanol has some advantages and disadvantages in comparison with conventional fuels. The methanol production technologies have already been proven in practice. They are reliable and widely used (alcohol production). The important advantages of methanol as a fuel are higher energy content per volume than other alternative fuels such as compressed natural gas (CNG) or LPG and the minimal changes needed in the existing fuel distribution network. Methanol can considerably reduce automotive emissions and may require some additives because of its fuel properties. It can be used directly as a replacement for gasoline in the gasoline and diesel blends. Higher octane number allows higher compression and following better effectiveness of the engine (De Alwis, Mohamad, and Mehrotra 2009; Enersol EU 2004).

The reason for research is how to use it in fuel cells as methanol has high ratio of hydrogen atoms and much higher energetic density than liquid hydrogen. Methanol can be made from corn cobs, wood or other forms of cellulose have lower emissions of all harming elements (20–70%). Methanol is furthermore cheaper than ethanol. Handling methanol is easier than gasoline, because it is safer in accidents and fire extinction can be done with water, because water dilutes it (Enersol EU 2004). Methanol is also favored over conventional fuels because of its lower ozone-forming potential, lower emissions of pollutants particularly benzene, PAHs, sulfur compounds, and low-evaporative emissions.

Methanol is highly corrosive and toxic when ingested or absorbed through the skin. Methanol causes stain of metal parts, removes oil from places where it is needed, and negatively affects plastic parts. Because of methanol's corrosive qualities, engine parts needed to be frequently replaced. Methanol vehicles produce a significant amount of formaldehyde. Compared to conventional fuels, its disadvantages include the possibility of higher CH_2O emissions, higher acute toxicity and lower cost-effectiveness (De Alwis, Mohamad, and Mehrotra 2009). Methanol vehicles produce a significant amount of formaldehyde.

Burning methanol is invisible. Starting the engines under the freezing point is also difficult. Warming the fuel similar to diesel helps to solve this problem. A further problem is the CH_2O smell when the engine is cold started. Catalysts help to shorten it by 2 minutes. Definitive elimination of these emissions can be assured by a system of warming the damp before inlet into the catalyst. The energy value of methanol is half of diesel. Because there is less energy in M85 than in gasoline, the driving range of vehicles operating on M85 is less. Therefore the vehicles need double the amount of fuel to the same place of destination (Enersol EU 2004).

4.9 Further Studies

These suggestions will provide additional information about the methanol fuel properties and about its effect on the engine performance, emissions, and combustion characteristics. Therefore, the following subject should be investigated for fundamental understanding if pure methanol or methanol–diesel blends were used in CI engines, especially in new technology CI engines. The main subjects may be written as:

- Long-term endurance tests and any engine modifications
- Cold weather startability and cold start emissions
- Lubricity properties of methanol–diesel blends
- Corrosive properties of methanol and wear effects on the engine and fuel system parts
- Engine tests at different engine operating parameters (effect of EGR, compression ratio, etc.)
- Engine tests at different injection parameters such as multiple injection and injection timing
- Methanol–diesel fuel blends spray characteristics using different injectors
- Knock production (noise level)
- Using cetane improvers for modifying ignition delay
- Optimum methanol–diesel blends at different engine operating conditions
- The effect of exhaust aftertreatment use on regulated emissions

In addition to a combination of the above subjects, to investigate the relationships between the fuel properties and combustion, the physical and chemical phenomena in the combustion chamber can be observed by using a high-speed camera system. This will provide valuable information on the nature of combustion process to help the researchers and engine designers for their future studies related to the fuel properties and methanol.

References

Abdel-Rahman, A. A. 1998. On the emissions from internal combustion engines: A review. *International Journal of Energy Research* 22:483–513.

Abu-Qudais, M., O. Haddad, and M. Qudaisat. 2000. The effect of alcohol fumigation on diesel engine performance and emissions. *Energy Conversion and Management* 41:389–99.

Adelman, H. 1979. Alcohols in diesel engines. *SAE* 790956.

Agarwal, A. K. 2007. Biofuels (alcohols and biodiesel) applications as fuels for internal combustion engines. *Progress in Energy and Combustion Science* 33:233–71.

Ajav, E. A., B. Singh, and T. K. Bhattacharya. 1998. Performance of a stationary diesel engine using vapourized ethanol as supplementary fuel. *Biomass and Bioenergy* 15:493–502.

Alla, G. H., H. A. Soliman, O. H. Badr, and M. F. Rabbo. 2002. Effect of injection timing on the performance of a dual fuel engine. *Energy Conversion and Management* 43:269–77.

Bayraktar, H. 2008. An experimental study on the performance parameters of an experimental CI engine fueled with diesel–methanol–dodecanol blends. *Fuel* 87:158–64.

Bhattacharyya, A., V. W. Chang, and D. J. Schumacher. 1998. CO_2 reforming of methane to syngas I: Evaluation of hydrotalcite clay-derived catalysts. *Applied Clay Science* 13:317–28.

Bradley, M. J. 2000. *Future wheels, Interviews with 44 global experts on the future of fuel cells and transportation and fuel cell infrastructure and a fuel cell primer.* Boston, MA: Northeast Advanced Vehicle Consortium (NAVC).

Bromberg, L., and D. R. Cohn. 2008. *Effective octane and efficiency advantages of direct injection alcohol engines.* MIT Laboratory for Energy and the Environment Report LFEE 2008-01. Online available at www.ethanolboost.com/LFEE%20 2008-01%20RP.pdf (accessed May 26, 2009).

Can, O., I. Celikten, and N. Usta. 2004. Effects of ethanol addition on performance and emissions of an IDI Diesel engine running at different injection pressures. *Energy Conversion and Management* 45:2429–40.

Canakci, M. 1996. Idealized engine emissions resulting from the combustion of isooctane supplemented with hydrogen. MSc Thesis, Vanderbilt University, Nashville, TN.

Canakci, M., C. Sayin, A. N. Ozsezen, and A. Turkcan. 2009. Effect of injection pressure on the combustion, performance and emission characteristics of a diesel engine fueled with methanol blended diesel fuel. *Energy and Fuels* 23:2908–20.

Canakci, M., C. Sayin, and M. Gumus. 2008. Exhaust emissions and combustion characteristics of a DI diesel engine fueled with methanol-diesel fuel blends at different injection timings. *Energy and Fuels* 22:3709–23.

Challen, B., and R. Baranescu. 1999. *Diesel engine reference book.* Oxford: Butterworth and Heinemann Publishing.

Chao, H.-R., T.-C. Lin, M.-R. Chao, F.-H. Chang, C.-I. Huang, and C.-B. Chen. 2000. Effect of methanol-containing additive on the emission of carbonyl compounds from a heavy-duty diesel engine. *Journal of Hazardous Materials* B73:39–54.

Chao, M.-R., C.-T. Lin, H.-R. Chao, F.-H. Chang, and C.-B. Chen. 2001. Effects of methanol-containing additive on emission characteristics from a heavy-duty diesel engine. *Science of the Total Environment* 279:167–79.

Chau, P. 2008. *Coal to methanol chemical plant report.* 2008. AIChE 2008 National Student Design Competition. Online available at www. maecourses.ucsd.edu/ ceng124/ rpts/gp9_final.doc (accessed May 27, 2009).

Chemlink Consultants. 2009. *Methanol (methyl alcohol).* Online available at http:// www.chemlink.com.au/methanol.htm (accessed May 27, 2009).

Cheng, C. H., C. S. Cheung, T. L. Chan, S. C. Lee, and C. D. Yao. 2008a. Experimental investigation on the performance, gaseous and particulate emissions of a methanol fumigated diesel engine. *Science of the Total Environment* 389:115–24.

Cheng, C. H., C. S. Cheung, T. L. Chan, S. C. Lee, C. D. Yao, and K. S. Tsang. 2008b. Comparison of emissions of a direct injection diesel engine operating on biodiesel with emulsified and fumigated methanol. *Fuel* 87:1870–79.

Cheng, W. H., and H. H. Kung. 1994. *Methanol production and use.* New York: Marcel Dekker.

Chu, W. 2008. The experimental study about the influence of methanol/diesel fuel mixture on diesel engine performance. Proceedings of the 2008 workshop on power electronics and intelligent transportation system; IEEE computer society: WA. August 2–3, 2008, Guangzhou, China, 324–27.

CMAI-Chemical Market Associates, Inc. 2009. *2009 world methanol analysis*. Online available at http: *www.cmaiglobal.com/WorldAnalysis/pdf/WMATOC.pdf* (accessed May 27, 2009).

De Alwis, H. P. N. S., A. A. Mohamad, and A. K. Mehrotra. 2009. Exergy analysis of direct and indirect combustion of methanol by utilizing solar energy or waste heat. *Energy and Fuels* 23:1723–33.

Ebbeson, B., H. C. Stokes, and C. A. Stokes. 2000. *Methanol—The other alcohol, A bridge to a sustainable clean liquid fuel.* WoodPro: The Pennsylvania Wood Products Productivity Program. Online available at http://www.deltastate.gov.ng/methanol.pdf (accessed May 27, 2009).

Enersol EU. 2004. *Biomass, Energy saving and renewable energy in vocational education.* Leonardo da Vinci Project ENERSOL EU (NL/01/B/F/PP-123143).

ESMAP. 2004. *Strategic gas plan for Nigeria*. Joint UNDP/World Bank Energy Sector Management Assistance Programme (ESMAP). Online available at http://wbln0018.worldbank.org/esmap/site.nsf/files/ESM27910paper.pdf/$FILE/ESM27910paper.pdf (accessed May 27, 2009).

Fishbein, L. 1997. *Methanol. Environmental health criteria 196.* Geneva: World Health Organization.

Ganesan, V. 1994. *Internal combustion engines.* New York: McGraw-Hill.

Ghojel, J., and D. Honnery. 2005. Heat release model for the combustion of diesel oil emulsions in DI diesel engines. *Applied Thermal Engineering* 25:2072–85.

Gumus, M. 2008. Evaluation of hazelnut kernel oil of Turkish origin as alternative fuel in diesel engines. *Renewable Energy* 33:2448–57.

Hamid, H., and M. A. Ali. 2004. *Handbook of MTBE and other gasoline oxygenates.* New York: Marcel Dekker.

Herz, R. 2001. *Methanol to gasoline conversion.* University of California, San Diego, CENG 176, Winter Quarter. Online available at http://chemelab.ucsd.edu/methanol01/project_proposal2.htm (accessed May 26, 2009).

Heywood, J. B. 1984. *Internal combustion engines.* New York: McGraw-Hill.

Hong, H., H. Jin, J. Ji, Z. Wang, and Z. Cai. 2005. Solar thermal power cycle with integration of methanol decomposition and middle-temperature solar thermal energy. *Solar Energy* 78:49–58.

Huang, Z. H., H. B. Lu, D. M. Jiang, K. Zeng, B. Liu, J. Q. Zhang, and X. Wang. 2004a. Engine performance and emissions of a compression ignition engine operating on the diesel-methanol blends. *Proceedings of the Institution of Mechanical Engineers, Part D: Journal of Automobile Engineering* 218:435–47.

Huang, Z. H., H. B. Lu, D. M. Jiang, K. Zeng, B. Liu, J. Q. Zhang, and X. Wang. 2004b. Combustion characteristics and heat release analysis of a compression ignition engine operating on a diesel/methanol blend. *Proceedings of the Institution of Mechanical Engineers, Part D: Journal of Automobile Engineering* 218:1011–24.

Huang, Z. H., H. B. Lu, D. M. Jiang, K. Zeng, B. Liu, J. Q. Zhang, and X. B. Wang. 2004c. Combustion behaviors of a compression-ignition engine fuelled with diesel/methanol blends under various fuel delivery advance angles. *Bioresource Technology* 95:331–41.

Huber, G. W., S. Lborra, and A. Corma. 2006. Synthesis of transportation fuels from biomass: Chemistry, catalysts, and engineering. *Chemical Reviews* 106:4044–98.

Ilhan, M. 2007. The effect of injection timing on the performance and emissions of a dual fuel diesel engine. MSc. Thesis, Marmara University, [in Turkish].

INEOS. 2009. *Paraform, Methanol price.* Online available at http://www.ineosparaform.com/55-Methanol_price.htm (accessed May 27, 2009).

Ingersoll, J. G. 1996. *Natural gas vehicles.* Lilburn: Fairmont Press

Jia, G. X., Y. S. Tan, and Y. Z. Han. 2006. A comparative study on the thermodynamics of dimethyl ether synthesis from CO hydrogenation and CO_2 hydrogenation. *Industrial and Engineering Chemistry Research* 45:1152–59.

Jothi, N. K. M., G. Nagarajan, and S. Renganarayanan. 2007. Experimental studies on homogeneous charge CI engine fueled with LPG using DEE as an ignition enhancer. *Renewable Energy* 32:1581–93.

Kaneko, S., Y. Kobayashi, T. Aruga, and T. Kabata. 2003. *Method and apparatus for producing methanol making use of biomass material.* U.S. Patent 6645442.

Kavet, R., and K. M. Nauss. 1990. The toxicity of inhaled methanol vapors. *Critical Reviews in Toxicology* 21:21–50.

Kowalecwicz, A. 1993. Methanol as a fuel for spark ignition engines: A review and analysis. *Proceedings of the Institution of Mechanical Engineers, Part D: Journal of Automobile Engineering* 207:43–52.

Kulakoglu, T. 2009. Effect of injection pressure on the performance and emissions of a diesel engine fueled with methanol-diesel blends. MSc. Thesis, Marmara University, [in Turkish].

Kumar, M. S., A. Kerihuel, J. Bellettre, and M. Tazerout. 2005. Experimental investigations on the use of preheated animal fat as fuel in a compression ignition engine. *Renewable Energy* 30:1443–56.

Kumar, M. S., A. Ramesh, and B. Nagalinman. 2003. An experimental comparison of methods to use methanol and Jatropha oil in a compression ignition engine. *Biomass and Bioenergy* 25:309–18.

Lange, J. P. 2001. Methanol synthesis: A short review of technology improvements. *Catalysis Today* 64:3–8.

Lee, J., G. Speight, and S. K. Loyalka. 2007. *Handbook of alternative fuel technologies.* New York: CRC Press.

Lee, S. 1997. *Methane and its derivatives.* New York: Marcel Dekker.

Liao, S. Y., D. M. Jiang, Q. Cheng, Z. E. Huang, and K. Zeng. 2006. Effect of methanol addition into gasoline on the combustion characteristics at relatively low temperatures. *Energy and Fuels* 20:84–90.

MERCK. 2006. *Product Specification*. Germany.

Methanol Fact Sheets. 1992. *Methanol: Reformulated natural gas.* American Methanol Institute: Washington, DC. Online available at http://www.afdc.energy.gov/afdc/pdfs/2475.pdf (accessed May 28, 2009).

Murayama, T., N. Miyamoto, T. Yamada, and J. I. Kawashima. 1982. A method to improve the solubility and combustion characteristics of alcohol–diesel fuel blends. *SAE* 821113.

National Research Council, Advisory Committee on Technology Innovation. 1983. *Alcohol fuels: Options for developing countries.* Washington, DC: National Academy Press.

Nouri, S., and A. M. Tillman. 2005. Evaluating synthesis gas based biomass to plastics (BTP) technologies. CPM—Centre for Environmental Assessment of Product and Material Systems, Chalmers University of Technology, Göteborg, CPM report 2005:6.

Nowell, G. P. 2009. *The promise of methanol fuel cell vehicles.* Prepared for the American Methanol Institute, State University of New York at Albany. Online available at http://www.methanol.org/fuelcell/special/amipromise.pdf (accessed May 28, 2009).

Nwafor, O. M. I. 2003. The effect of elevated fuel inlet temperature on performance of diesel engine running on neat vegetable oil at constant speed conditions. *Renewable Energy* 28:171–81.

Nwafor, O. M. I., G. Rice, and A. I. Ogbonna. 2000. Effect of advanced injection timing on the performance of rapeseed oil in diesel engines. *Renewable Energy* 21:433–44.

Olah, G. A., A. Goeppert, and G. K. Prakash. 2006. *Beyond oil and gas: The methanol economy.* Weinheim, Germany: Wiley-VCH Verlag GmbH & Co.

Olah, G. A., and G. K. S. Prakash. 2006. *Selective oxidative conversion of methane to methanol, dimethyl ether and derived products.* WIPO Patent Application WO/2006/113294.

Olah, G. A., and G. K. S. Prakash. 2009. *Conversion of carbon dioxide to dimethyl ether using bi-reforming of methane or natural gas.* U.S. Patent Application 2009/0030240.

OPPT. 1994. *Chemical summary for methanol.* Office of Pollution Prevention and Toxics, U.S. Environmental Protection Agency. Online available at http://www.epa.gov/opptintr/chemfact/s_methan.txt (accessed May 28, 2009).

Ozaktas, T., M. Ergeneman, F. Karaosmanoğlu, and E. Arslan. 2000. Ignition delay and soot emission characteristics of methanol-diesel fuel blends. *Petroleum Science and Technology* 18:15–32.

Ozsezen, A. N., M. Canakci, and C. Sayin. 2008. Effects of biodiesel from used frying palm oil on the performance, injection, and combustion characteristics of an indirect injection diesel engine. *Energy and Fuels* 22:1297–1305.

Pirnie, M. 1999. *Evaluation of the fate and transport of methanol in the environment.* Washington, DC: Technical memorandum, prepared for American Methanol Institute.

Pons, J. H. 2009. *Methanol—An alternative fuel.* University of Southwestern Louisiana. Online available at http://www.dnr.louisiana.gov/sec/execdiv/techasmt/ecep/auto/m/m.htm (accessed May 26, 2009).

Popa, M. J., N. Negurescu, and C. Pana. 2001. Results obtained by methanol fuelling diesel engine. *SAE* 2001-01-3748.

Pukrakek, W. W. 1997. *Engineering fundamentals of the internal combustion engines.* New York: Simon and Schuster Co.

Ristinen, R., and J. Kraushaar. 2006. *Energy and environment.* New York: John Wiley and Sons.

Roan, V., D. Betts, A. Twining, K. Dinh, P. Wassink, and T. Simmons. 2004. *An investigation of the feasibility of coal-based methanol for application in transportation fuel cell systems.* USA: Internal report of University of Florida.

Sayin, C., H. M. Ertunc, M. Hosoz, I. Kilicaslan, and M. Canakci. 2007. Performance and exhaust emissions of a gasoline engine using artificial neural network. *Applied Thermal Engineering* 27:46–54.

Sayin, C., and K. Uslu. 2008. Influence of advanced injection timing on the performance and emissions of CI engine fueled with ethanol-blended diesel fuel. *International Journal of Energy Research* 32:1006–15.

Sayin, C., M. Ilhan, M. Canakci, and M. Gumus. 2009. Effect of injection timing on the exhaust emissions of a diesel engine using diesel–methanol blends. *Renewable Energy* 34:1261–69.

Schrader, J., M. Schilling, D. Holtmann, D. Sell, M. V. Filho, A. Marx, and J. A. Vorholt. 2009. Methanol-based industrial biotechnology: Current status and future perspectives of methylotrophic bacteria. *Trends in Biotechnology* 27:107–15.

Senthil, K. M., A. Kerihuel, J. Bellettre, and M. Tazerout. 2005. Experimental investigations on the use of preheated animal fat as fuel in a compression ignition engine. *Renewable Energy* 30:1443–56.

Shi, S. X., K. H. Zhao, M. L. Fu, S. K. Wang, and Z. Y. Sun. 1983. An investigation of using methanol as an alternative fuel for diesel engines. In *Proceedings of the 15th International Congress on Combustion Engines* Paris: CIMAC.

Song, R., J. Liu, L. Wang, and S. Liu. 2008. Performance and emissions of a diesel engine fuelled with methanol. *Energy and Fuels* 22:3883–88.

TUPRAS. 2005. *Product Specification.* Turkey.

Tzeng, G. H., C. W. Lin, and S. Opricovic. 2005. Multi-criteria analysis of alternative-fuel buses for public transportation. *Energy Policy* 33:1373–83.

Udayakumar, R., S. Sundaram, and S. Sivakumar. 2004. Engine performance and exhaust characteristics of dual fuel operation in DI diesel engine with methanol. *SAE* 2004-01-0096.

Vucins, P. A. 2006. *The global synthetic fuels alliance.* MSc. Thesis, Delft University of Technology, Delft, The Netherlands.

Wagner, T. O., D. S. Gray, B. Y. Zarah, and A. A. Kozinski. 1979. Practicality of alcohols as motor fuel. *SAE* 790429.

Wang, L. J., R. Z. Song, H. B. Zou, S. H. Liu, and L. B. Zhou. 2008. Study on combustion characteristics of a methanol–diesel dual-fuel compression ignition engine. *Proceedings of Instrumentation of Mechanical Engineers, Part D: Journal of Automobile Engineering* 222:619–27.

Watkins, C. 2008. *Carbon monoxide reduction to methanol. Organo-metallic chemistry.* Online available at http://bama.ua.edu/~kshaughn/ch609/paper/paper-s08/watkins-CO-red.pdf (accessed May 27, 2009).

Weissermel, K., H. J. Arpe, and C. R. Lindley. 2003. *Industrial organic chemistry.* Weinheim: Wiley-VCH Verlag GmbH & Co.

Yanfeng, G., L. Shenghua, and L. Yu. 2007. Investigation on methanol spray characteristics. *Energy and Fuels* 21:2991–97.

Yao, C., C. S. Cheung, C. Cheung, Y. Wang, T. L. Chani, and S. C. Lee. 2008. Effect of diesel/methanol compound combustion on diesel engine combustion and emissions. *Energy Conversion and Management* 49:1696–1704.

Yao, C. D., C. S. Cheung, C. H. Cheng, and Y. S. Wang. 2007. Reduction of smoke and NOx from diesel engines using a diesel/methanol compound combustion system. *Energy and Fuels* 21:686–91.

Zhu, Y. 1997. Alternative fuel engine control system. MSc. Thesis, Texas Tech University, Lubbock, TX.

5

Ethanol

Alan C. Hansen, Carroll E. Goering, and Arumugam Sakunthalai Ramadhas

CONTENTS

5.1 Introduction

Ethanol is an oxygenated fuel, produced from fermentation of biological renewable resources such as molasses, sugar cane, or starch. Because of benefits associated with ethanol, some countries started using ethanol as a substitute for gasoline or diesel. Sustaining a clean environment has become an important issue in an increasingly industrialized society. Ethanol was deemed the "fuel of the future" by Henry Ford and has continued to be the most popular alcohol-based fuel for several reasons including its production from renewable agricultural products and also because the incomplete oxidation by-products of ethanol (acetic acid and acetaldehyde) are less toxic than the incomplete oxidation by-products of other alcohols (Minteer 2006).

The use of ethanol for internal combustion engines was first investigated in 1897. However, ethanol utilization again picked up during fuel crises arising from World War I and II in the twentieth century. In the United States, a federal ethanol program was started during the energy crisis in the 1970s. The clean air act of 1970 allowed the Environmental Protection Agency (EPA) to set standards for vehicular emissions. This led to a requirement of oxygenated fuels to reduce the vehicular emissions from gasoline engines. The requirement corresponded to approximately 7.5% ethanol or 15% methyl tertiary butyl ether (MTBE) in gasoline by volume.

Gasoline powered vehicles typically emit less particulate matter (PM) compared to diesel powered vehicles. Research has shown that ethanol blended with gasoline could reduce environmental pollution and also slow depletion of petroleum reserves. Because of these benefits, many countries started using ethanol as a substitute or partial replacement for gasoline or diesel. If ethanol can be produced abundantly and economically, it will be an attractive alternative fuel for spark ignition (SI) engines. It can be used either as neat fuel or as a gasoline blend. Both options provide some advantages for engine performance, fuel economy, and exhaust emissions. The predominant use of ethanol as a fuel has been in the form of a gasoline-blending component to power SI engines. Its high octane number makes it attractive for use in such engines.

With the rapidly expanding production of ethanol and its lower cost relative to petroleum-based fuel in countries such as Brazil, interest in using ethanol as a blending component in diesel fuel has naturally occurred. Alternative methods of introducing ethanol into a compression-ignition engine such as fumigation have also been explored (Shropshire and Goering 1982). In the 1970s such interest was driven by a global fuel crisis and ethanol–diesel blends became an important topic of research (Hansen, Zhang, and Lyne 2005). From the late 1990s onward, renewed efforts in establishing the viability of ethanol–diesel blends as a commercial fuel took place. The U.S. Department of Energy published a study concerning the technical barriers to the use of ethanol in diesel fuel (McCormick and Parish 2001). Studies into creating stable ethanol–diesel blends with properties that were a closer match to diesel fuel were conducted. Moreover, laboratory and field tests have been performed to evaluate the impact of ethanol on compression-ignition engine performance, emissions, and durability.

The phase out of MTBE as a gasoline additive in California and several other U.S. states has generated renewed interest in the use of ethanol as a gasoline oxygenate. Worldwide, ethanol can be used at approximately 8%v in oxygenated gasoline or approximately 6%v in reformulated gasoline (RFG). The most significant use of ethanol in vehicles started in Brazil in 1970s. In that country, the National Alcohol Program was created to cope with the high oil prices of the 1970s and 1980s. Federal incentives, in combination with the participation of the automobile industry and the strong environmental appeal, made the program a success (EPA 2001). Linked with production and utilization is the need for fuel quality regulation to ensure engine compatibility and safety.

5.2 Ethanol Production

Fuel ethanol production is rapidly expanding all over the world. However, Brazil is the world leader of the production and commercialization of ethanol as engine fuel. In Brazil, out of the total sugar cane available for crushing, 45% goes for sugar cane production and the balance, 55%, for production of ethanol directly from sugar cane. Raw materials for producing ethanol vary from country to country depending upon the availability. As examples, sugar cane is used in Brazil, cereals in the United States, sugar beets in Europe, and molasses in India. Brazil uses ethanol as the sole fuel in about 20% of their vehicles and 25% ethanol and 75% gasoline blends in the remaining vehicles. In 1999, approximately 75% of all automobiles in Brazil ran on gasoline containing 24% ethanol, with a total fuel alcohol consumption of 13.8×10^6 m^3/year (EPA 2001). With the rapidly expanding production of ethanol and its lower cost relative to petroleum-based fuel in countries such as Brazil, there is increased interest in using ethanol as a blending component. The world's ethanol production in the year 2008 was 17 billion gallons as compared to 13 billion gallons in the year 2007. The break-up of world ethanol production is shown in Figure 5.1.

In addition to sugar cane and starch, ethanol can also be produced from natural gas, shale oil, and cellulosic biomass. A typical estimate of greenhouse gas reduction potential with respect to sources of ethanol is depicted in Figure 5.2 (www.eere.energy.gov).

In the United States, the most popular feedstock for the production of ethanol fuel is starch obtained from corn (maize). Conversion of corn to ethanol requires a multiple-step process. The corn is ground to remove the starch

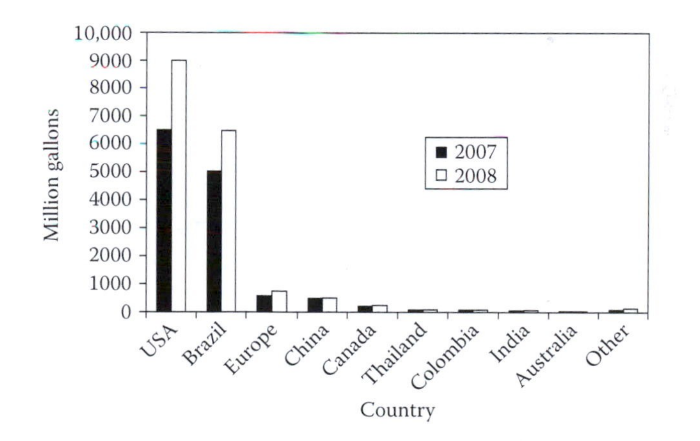

FIGURE 5.1
Ethanol production in the world. (From Licht, F. O., cited in Renewable Fuels Association, Ethanol Industry Outlook 2008 and 2009, www.ethanolrfa.org.)

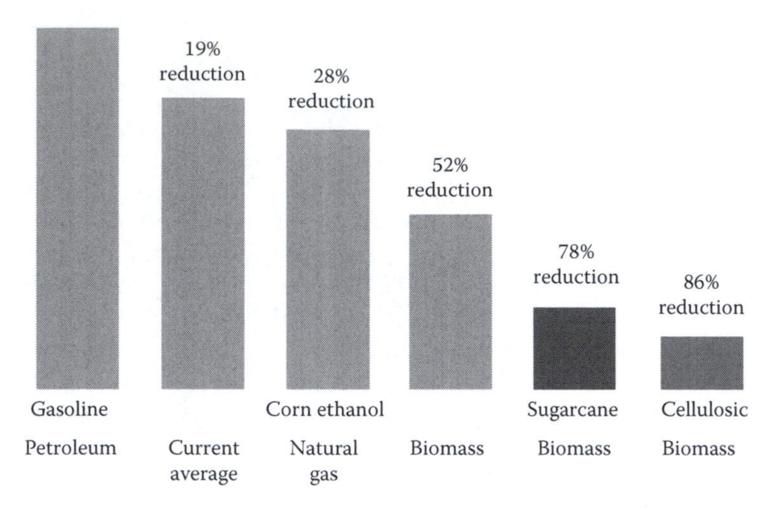

FIGURE 5.2
Greenhouse gas emission reduction potential with respect to source. (From Life-Cycle Energy and Greenhouse Gas Emission Impacts of Different Corn Ethanol Plant Types, www.eere.energy.gov/biomass/environmental.html.)

from the corn kernels. The starch comprises approximately 72% of the corn kernels. The starch is converted to glucose. The glucose is fermented to an ethanol beer. The beer is distilled to remove most of the water and, finally, a molecular sieve is used to obtain anhydrous ethanol. Saccharification, the process of converting starch to sugar, occurs as follows:

$$\begin{array}{ccc} \text{Starch} & \text{Water} & \text{Glucose} \\ (C_6H_{10}O_5)_n + & n(H_2O) \rightarrow & n(C_6H_{12}O_6). \end{array}$$

The process theoretically produces 1.11 kg of glucose per kg of corn starch. The fermentation of glucose to ethanol proceeds as follows:

$$\begin{array}{ccc} \text{Glucose} & \text{Yeast} & \text{Ethanol} \\ C_6H_{12}O_6 & \rightarrow & 2(C_2H_5OH) + 2\,CO_2. \end{array}$$

The process theoretically produces 0.511 kg of ethanol per kg of glucose. In English units, 19.46 pounds or 2.97 gallons of anhydrous ethanol are theoretically produced per bushel of corn. In metric units, the process produces 250 kg or 319 liters of anhydrous ethanol per cubic meter of corn. At least 85% of the theoretical yield has been achieved in practice. The yeast is not a reactant in the process; the living yeasts consume the glucose and produces ethanol. The process cannot progress beyond the point at which the concentration of ethanol in the beer becomes toxic to the yeast, at an ethanol concentration of approximately 15%. The beer is at least 85% water, and is distilled to separate out the ethanol. However, as the ethanol boils off and is condensed into liquid, some water accompanies the ethanol. Distillation cannot remove the last 5% of the water. Usually, a molecular sieve is used to remove this water to produce anhydrous ethanol. The corn-to-ethanol process also produces a high protein material that, after drying, is called distillers dried grains (DDG) and is used as animal feed.

For sugary crops such as sugar cane or sugar beets, the conversion of starch to glucose is not required. The sugar is extracted directly from the crop and the remainder of the process is the same as for converting corn starch to ethanol. Sugar cane is the primary feedstock for production of fuel ethanol in Brazil.

If the ethanol production feedstocks are limited to corn, sugar cane, or sugar beets, not enough ethanol can be produced to meet the total demand for engine fuel. For example, corn-based ethanol can meet less than 10% of the total demand for transportation fuel. Thus, there is great interest in expanding the feedstock pool to include cellulosic materials. The pool of potential cellulosic feedstocks is large enough that, when converted into ethanol, could potentially supply most if not all of the transportation fuel needs.

The technology for cellulose conversion is still in its infancy. A difficult key problem is to deconstruct the waxy lignin that provides the plant structural support while containing and protecting the cellulose and hemicellulose within. The lignin itself contains no sugars but it must be deconstructed to allow access to the cellulose and hemicelluloses. The growing interest in cellulosic-based fuel ethanol has sparked a great amount of research on lignin deconstruction. Three different approaches (physical, chemical and biological, and combinations of those) are being studied. Pulverization is an example of the physical approach. Treatment with concentrated acid is one chemical approach. Genetic modification of plants to produce more easily deconstructed lignin is a biological approach. A successful approach must be able to free up a high percentage of the sugars, be reasonably fast, and not consume excessive energy. Some of the sugars in cellulose and hemicellulose are not directly fermentable to ethanol. Thus, after the lignin is deconstructed and removed, enzymes are used to convert these unfermentable sugars to a fermentable form. The remainder of the process is then similar to the production of ethanol from corn.

Beyond the conversion of cellulose to ethanol, other problems must be solved before cellulose can become a useful feedstock for ethanol production. These include a selection of the best plant species and development of procedures for harvesting, densification, transport, and storage of the cellulosic crops. Among the species being considered are grasses such as switch grass or miscanthus. Currently unutilized crop residues such as corn cobs, corn stalks, or sugar cane bagasse are also possibilities. Products such as waste paper or wood wastes could also be sources of cellulose. For each such cellulose source, it would be necessary to develop economical procedures for harvesting or collecting the material and perhaps increasing its density to aid in transport from the point of collection/harvest to the ethanol processing plant. Also, since crop production is seasonal but ethanol processing plants require a steady input of feedstocks, ways must be found to temporarily store the cellulosic materials without subjecting them to excessive degradation.

The entire processes for converting starchy or sugary crops to fuel ethanol are now well established. Sustained research in this area has led to the processes becoming much more efficient. Many of the processes for production of cellulosic fuel ethanol are still in the research stage. Plans for constructing cellulosic ethanol plants are under consideration. In May 2008, the Verenium Company of Cambridge, Massachusetts, opened a cellulosic ethanol plant in Jennings, Louisiana, to convert sugar cane bagasse to ethanol (Bullis 2008). Over time, as more problems are worked out through research, the processes to convert cellulosic materials to fuel ethanol will likely become more efficient, just as they did for starchy or sugary crop conversion to ethanol.

5.3 Ethanol Properties

The quality of ethanol as a fuel has a major influence on factors such as blend stability, materials compatibility in the engine and corrosion of engine components. Quality standards for ethanol fuel blended with gasoline have been established in a number of countries. ASTM D4806 and D5798 standards in the United States are two specifications addressing the quality of ethanol blended with gasoline. D4806 was first published in 1999 and covers anhydrous denatured fuel ethanol intended to be blended with unleaded or leaded gasoline at 1–10 volumetric percentages for use as a SI automotive engine fuel. D5798 was also published in 1999 to regulate the quality of E85, a fuel blend specified as 75–85%v denatured fuel ethanol and 25–15%v hydrocarbons for use in automotive SI engines. ASTM D4806 has been used as the basis for development of standards in countries such as Australia, Canada, and China. In Europe, the EN 15376 standard was finalized in 2008 for use of ethanol as a blending component for gasoline up to 5%v.

Gasoline is composed of C_4–C_{12} hydrocarbons, and therefore has wider transitional properties than ethanol. The alcohol contains an oxygen atom so that it can be viewed as a partially oxidized hydrocarbon. Ethanol is isomeric with dimethyl ether (DME) and both ethanol and DME can be expressed by the chemical formula C_2H_6O. The oxygen atom in ethanol possibly induces three hydrogen bonds. Although, they may have the same physical formula, the thermodynamic behavior of ethanol differs significantly from that of DME on account of the stronger molecular association via hydrogen bonds in ethanol. Ethanol contains about 35% of oxygen, which improves the combustion and reduces the partial combustion of fuel. Ethanol is the preferred ingredient for the biodiesel transesterification process as it is produced from agricultural products and renewable in nature and it is less objectionable to environment. The properties of ethanol in comparison with gasoline and diesel are shown in Table 5.1.

Ethanol has higher research octane numbers (RON; > 100) than gasoline. When ethanol is blended with gasoline, the RON is boosted but there is minimal or no increase in motor octane number (MON). Ethanol–gasoline blends of the same RON have given better performance during low-speed and accelerating conditions, but were inferior in performance at higher speed conditions in comparison with gasoline. The MON can be increased by adding suitable components in the refinery itself and hence the risk of engine damage can be minimized.

High octane helps prevent engine knocking and is extremely important in engines designed to operate at a high compression ratio to generate more power. High compression ratios result in higher energy efficiency. Low-level blends of ethanol, such as E10 ($10\%_v$ ethanol, $90\%_v$ gasoline), generally have a higher octane rating than unleaded gasoline. Low-octane gasoline can be blended with 10% ethanol to attain the standard $(M + R)/2$ requirements of 87 for regular gasoline. $(M + R)/2$ is the numerical average of the MON and RON. Abdel-Rahman and Osman (1997) tested 10%, 20%, 30%, and 40% volume blends of ethanol in gasoline for use in a variable compression ratio engine. They found that increased ethanol content increased the octane number, but decreased the heating value. They reported that E10 is an optimum blend for gasoline engine applications.

Ethanol is an effective solvent and can be considered a fuel detergent. Thus ethanol use as a gasoline additive helps to remove gum and deposits from fuel systems. Deposits in fuel tanks and carburetor bowls will eventually cause problems in engines running on straight gasoline; however, ethanol blends may accelerate the release of deposits. It has been reported that ethanol has some detergent properties that reduce buildup, which keeps engines running smoothly and fuel injection systems clean for better performance (EPA420-F-00-035 2002).

Blends of gasoline and ethanol form azeotropes that cause a disproportionate increase in vapor pressure and a reduction in front-end distillation temperature. This effect varies with the amount of ethanol content but becomes significant at ethanol concentrations around $10\%_v$. The increase in vapor pressure could cause hot drivability problems in vehicles. Hence, random mixing of gasoline and ethanol should be avoided for hot drivability problems. To tackle this problem, it is necessary to remove high volatility components such as butane from the gasoline. Moreover, high ethanol concentrations in blends can cause cold drivability problems because ethanol has a higher latent heat of vaporization than gasoline.

Though the Reid vapor pressure of pure ethanol is low in comparison with gasoline, the RVP of gasoline-ethanol blends rises depending on the ethanol

TABLE 5.1

Ethanol Fuel Properties in Comparison with Gasoline and Diesel

Property	Ethanol	Gasoline	No. 2 Diesel
Chemical Formula	C_2H_5OH	C_4 to C_{12}	C_3 to C_{25}
Molecular Weight	46.07	100–105	≈200
Carbon, %w	52.2	85–88	84–87
Hydrogen, %w	13.1	12–15	33–16
Oxygen, %w	34.7	0	0
Specific gravity, 15.5°C/15.5°C	0.796	0.72–0.78	0.81–0.89
Density, kg/m^3	735	719–779	848
Boiling temperature, °C	78	27–225	180–340
Reid vapor pressure, bar	0.16	0.55–1.03	<0.01
Research octane no.	108	90–100	–
Motor octane no.	92	81–90	–
(R + M)/2	100	86–94	N/A
Fuel in water, volume %	100	Negligible	Negligible
Water in fuel, volume %	100	Negligible	Negligible
Freezing point, °C	–114	–40	–40 to –1 (pour point)
Viscosity, cSt @ 20°C	1.50	0.5–0.6	2.8–5.0
Flash point, closed cup, °C	13	–43	60–80
Autoignition temperature, °C	423	257	316
Flammability limits, v%	4.3–19.0	1.4–7.6	1.0–6.0
Higher heating value, MJ/kg	29.84	46.53	45.76
Stoichiometric ratio	9.0	14.7	14.7

Source: U.S. Department of Energy, Office of Energy Efficiency and Renewable Energy, Alternative Fuels Data Center, http://www.eere.energy.gov/afdc/altfuel/fuel_properties.html, 1999.

proportion in the blend. Low RVP can cause cold starting problems; therefore, volatile additives should be used when pure ethanol is used (He et al. 2003a and 2003b).

Alcohol is completely miscible with water in all proportions, while gasoline and water are immiscible. Ethanol is completely miscible with gasoline. But even a small percentage of water in gasoline or ethanol leads to phase separation. Such separation can result in poor engine performance and damage to the engine. Hence, it is necessary to ensure that ethanol to be blended with gasoline is anhydrous (CONCAWE 1995).

If ethanol-gasoline blends contain water, there may be corrosion of mechanical components, especially components made of copper, brass, or aluminum. Hence, these materials should be avoided in the fuel delivery systems. Alcohol can react with most rubber, causing deterioration that can jam in the fuel lines. Therefore, when ethanol is included in the fuel, it is better to use fluorocarbon rubber as a replacement for rubber (Naegeli et al. 1997).

The autoignition temperature and flash point of alcohol are higher than those of gasoline, which make it safer for transportation and storage. The latent heat of evaporation of alcohol is more than three times higher than that of gasoline. This leads to a reduction in the intake temperature in the manifold and hence increases the volumetric efficiency. The calorific value of ethanol is lower than that of gasoline and hence more ethanol is required to achieve the same energy output. Since the ethanol itself contains oxygen, the amount of air required for complete combustion is lesser for alcohol. The stoichiometric air–fuel ratio (AFR) of alcohol is about 9 as compared to gasoline, which is 14.7.

Gasoline-alcohol mixtures may be prepared by addition of a certain amount of ethanol to the gasoline. Gasoline-ethanol mixtures, which contain up to $20\%_v$ ethanol, can be safely used without causing any damage to the engine. To use gasoline-ethanol mixtures as a motor fuel, the mixture must be stable and a separation of phases should not occur. In gasoline-ethanol-water systems, the phase separation depends on the ethanol and water content of the blend, the environmental temperature, and the composition of gasoline. In order to lower the phase separation temperature, higher aliphatic alcohols such as tertiary butyl alcohol, benzyl alcohol, cyclohexanol, or toluene are usually added to the gasoline-alcohol blends (Ferfecki and Sorenson 1983; Fikret and Yuksel 2004).

Environmentalists prefer low-carbon fuels to help reduce the emission of carbon into the atmosphere. Ethanol fuel contains a lower percentage of carbon than gasoline or diesel fuel. However, much of that advantage is lost because the lesser heating value of ethanol, compared to gasoline or diesel fuel, requires the ethanol to be consumed at a higher rate to achieve the same power output as from the petroleum fuel.

Biodegradability is a useful fuel property when fuel is inadvertently spilled. Speidel and Ahmad (1999) studied the biodegradability of a number of fuels. Their technique was to inoculate the fuel with microbes, seal the fuel in a closed chamber, and monitor the CO_2 production as the microbes consumed the fuel. Higher CO_2 production indicates a higher level of biodegradation of the fuel. Relevant results of their study are shown in Table 5.2. Sucrose is readily biodegradable and was assigned a value of 100%. On this relative scale, ethanol was twice as biodegradable as petroleum diesel fuel. Oxy-diesel consisted of 80% petroleum diesel, 15% ethanol, and 5% of PEC,

TABLE 5.2

Biodegradability of Selected Fuels

Fuel	Percent
Sucrose	100
Ethanol	58
Oxydiesel	50
Diesel	28

Source: From Speidel, H. K. and Ahmed, I., *SAE* 1999–01-3518, in *SAE* SP-1482, 1999.

a proprietary additive of the Pure Energy Corporation. The oxy-diesel contained both ethanol and diesel fuel and thus, not surprisingly, fell between the two in biodegradability.

5.4 Ethanol–Gasoline Engine Tests

Ethanol can be used either as a pure fuel or as a gasoline additive. Both options can provide some advantages for engine performance, fuel economy, and exhaust emissions. Ethanol can be blended with gasoline at any percentages in pure form and used in SI engines. Gasoline-ethanol blends at low proportions can be used without any engine modification but pure ethanol requires major modifications to the engine design and fuel system.

5.4.1 Engine Dynamometer Tests

Palmer (1986) used various blend rates of ethanol-gasoline fuels in engine tests. Results indicated that 10%v ethanol addition increases the engine power output by 5%, and the octane number can be increased by 5% for each 10%v ethanol added. The reduction of CO emission is apparently caused by the wide flammability and oxygenated characteristics of ethanol. Therefore, the ethanol blends provided improvements in power output, efficiency, and fuel economy. Al Hasan (2003) conducted tests using a four cylinder, four stroke SI engine with a swept volume of 1452 c.c., a compression ratio of 9:1 and a maximum power of 52kW at 5600 rpm.

Brake thermal efficiency of the engine increased with increased ethanol concentration in the blends. The maximum brake thermal efficiency was achieved with an E20 blend. Torque and power increased with ethanol content up to $20\%_v$ but decreased with higher ethanol concentrations (Figure 5.3).

During the compression stroke, vaporization of fuel tends to decrease the temperature of the fuel–air mixture and increase the quantity of vapor. However, when a low-latent heat fuel such as gasoline is used, the effect of cooling is not sufficient to overcome the effect of additional vapor. Increasing the latent heat of the fuel blend used by increasing the ethanol concentration increases the effect of cooling, which reduces the compression work. As the ethanol concentration increases in the fuel blend, the pressure and temperature decrease at the beginning of combustion. However, increasing the ethanol concentration increases the AFR; that is, decreases the heat transfer to the cylinder walls due to incomplete combustion, and therefore, increases the value of maximum pressure. Hence, increasing the ethanol concentration in the fuel blend increases the indicated efficiency remarkably.

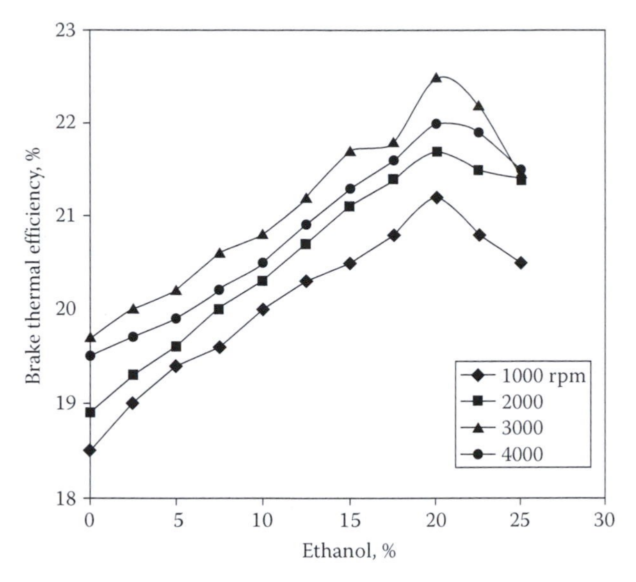

FIGURE 5.3
Brake thermal efficiency of ethanol-gasoline blends. (From Al-Hasan, M., *Energy Conversion and Management*, 44, 1547–61, 2003. Reprinted with permission from Elsevier Publication.)

Figures 5.4 through 5.6 show the CO, HC, and CO_2 emissions of ethanol-gasoline blends. Hydrocarbon and carbon monoxide emissions decrease with increased ethanol concentration in blends up to 20%v. In contrast, carbon dioxide emissions increase with increased ethanol concentration up to 20%v.

The CO emissions for ethanol–gasoline blends are reduced due to oxygen enrichment coming from ethanol. This result can be regarded as a "premixed oxygen effect" to make the reaction go to a more complete state. Ethanol molecules are polar, and cannot be absorbed easily by the unpolar molecules in the lubricating oil layer; and therefore ethanol can lower the possibility of producing HC emissions. In general, unburned hydrocarbons in the exhaust are mainly caused by three mechanisms:

1. Misfiring or incomplete combustion, which occurs in highly rich or lean situations, or when the air–fuel mixture contains excess amounts of recirculated exhaust gases.
2. Flame quenching, which takes place near combustion chamber surfaces.
3. Formation of deposits on combustion chamber inner walls surfaces that absorb fuel during the intake stroke and release fuel during the exhaust stroke. Minimum HC emissions occur in the condition of stoichiometric to slightly lean combustion.

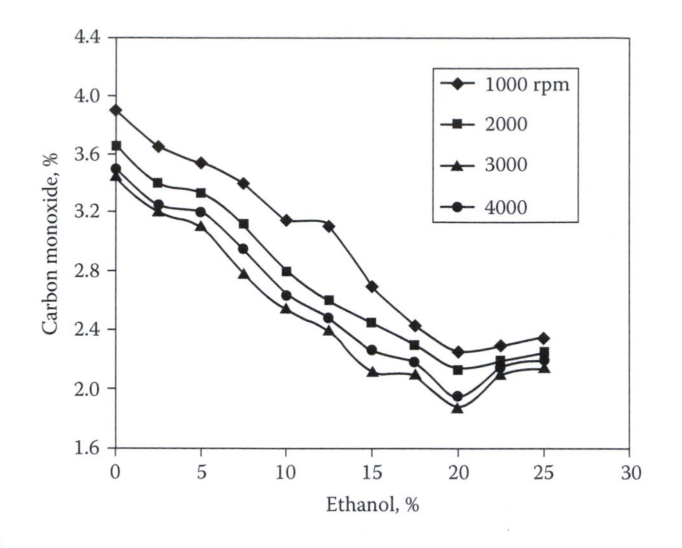

FIGURE 5.4
CO emissions with ethanol–gasoline blends. (From Al-Hasan, M., *Energy Conversion and Management*, 44, 1547–61, 2003. Reprinted with permission from Elsevier Publication.)

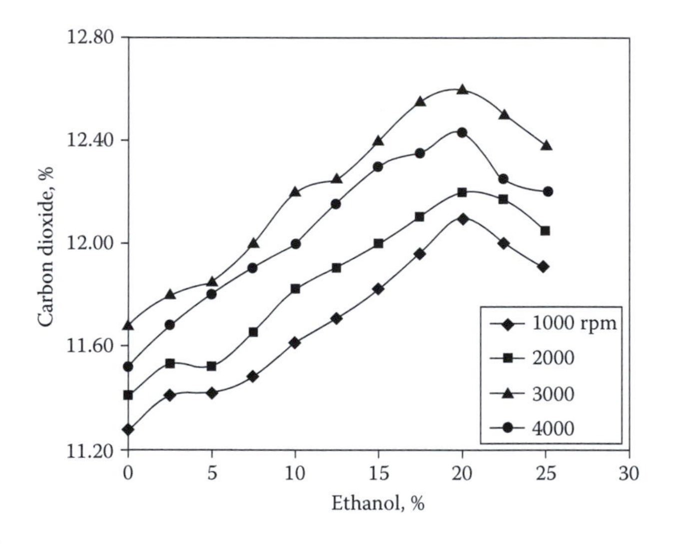

FIGURE 5.5
CO_2 emissions with ethanol–gasoline blends. (From Al-Hasan, M., *Energy Conversion and Management*, 44, 1547–61, 2003. Reprinted with permission from Elsevier Publication.)

The exhaust emissions of various blends of gasoline–ethanol with respect to gasoline at idle speed is shown in Figure 5.7. Ethanol emissions in the engine exhaust is shown in Figure 5.8.

Unregulated emissions of a gasoline–ethanol fueled, 66 kW, multipoint fuel injection system (MPFI) engine were analyzed by He et al. (2003a and

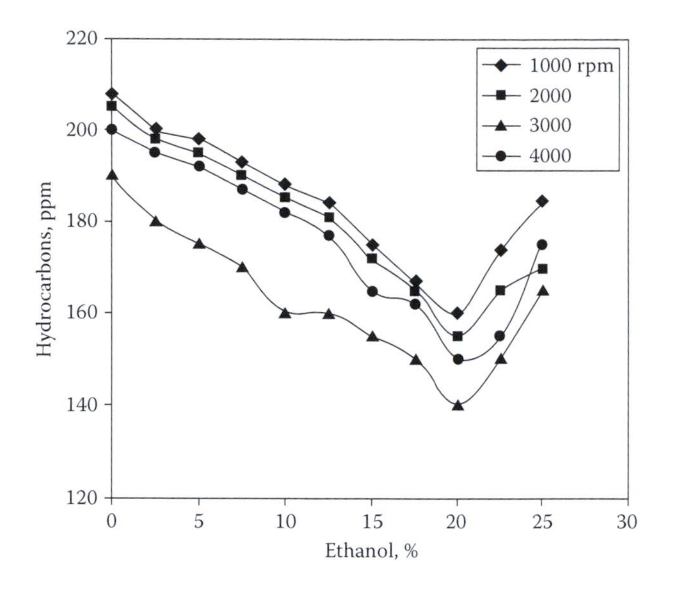

FIGURE 5.6
HC emissions with ethanol–gasoline blends. (From Al-Hasan, M., *Energy Conversion and Management*, 44, 1547–61, 2003. Reprinted with permission from Elsevier Publication.)

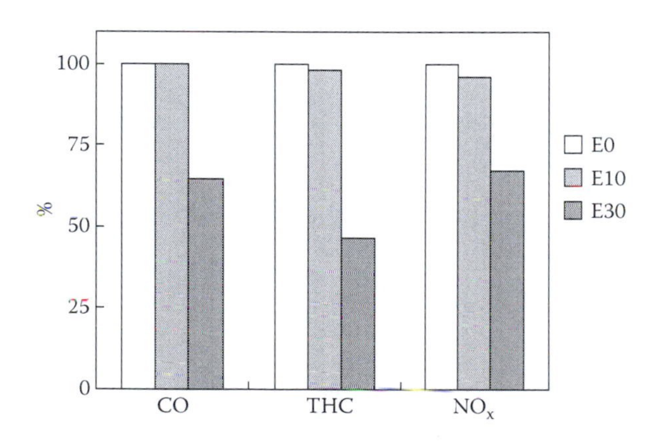

FIGURE 5.7
Exhaust emissions at idle speed. (From He, B. Q., Wang, J. X., Hao, J. M., Yan, X. G., and Xiao, J. H., *Atmospheric Environment*, 37, 949–57, 2003. Reprinted with permission from Elsevier Publications.)

2003b). It has been reported that engine-out unburned ethanol emissions of E30 are more than two times those of E10. The engine-out acetaldehyde emissions increased as the proportion of ethanol increased. Engine-out acetaldehyde emissions of E0 were quite low relative to those of the blended fuels, indicating that more acetaldehyde emissions are formed due to the oxidation of ethanol (Figure 5.9).

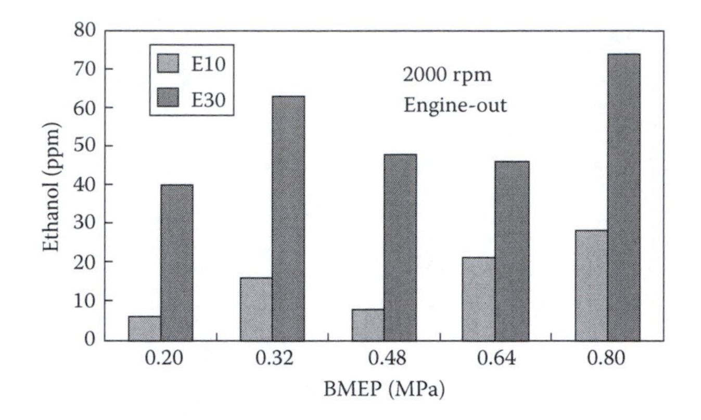

FIGURE 5.8
Unburned ethanol emission in exhaust of an ethanol-gasoline blend engine. (From He, B. Q., Wang, J. X., Hao, J. M., Yan, X. G., and Xiao, J. H., *Atmospheric Environment*, 37, 949–57, 2003. Reprinted with permission from Elsevier Publications.)

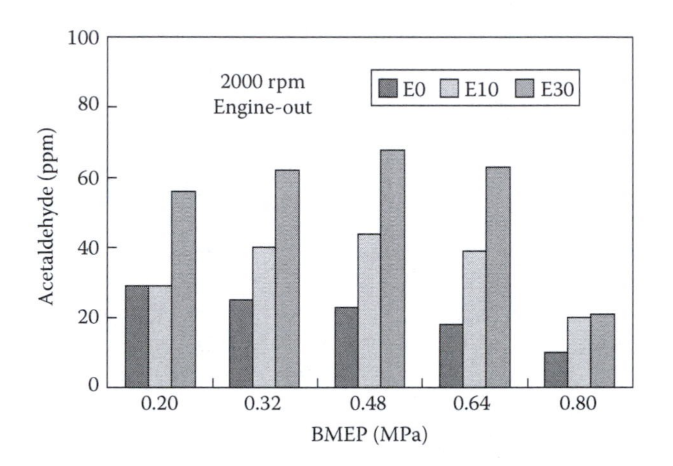

FIGURE 5.9
Acetaldehyde emission in exhaust of an ethanol–gasoline blend engine. (From He, B. Q., Wang, J. X., Hao, J. M., Yan, X. G., and Xiao, J. H., *Atmospheric Environment*, 37, 949–57, 2003. Reprinted with permission from Elsevier Publications.)

5.4.2 Engine Chassis Dynamometer Tests

Jia et al. (2005) conducted experiments using blends of gasoline–ethanol on a single-cylinder, four-stroke, air-cooled gasoline engine. They analyzed the engine emission characteristics of ethanol–gasoline blends by conducting ECE15 emission cycles on chassis dynamometer. The emission results are depicted Figure 5.10.

The exhaust emissions for base and E10 test fuels were 6.85 and 4.74 g/km. CO emissions decreased by about 30.8% for E10 fuel, compared to base fuel. This is due to improving the combustion process as a result of the oxygen

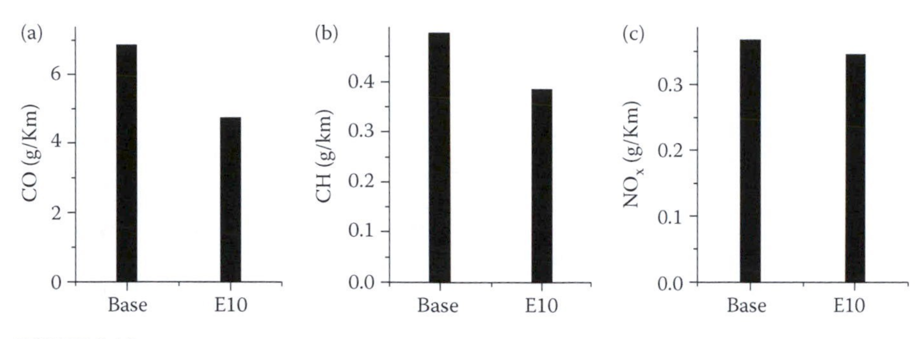

FIGURE 5.10
Exhaust emissions of an ethanol–gasoline vehicle. (From Jia, L. W., Shen, M. Q., Wang, J., and Lin, M. Q., *Journal of Hazardous Materials A123*, 29–34, 2005. Reprinted with permission from Elsevier Publications.)

content in ethanol fuels. HC emissions were reduced by nearly 31.7% with E10 fuel for one whole cycle. The oxygenate characteristics of ethanol in blend fuels are more effective in enhancing oxidation of hydrocarbons. The reduction of NO_x emissions was relatively small, about 5.9% for E10 fuel. The ethanol blends produced minor effect on the decrease of NO_x emissions.

Blending ethanol with gasoline positively affects the geometric properties of the flame and the mass burning rate, leading to faster burning. It also produces higher cylinder pressures and temperatures compared with gasoline. Higher combustion temperatures results in higher dissociation rates that tend to increase the NO_x emissions. Rising pressure and temperature can cause damage to engine structural components such as pistons, cylinders, and valves. For the use of ethanol blends, engine components should be manufactured to resist higher pressure and temperature.

Ethanol has been used for racing purposes because it has desirable properties that increase the power and torque outputs. Since the stoichiometric air–fuel ratio of ethanol is less than gasoline, an engine operating on E85 can use about 1.48 times more E85 for the same amount of air. About 1.40 times more E85 is required to equal the energy of gasoline on volumetric basis, leading to about 6 to 7% increase in power. Ethanol blends can burn cleaner and the engine spark timing can be advanced because of its higher octane number and hence higher power can be achieved (Davis 2006).

During the suction stroke, the engine draws in air or an air–fuel mixture. The high temperature in the engine evaporates the mixture and reduces the volumetric efficiency of the engine. The latent heat of vaporization of gasoline is lower than that of ethanol and hence ethanol requires more heat to evaporate than gasoline. This leads to an increase in volumetric efficiency with ethanol–gasoline blends and hence an improvement in power also.

Acetaldehyde emissions increase slightly when ethanol–gasoline blended fuels are used, since acetaldehyde may be produced through the partial oxidation of the ethanol in E10 fuel. It is well known that humans develop

TABLE 5.3

E85 Emission Reduction in comparison with Gasoline

CO	VOC	PM	NO_x	Sulfate
40	15	20	10	80

irritation of the eyes, skin, and the respiratory tract when exposed to acetaldehyde vapors. Ethylene emission amounts were higher for E10 fuel than for the base fuel at different cruising stages. Ethanol in fuel blends favors combustion of methane, which was mainly produced by the decomposition of other long-chain hydrocarbons. Moreover, the amounts of butane, pentane, and hexane emission at each driving mode were decreased to some extent when E10 fuel was used.

A U.S. EPA420-F-00-035 report on clean alternative fuels, ethanol highlights the emission characteristics of E85. It was reported that E85 produced fewer total toxic emissions, higher ethanol and aldehydes emissions and low-reactivity hydrocarbon emissions. The comparison emission characteristics of E85 with conventional gasoline are depicted in Table 5.3.

5.5 Ethanol–Diesel Engine Tests

Combustion in a CI engine is fundamentally different from that in a SI engine. Unlike a SI engine that has a spark plug to ignite the air-fuel mixture, the CI engine relies on high compression pressure to raise the temperature in the combustion chamber above self ignition temperature (SIT) of the fuel. However, the fuel does not ignite immediately upon reaching its SIT. Instead, there is an ignition delay period that starts upon injection of fuel into the combustion chamber and ends after certain preflame reactions are completed. Figure 5.11 is a conceptual illustration of combustion in a CI engine. All of the fuel that vaporizes and mixes with air during the ignition delay period combusts very quickly upon ignition, resulting in the triangular peak of fast energy release known as premixed combustion. The subsequent, slower combustion rate is limited by the rate at which fuel vapor and air can diffuse into each other and is known as diffusion combustion.

The cetane rating of CI engine fuels is closely related to the ignition delay; that is, a fuel with a high cetane rating has a short ignition delay and vice versa. Lowering the cetane rating of the fuel lengthens the ignition delay and thus shifts more combustion into the premixed mode. However, lowering the cetane rating too much can result in ignition being delayed until after the piston has moved far enough into the power stroke to lower the chamber temperature enough to prevent ignition entirely. Ethanol has a very

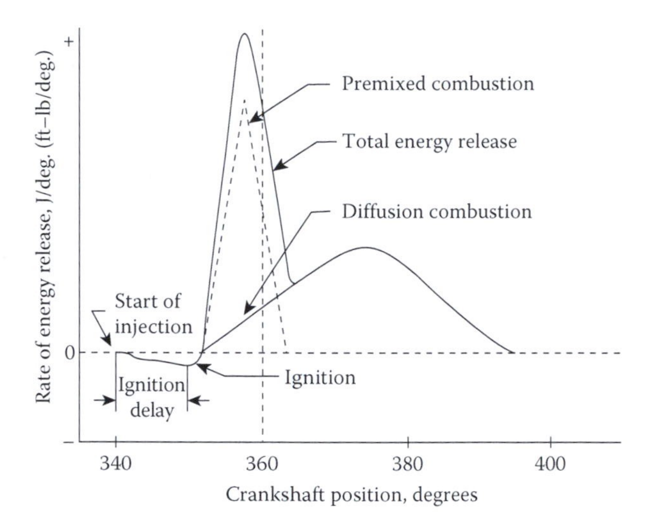

FIGURE 5.11
A conceptual diagram of heat release in a compression-ignition engine. (From Goering, C. E., Stone, M. L., Smith, D. W., and Turnquist, P. K., *Off-Road Vehicle Engineering Principles*, 2003. Reprinted with permission from the American Society of Agricultural and Biological Engineers.)

low-cetane rating and thus, when used as the complete fuel, will not self-ignite in an ordinary CI engine.

Consideration of the heat release diagram of a CI engine provides insights into the effect of ethanol on engine performance and durability. Premixed combustion is more efficient than diffusion combustion because premixed combustion releases energy near the start of the power stroke. Conversely, energy release from diffusion combustion occurs later in the power stroke and thus has less of the stroke in which to exert its influence. Up to a certain point, blending ethanol into diesel fuel increases the ignition delay, thereby increasing the indicated thermal efficiency of the CI engine. However, blending more than the optimum amount of ethanol into diesel fuel can cause ignition to be delayed well into the power stroke, thereby reducing the indicated thermal efficiency. Pressures and temperatures are higher in premixed combustion. Therefore premixed combustion puts more mechanical stress on the engine than diffusion combustion.

The heat release diagram gives insights into CI engine exhaust emissions. Most of the NO_x emissions occur in premixed combustion, where oxygen and nitrogen are relatively plentiful, while pressures and temperatures are high. Air is relatively scarcer in diffusion combustion and, consequently, most of the CO and smoke emissions are generated there. Thus, when more of the combustion is shifted into the premixed mode due to blending ethanol into the diesel fuel, the NO_x emissions are likely to increase, while CO and smoke emissions decrease. The increase in NO_x emissions can be moderated or eliminated by retarding the start of fuel injection into the cylinders.

Retarding the start of injection causes injection to begin when combustion chamber pressures and temperatures are higher, thus shortening the ignition delay. However, shortening the ignition delay increases the amount of diffusion combustion, with a consequent increase in CO and smoke emissions. Since the effect of ignition delay on NO_x emissions is opposite the effect on CO and smoke, choice of the desired ignition delay is a compromise.

As previously mentioned, when used as the complete fuel, ethanol will not self-ignite in an ordinary CI engine. However, there are a number of techniques available to use ethanol fuel in a CI engine. These techniques can be grouped into those involving modification or adaptation of the engine to be able to combust ethanol and those that rely on the fuel being modified to achieve CI engine compatibility, including blending of ethanol with diesel fuel. The most common methods for achieving dual fuel operation are

1. Alcohol fumigation: The addition of alcohols to the intake air charge, displacing up to 50% of the diesel fuel demand.
2. Dual injection: Separate injection systems for each fuel, displacing up to 90% of diesel fuel demand.
3. Alcohol-diesel fuel blend: Mixture of the fuels just prior to injection, displacing up to 25% of the diesel fuel demand.
4. Alcohol-diesel fuel emulsion: Using an emulsifier to mix the fuels to prevent separation, displacing up to 25% diesel fuel demand.

5.5.1 Fumigation

Fumigation is a method by which alcohol is introduced into the engine by carbureting, vaporizing, or injecting the alcohol into the intake air stream. This requires the addition of a carburetor, vaporizer or injector, along with a separate fuel tank, lines, and controls.

The intake manifold and intake valves of a conventional CI engine handle only air but, in a technique known as fumigation, ethanol can be blended with the air to supply partial fueling to the engine, while the primary fuel of adequate cetane rating is injected through the usual fuel injection system. While carburetors have been investigated as an inexpensive way of introducing ethanol into the intake airstream (Ajav, Singh, and Bhattacharya 1998; Goering and Wood 1982), manifold injection is regarded as the preferred approach in keeping with gasoline injection systems used in most applications at present.

Sullivan and Bashford (1981) sprayed ethanol into the air inlet of a turbocharged CI engine. Impingement of the ethanol on the compressor blades caused blade erosion after only 30 hours of running. Injection of the ethanol separately into each engine cylinder is preferable to insure equal distribution of the ethanol to each cylinder. Otherwise, the maximum rate of fumigation

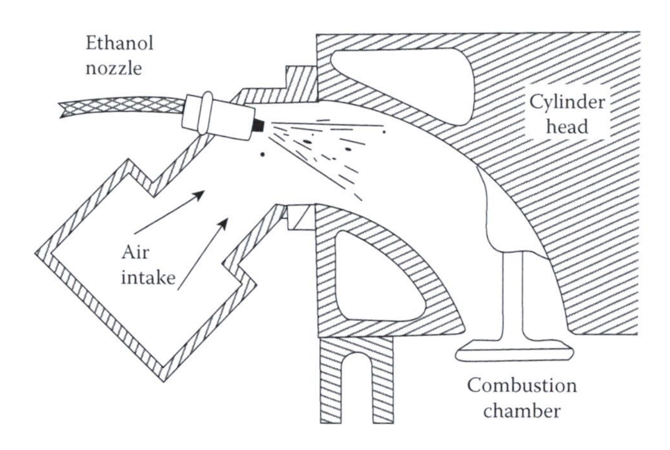

FIGURE 5.12
Nozzle placement to fumigate ethanol into a compression-ignition engine. (From Shropshire, G. J. and Goering, C. E., *ASABE Transactions*, 25, 570–75, 1982. Reprinted with permission.)

will be limited by the cylinder receiving the disproportionately larger share of the ethanol, while the other cylinders are underutilized. Figure 5.12 illustrates a fumigation system used by Shropshire and Goering (1982). They replaced the intake manifold with a segmented manifold designed for use on a SI engine. The segmented manifold kept the fumigated ethanol near the appropriate intake port to assist with equal distribution to all cylinders. The fumigated ethanol can be sprayed continuously or timed to be sprayed only when the intake valve is open. Goering et al. (1987) found that spraying the ethanol only when the intake valve was open produced excessive rates of cylinder pressure rise when the engine was at full load. Better results were obtained by spraying a portion of the ethanol when the intake valve was closed. Use of continuous spraying also permits the use of a simpler fumigation control system.

Depending on the technique used, the fumigated ethanol can replace part of the primary fuel or can be added to the primary fuel to over fuel the engine. Figure 5.13 illustrates the possibilities. Consider when the engine is operating at Point A in the governor-controlled range before fumigation begins. When the fumigated ethanol enters the engine its speed increases, causing the governor to reduce the primary fuel input. Operation now moves to Point B, at a slightly higher torque and speed than without fumigation. The situation in the load-controlled range is different. Here, the governor is unable to reduce the primary fuel consumption. Rather than replacing primary fuel, the fumigated ethanol only increases the power output of the engine, possibly to levels the engine is not designed to handle. Thus, when adding a fumigation system to a CI engine, provision must be made to prevent heavy over fueling in the load-controlled range of engine operation.

The maximum percentage of the total engine energy that can be supplied by fumigated ethanol varies depending on the situation. In a study by

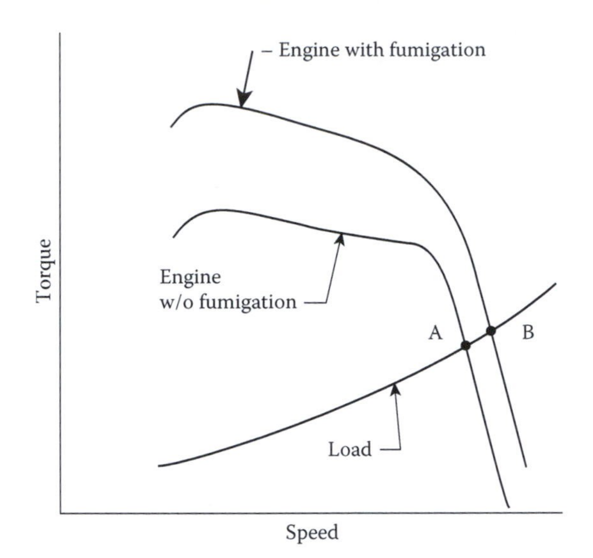

FIGURE 5.13
Engine torque output with and without fumigation. (From Goering, C. E. and Wood, D. R., *ASABE Transactions*, 25, 576–80, 1982. Reprinted with permission.)

TABLE 5.4
Maximum Percentage of Fuel Energy Supplied by Fumigated Ethanol to a CI Engine

		Speed Rev/Min			
Proof	**Load**	**1600**	**1800**	**2000**	**2200**
190	Full	55	48[a]	62	56
190	2/3	48	45	39	48[b]
190	1/3	37	31	51	44[b]
150	Full	56	66	59	54
150	2/3	47	42	55	49
150	1/3	37	34	47	43
100	Full	17	19*	19	
100	2/3	25	21	21	17
100	1/3	35	31	29	25

Source: From Shropshire, G. J. and Goering, C. E., *ASABE Transactions*, 25, 570–75, 1982.
[a] Knock limited.
[b] Efficiency limited.

Shropshire and Goering (1982), 190-proof fumigated ethanol supplied 62% of the total energy to a four-cylinder, naturally aspirated (NA), direct-injected (DI), CI running at full load, as shown in Table 5.4. Lowering the ethanol proof greatly reduced the percentage of energy that could be supplied by fumigated ethanol.

In a study by Goering et al. (1987), 200-proof fumigated ethanol supplied 35% of the total energy to a six-cylinder, turbocharged (TC), DI, CI engine running at full load and 90% under light loads. The latter investigators found that engine knock limited the maximum fumigation rate at full load, while engine misfire limited the maximum rate at light load. Barnes, Kittelson, and Murphy (1975) were able to supply 40% of the total energy to a multicylinder, TC engine at full load. Note that, if the rate of ethanol fumigation is not varied with engine load, the engine governor will cause the percentage of energy supplied by fumigated ethanol to be highest under light load and to decrease progressively as engine load is increased in the governor-controlled range. Provision must be made for limiting or stopping the flow of fumigated ethanol when all loads are removed from the engine. The engine will accelerate to the point of destruction otherwise.

When fumigated ethanol is used to replace a portion of the primary fuel rather than being used to over fuel the engine, most investigators have found a substantial reduction in exhaust smoke, little change in carbon monoxide emissions but a large increase in unburned hydrocarbons, especially at lighter loads. Goering et al. (1987) found that fumigation greatly reduced NO_x emissions at light loads but slightly increased them at moderate to full loads. Most investigators found that brake thermal efficiency was slightly reduced by fumigation under light loads but slightly increased under a full load.

5.5.2 Glow Plug Assisted Ignition

Kroeger (1986) modified a Caterpillar CI engine to run on methanol by installing a glow plug in each cylinder to ignite the fuel. An injection system of greater capacity was installed in the modified engine to offset the lesser heating value of the methanol.

Goering, Parcell, and Ritter (1998) experimented with a Detroit Diesel DI, CI, V6, 2-cycle bus engine that had been modified by Detroit Diesel to run on ethanol fuel. The Roots blower in the engine was bypassed at higher loads and a turbocharger was used to boost air consumption. Glow plugs were used to assist autoignition of the ethanol at start-up, during warm-up and for low-speed, light-load operations. Exhaust scavenging control was used to increase the residual fraction of exhaust gases in the combustion chambers to aid autoignition. The engine was equipped with a catalytic converter for exhaust gas treatment. The engine was calibrated to provide 207 kW of brake power while running on 200-proof ethanol denatured with 5% gasoline as the baseline fuel. Lubrizol 9520A fuel additive was added to the baseline fuel in 0.06% concentration. In the experiments by Goering, Parcell, and Ritter (1998) the engine was run on the baseline fuel and on the baseline fuel with enough distilled water added to simulate 190-proof ethanol. The engine was run for 454 hours on a load–speed cycle designed to simulate operation of Chicago Transit Authority buses. Test results are summarized in Figure 5.14. Except for an incident of injector needle sticking that caused power to decline, the

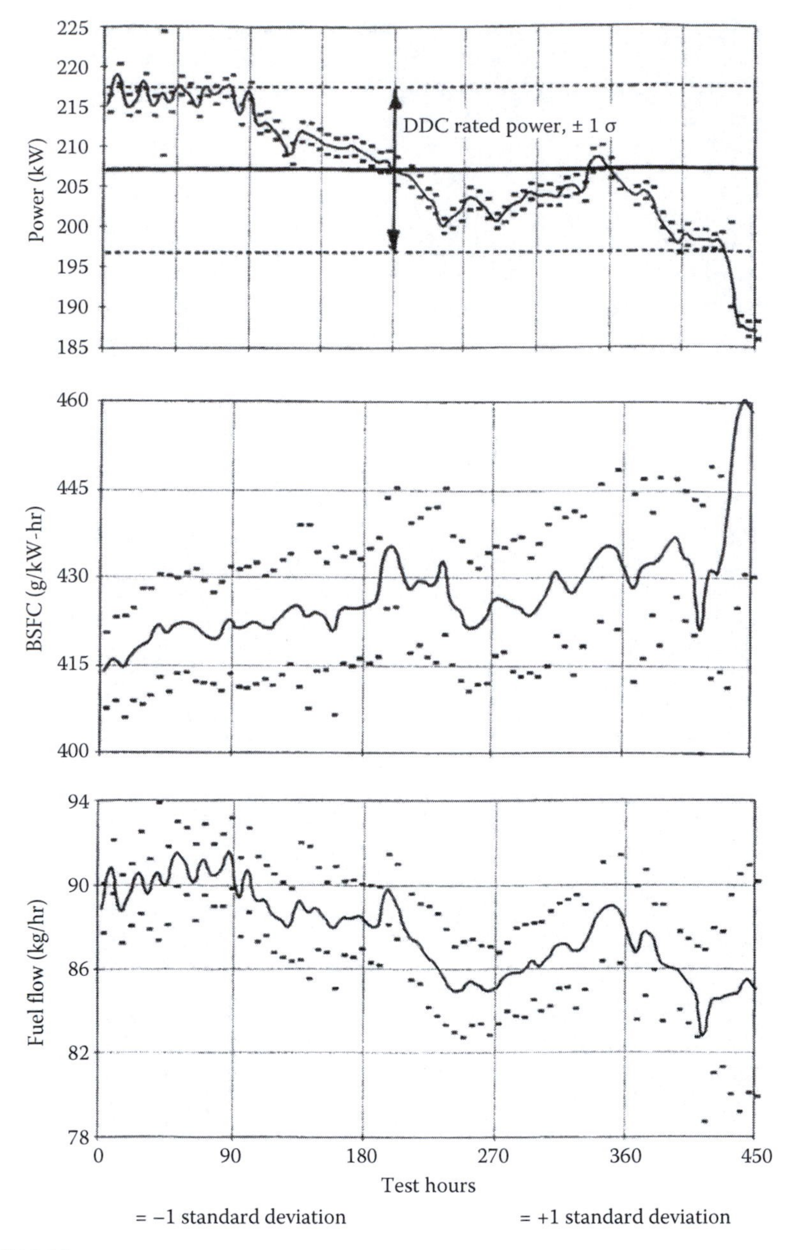

FIGURE 5.14
Test results for a Detroit Diesel engine modified to run on ethanol. (From Goering, C. E., Parcell, R. T., and Ritter, C. P., *ASABE Transactions*, 41, 1255–60, 1998. Reprinted with permission.)

TABLE 5.5
Emissions from a Detroit Diesel Engine Adapted to Run on Ethanol

Emissions (ppm)	Anhydrous, 0 h		Hydrated, 0 h		Hydrated, 454 h	
	Before Cat	After Cat	Before Cat	After Cat	Before Cat	After Cat
30% Throttle						
CO	995	125	133	35	1167	83
HC	519	75	677	76	6714	90
NO_x	55	60	9	56	49	50
100% Throttle						
CO	479	71	232	12	147	42
HC	203	103	149	50	186	69
NO_x	82	76	75	72	78	73

Source: From Goering, C. E., Parcell, R. T., and Ritter, C. P., *ASABE Transactions*, 41, 1255–60, 1998. With permission.
Note: An AC Rochester model HC-1590 HN-6941 catalytic converter was used.

engine ran well on the 190-proof ethanol. As shown in Table 5.5, the water in the 190-proof ethanol reduced the concentrations of NO_x, CO, and unburned hydrocarbons in the exhaust gases. The catalytic converter was also successful in reducing exhaust emissions.

5.5.3 Cetane Enhanced Ethanol

Attempts to burn ethanol in unmodified CI engines have highlighted the need for the use of fuel additives, both when burning ethanol on its own and when blended with diesel fuel. Special ignition improvers are available that, when blended with ethanol, can raise the cetane rating of ethanol high enough for combustion in a CI engine. However, a limited amount ethanol can be blended with a compression-ignition fuel to achieve a fuel capable of self-ignition.

Additives have been developed that will promote the self-ignition of ethanol in a CI engine. For example, Imperial Chemicals International developed an additive called Avocet in the early 1980s that contained an ignition improver, a lubricant to prevent excessive wear of the injection system, and a corrosion inhibitor. The mechanism by which ignition improvers cause ethanol to ignite in a CI engine is not well understood. Hardenburg and Schaefer (1981) believed it was a temperature-induced decomposition of nitrate molecules, resulting in formation of radicals that react with the ethanol to initiate and accelerate the combustion chain reaction. Goering et al. (1992) experimented with the use of Avocet-enhanced ethanol as the fuel for an International Harvester (IH) tractor equipped with a 128 kW, six-cylinder, DI, TC, CI engine. Avocet concentrations of 3, 3.5, and 4% were evaluated. The engine misfired badly with the 3% Avocet concentration. The engine also misfired under light load when 3.5% Avocet was used. The engine ran well under all loading conditions when the Avocet concentration was 4%.

TABLE 5.6

Maximum Power Engine Performance on Diesel Fuel and Enhanced Ethanol

Parameter	D2[a]	EE[b]
Engine speed, rpm	2113	2149
Power, kW	107	63
Fuel consumption, kg/h	28.2	29.3
A/F ratio	24.4	21.1
Indicated thermal efficiency, %	39	44
Brake thermal efficiency, %	32	29
BSFC, g/kW.h	265	466
Ignition delay, crankshaft degrees	3.50	7.00
Start of combustion, deg. before HDC	14	8
Peak combustion pressure, MPa	11.2	8.6
Max. rate of pressure rise, kPa/degrees	466	487
Maximum rate of heat release, kJ/degrees	80.9	118.1
Bosch smoke number	2.4	0

Source: From Goering, C. E., Crowell, T. J., Savage, L. D., Griffith, D. R., and Jarrett, M. W., *ASABE Transactions*, 35, 423–28, 1992.

[a] Original setting of injector pump.

[b] Injector pump reset for increased delivery.

Air temperature, 22–23°C.

Barometric pressure, 98.4–99.2 kPa.

Goering et al. (1992) used the IH tractor to evaluate the performance of ethanol with 4% Avocet compared to the performance of No. 2 diesel fuel. Test results are summarized in Table 5.6. The greater latent heat of vaporization of the ethanol lowered the temperatures in the combustion chamber, such that ignition delay increased. At maximum power, the ignition delay increased from 3.5° for No. 2 diesel to 7.0° for the ethanol fuel. Because the piston was already moving in the power stroke when the main combustion of the ethanol fuel occurred, the peak pressure diminished compared to No. 2 diesel fuel. The ethanol fuel has 39% less energy per unit volume than No. 2 diesel. Therefore, while the engine consumed the ethanol fuel at a slightly higher rate, it produced 41% less peak power compared to No. 2 diesel fuel. Indicated thermal efficiency was 39% for the No. 2 diesel and 44% for the ethanol fuel. However, the reduced power output lowered the mechanical efficiency and thus, the brake thermal efficiency was 39% for the No. 2 diesel fuel, but only 29% for the ethanol fuel. The engine produced modest smoke (Bosch smoke No. = 2.4) when running on No. 2 diesel but generated zero smoke when running on the ethanol fuel.

5.5.4 Ethanol Blended with Diesel

Adapting ethanol for combustion in a CI engine necessitates modifying its properties with the aid of additives, some of which have been mentioned earlier. Also it is necessary to overcome the immiscibility of ethanol in diesel fuel using suitable additives. E-diesel fuel refers to blends of ethanol with diesel fuel, which typically contain an additive to ensure diesel-like combustion and lubricity, as well as providing a stable blend. Commercial additives have been developed that allow anhydrous ethanol to be splash-blended with diesel fuel (Hansen, Zhang, and Lyne 2005). The amount of additive required varies in proportion to the percentage of ethanol in the blend.

The physical properties of diesel fuel are changed when ethanol is added into the solution (blend). The addition of ethanol causes the viscosity of diesel fuel to decrease. Also, the addition of ethanol in solutions with diesel fuel causes the cetane rating to drop and the heating values to be lower.

Blend stability is a key aspect in the formulation of ethanol–diesel fuel blends. This stability is affected primarily by two factors; namely, fuel temperature and water content. Ethanol and diesel fuel will separate when any water is added to the blend, or when the blend is cooled below about 10°C in the case of dry ethanol blends (Hansen, Zhang, and Lyne 2005). Prevention of this separation can be accomplished through the use of either emulsifiers or cosolvents. Figures 5.15 and 5.16 illustrate phase separation as functions of temperature and water addition (Letcher 1980, 1983).

Wrage and Goering (1980) investigated diesohol, a blend of 10% anhydrous ethanol and 90% No. 2 petroleum diesel fuel. Their concern as to whether the blend would have sufficient viscosity to lubricate the injection system led them to develop Figure 5.17. The blend of 10% anhydrous ethanol and 90% No. 2 diesel had viscosity well above the minimum for No. 2 diesel. In addition to measuring fuel properties, Wrage and Goering tested the fuel in a John Deere 37 kW, 3-cylinder, DI, CI engine equipped with a distributor type injection pump. The ethanol in the diesohol lowered the viscosity and cetane rating but both properties were above the lower limit for No. 2 diesel fuel. When used in the engine, the ethanol vaporized in the injection pump, causing the engine to lose power and stall. That problem was solved by pressurizing the diesohol before it reached the injector pump. Chilling the diesohol also solved the vapor lock problem but was considered less practical than pressurizing the fuel. The engine produced approximately the same power output on both fuels, but consumed the diesohol at a higher rate. Finally, the engine produced 30% less smoke when running on diesohol.

Boruff et al. (1982) used microemulsions to solve the phase separation problem. Microemulsions are transparent, thermodynamically stable colloidal dispersions in which the diameter of the dispersed-phase particles is less than one-fourth the wave length of visible light. A surface-active agent (i.e., a surfactant) is added to the ethanol–diesel blend to form the microemulsion. Boruff et al. evaluated a fuel blend containing 66.7% of No. 2 diesel fuel, 16.7% of 190-proof ethanol, 12.5% of Emersol 315, a commercial surfactant containing a blend of soy oil fatty acids, and 4.1% of N,N-dimethylethanolamine.

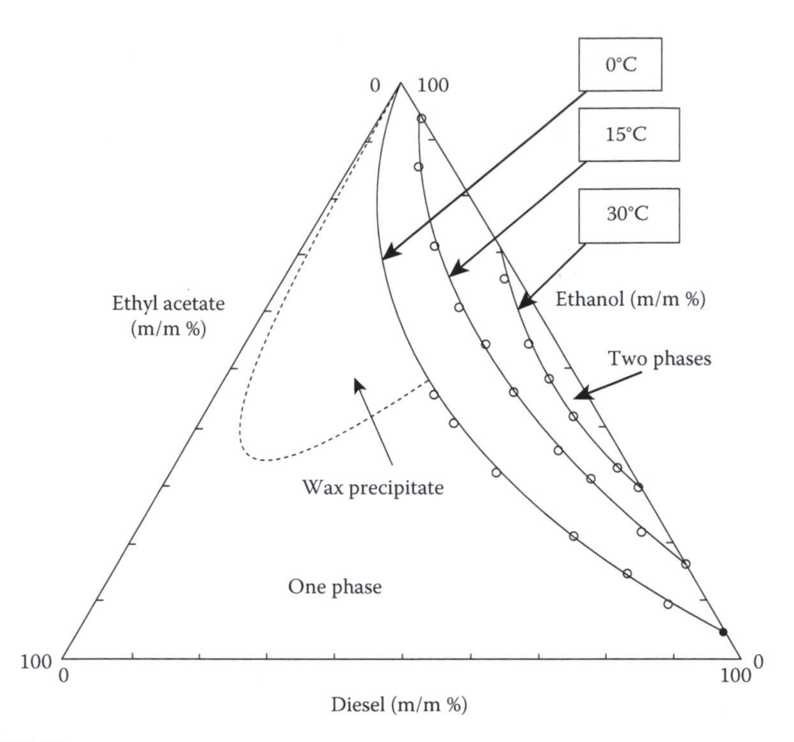

FIGURE 5.15
Phase separation diagram of diesel–ethanol–ethyl acetate blends as affected by temperature. (From Letcher, T. M., *South Africa Journal of Science,* 79(1), 4–7, 1983. Reprinted with permission from Academy of Science of South Africa.)

The microemulsion fuel was comparatively tested in a Ford three-cylinder, DI, CI engine equipped with a distributor-type injection pump. Compared to No. 2 diesel, the microemulsion fuel produced lower exhaust temperatures and 4–5% higher brake thermal efficiency. The microemulsion fuel also reduced exhaust smoke and CO emissions, but increased the level of unburned hydrocarbons. A concern for fuel safety led Boruff et.al. (1982) to develop Figure 5.18. In a closed fuel container, either for fuel storage or on a vehicle, the flammability of the gases above the liquid is a concern. Over a range of typical environmental temperatures, the mixture above either No. 1 or No. 2 diesel is usually too lean to burn if ignited. The mixture above gasoline is usually too rich to burn. However, the fuels containing ethanol or butanol are flammable at common environmental temperatures and therefore must be handled with caution.

More recent research has examined the use of biodiesel as an additive to offset all the blending deficiencies created by ethanol. Depending on the vegetable oil or animal fat from which the biodiesel is produced, it will typically have a higher cetane number, higher viscosity, and higher lubricity than diesel fuel, thereby being able to restore these properties in the case

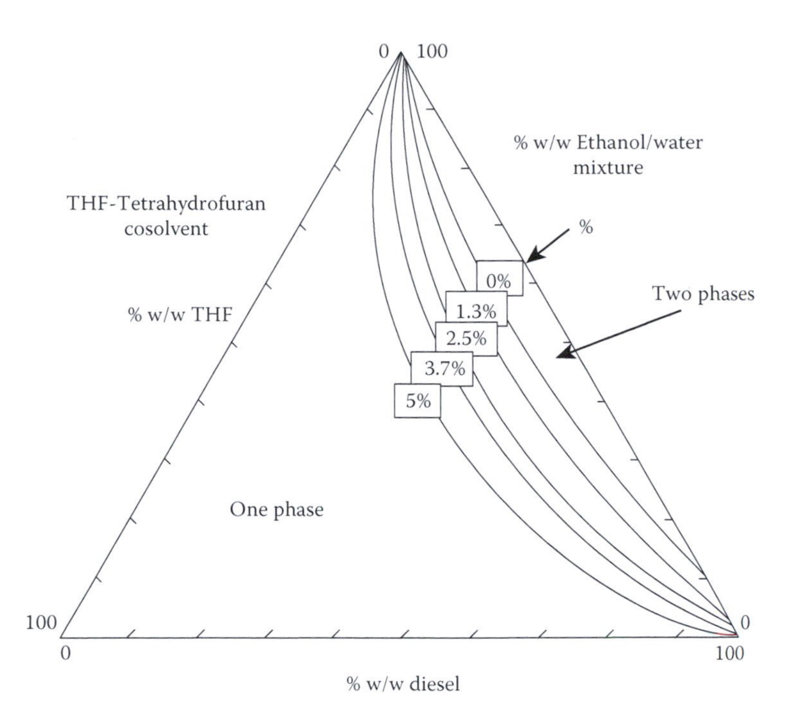

FIGURE 5.16
Phase separation of diesel-ethanol-cosolvent blends as affected by ethanol water content. (From Letcher, T. M., *South Africa Journal of Science*, 76(1), 130–32, 1980. Reprinted with permission from Academy of Science of South Africa.)

of ethanol blends to the levels specified in the ASTM diesel fuel standard. Furthermore, methyl esters tend to have higher energy content than diesel fuel, thereby offsetting the reduction in energy content caused by the addition of ethanol. Another very important attribute of methyl esters is that they have been shown to be effective surfactants in creating microemulsions. Fernando and Hanna (2005) explored the addition of biodiesel in ethanol–diesel blends and concluded that it could be used effectively as an amphiphile.

Shi et al. (2005) investigated the impact of methyl soyate–ethanol–diesel blends on exhaust emissions. For three different blends comprising 20% methyl soyate-80% diesel, 3% ethanol 12% methyl soyate 85% diesel, and 4% ethanol 16% methyl soyate 80% diesel, they found that the blends remained very stable over a period of three months even when exposed to air. Also the blends containing ethanol were able to tolerate water content of 1%–2% at a temperature of 20°C.

In a study by Kwanchareon, Luengnaruemitchai, and Jai-In (2007) it was concluded that diesel–biodiesel–ethanol (diesohol) blends containing 5% ethanol had properties that were similar to diesel fuel. They also stated that even though the physical characteristics of diesohol made it technically

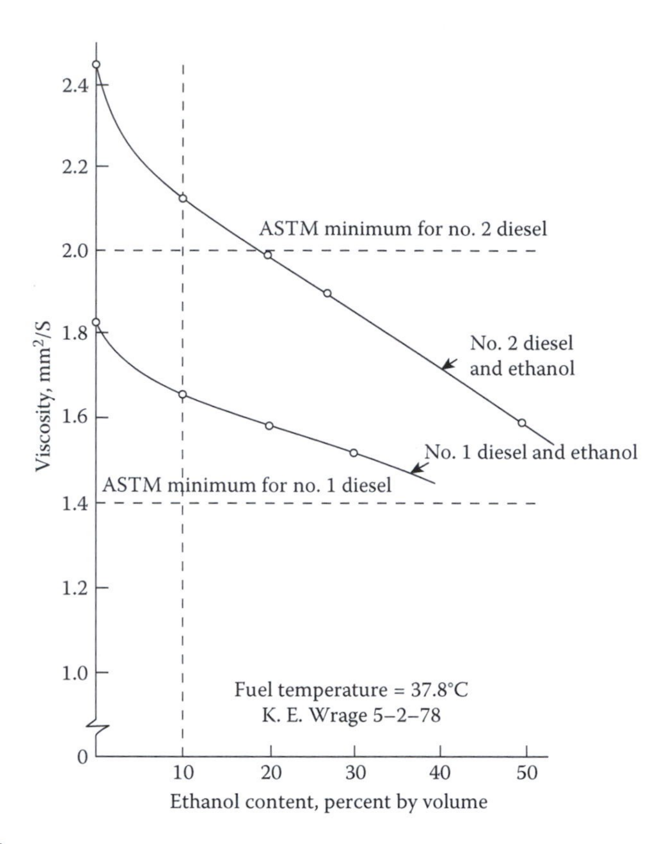

FIGURE 5.17
Viscosity of diesel-ethanol blends. (From Wrage, K. E. and Goering, C. E., *ASABE Transactions*, 23, 1338–43, 1980. Reprinted with permission.)

impossible to meet the diesel fuel standard, the operational integrity of an engine adjusted to run on conventional diesel fuel was not compromised when run on diesohol. They pointed to the need for additional fuel property specifications applicable to such blends to ensure they were compatible for fueling diesel engines.

5.5.5 Exhaust Emissions with E-Diesel

Exhaust emissions play a major role in the assessment of engine performance and the effect of changes in fuel properties. How biofuels alter the exhaust emissions from diesel engines therefore is of great importance. A number of studies have been carried out on E-diesel regarding both regulated and unregulated emissions. Earlier tests focusing on soot, carbon monoxide (CO) and unburned hydrocarbon (HC) emissions showed reductions in soot (30–40%) and in CO, but HC emissions increased (Boruff et al. 1982; Wrage and Goering 1980). Major reductions in PM emissions were obtained by Shi

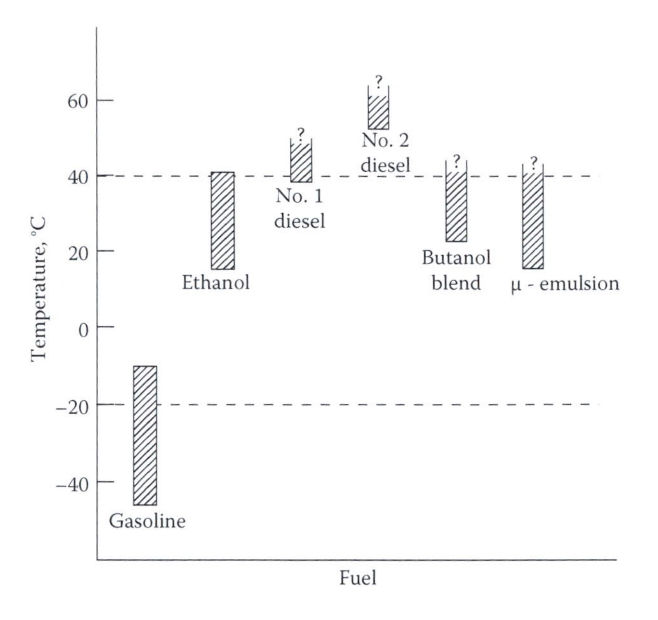

FIGURE 5.18
Flammability ranges of several fuels in the absence of forced ventilation. (From Boruff, P. A., Schwab, A. W., Goering, C. E., and Pryde, C. E., *ASABE Transactions*, 25, 47–53, 1982. Reprinted with permission.)

et al. (2005) with the greatest being for the blend with the highest amount of biofuel because of the high oxygen content. The NO_x emissions tended to increase while CO emissions were lower with the blends. In another study by Merritt et al. (2005) involving three engines fitted with a rotary pump-line nozzle, unit injector, and high pressure common rail fuel injection systems respectively, smoke and PM emissions decreased with increasing ethanol concentration in diesel fuel, but emissions of acetaldehyde increased. NO_x emissions improved with ethanol use in two of the engines, but not the engine fitted with high pressure common rail injection system. Also CO emissions were reduced in two of the three engines, the engine fitted with the rotary pump-line nozzle system being the exception. The latter two results highlighted the impact of fuel injection technology on engine response to E-diesel blends.

Hansen, Zhang, and Lyne (2005) summarized the results of emissions tests for CI engines running on ethanol–diesel blends of 10% and 15% ethanol concentration, as shown in Table 5.7. The test engines varied considerably, as shown in the table, and two different test procedures were used. Increased concentration of ethanol reduced the PM, carbon monoxide (CO), and unburned hydrocarbon (HC) emissions but had no effect on the NO_x emissions.

Rakopoulos, Antonopoulos, and Rakopoulos (2007) and Rakopoulos et al. (2008) carried out in-depth experimental investigations of the heat release

TABLE 5.7

Summary of Emissions Test Results for 10% and 15% Ethanol–Diesel Blends

Reference	Spreen (1999)		Schaus et al. (2000)		Kass et al. (2001)	
Test engine	1991 DDc series 60 6-cyl, 12.7 L DI with turbocharger and inter cooler		1997 VW TDI 4-cyl, 1.9 L DI with turbocharger, EGR and oxidation catalyst		1999 cummins ISB 6-cyl, 5.9 L DI with turbocharger and inter cooler	
Test procedure	Hot start transient tests based on EPA FTP procedure (CFR 40 86N)		Steady state 5 × 5 speed torque test matrix (SAE J1003)		Steady state AVL 8-mode test cycle	
Reference fuel	Emission grade fuel meeting 1998 EPA		No. 2 diesel		Philips petroleum certification fuel containing 350 ppm sulfur	
Test fuel (%v)						
Ethanol	10	15	10	15	10	15
Additive	2.35 PEC	2.35 PEC	2 GE	2 GE	2 GE	2 GE
Diesel	87–65	82.65	88	83	88	83
Ave. emissions (blend/ref. fuel ratio %)						
PM	73	59	27–159	25–157	80	70
NO_x	96	95	80–125	40–125	80	70
CO	80	73	—	—	160	140
HC	171	210	—	—	200	175

Source: From Hansen, A. C., Zhang, Q., and Lyne, P. W. L., *Bioresource Technology*, 96, 277–85, 2005. Reprinted with permission from Elsevier Publications.

Note: PEC: Pure Energy Corporation additive. GE: GE Betz additive.

and emissions characteristics of ethanol–diesel blends with 5%, 10%, and 15% ethanol content. Their results also showed a substantial reduction in smoke density with the reduction being higher with higher ethanol content. CO and NO_x emissions were also reduced with the same trend relative to ethanol content. However, HC emissions increased and the increase was higher with higher ethanol content. They suggested possible reasons for the HC increase being higher heat of evaporation and therefore slower fuel–air mixing, as well as increased spray penetration and possible unwanted fuel impingement on the piston.

Measurements of unregulated emissions have indicated increases of aldehydes with ethanol–diesel blends, but the results depended on engine load and speed (Cheung, Di, and Huang 2008; He et al. 2003a and 2003b; Merritt et al. 2005; Shi et al. 2005). Acetaldehyde is a potential intermediate product from partial oxidation of ethanol. The studies showed that acetaldehyde emissions increased particularly at high loads. Also Shi et al. (2006) measured increases in acetone emissions at all test modes.

5.5.6 Engine Durability with E-Diesel

The impact of a biofuel on engine durability is of particular concern to the engine manufacturers especially with reference to covering engine warranties and ensuring the longevity of their engines. In the early 1980s in-field durability tests by Hansen et al. (1982) and Meiring et al. (1983) on blends with diesel fuel containing 10% and 15% dry ethanol showed no abnormal wear taking place. Corkwell and Jackson (2002) carried out the Scuffing Load Ball-on-Cylinder Lubricity Evaluator (SBOCLE) test and 576 hour pump tests on blends containing 10% anhydrous ethanol in low-sulfur, No. 2 diesel fuel. The various blends also contained from 300 to 2000 ppm of water and various levels of a lubricity improver. Except for the fuel without lubricity improver, all of the diesel blends showed very good performance in the SBOCLE tests and the injector pump tests. The fuel without lubricity improver passed the SBOCLE test but failed the pump test. Further work was required to fully understand the effects of impurities on lubricity.

Frame and McCormick (2005) tested six elastomers by soaking them in certification diesel and in certification diesel containing 15% ethanol. The elastomers, which are commonly used in fuel systems, included N674 general purpose nitrile rubber, N0497 high aceto-nitrile content rubber, N1059 peroxide-cured nitrile rubber, V747 flourocarbon filled with carbon black, and V884 flourocarbon without carbon black. The following parameters were evaluated after soaking: break load, break stress, O-ring inside diameter, O-ring thickness, and O-ring volume. Samples soaked in the ethanol–diesel blend exhibited a significant reduction in break load for all elastomers except N0479. The largest increases in O-ring dimensions and volume were for the elastomers soaked in the ethanol–diesel blends.

Wrage and Goering (1980) soaked the seals from a distributor-type injector pump of a John Deere 830 diesel tractor in No. 2 diesel fuel blends containing from 0 to 50% anhydrous ethanol. After one year of soaking, there was no visible deterioration of any of the seals. However, Wrage and Goering did not do strength tests on the soaked seals.

Recent test results reported by Marek and Evanoff (2001) concerning two trucks accumulating over 400,000 km and a fleet of 15 buses amassing 434,500 km while operating on a 15% ethanol blend of E-diesel, also indicated that no abnormal deterioration in condition had taken place. Also in a farm demonstration project with four tractors and two combine harvesters running on a 10% ethanol blend of E-diesel, Hansen et al. (2001) reported that no abnormal wear patterns could be detected according to oil analyses after at least two seasons of operation.

In a laboratory-based 500 h durability test on an engine running on a 15% ethanol blend of E-diesel, Hansen et al. (2000) found that with the exception of the fuel injection system no abnormal deterioration in engine condition could be detected. One nonmetal component in the fuel injection system failed because of a possible chemical corrosion effect and there was heavy wear and signs of erosion in the fuel injectors.

5.6 Flexible Fuel Vehicles

Brazilian car makers modified gasoline engines to support ethanol characteristics and changes included compression ratio, amount of fuel injected, replacement of materials that would be corroded by contact with ethanol, use of colder spark plugs suitable for dissipating heat due to higher flame temperatures, and an auxiliary cold-start system that injects gasoline from a small tank in the engine compartment to help starting when cold.

A desire to increase the use of ethanol fuel and a concern that high ethanol concentrations could cause material compatibility problems led to the development of a new type of vehicle. The flexible fuel vehicle (FFV) can operate using various blends of alcohol in the range of 0 to 85% ethanol. A vehicle optimized for burning E85 could have been produced at lower cost than a FFV. However, the stoichiometric AFR is 9.8 for E85 and 15 for gasoline. These ratios are too far apart to allow an engine optimized for E85 to run on gasoline. The ability of the FFV to automatically adjust its AFR allows it to use any combination of gasoline–ethanol blends. Thus, the FFV can run on gasoline in locations where E85 is not available but helps to reduce the demand for petroleum fuels where E85 is available.

Automotive manufacturers produced the FFV vehicles compatible with E85. Further, due to the different AFR mixture requirement of neat gasoline and E85, the fuel delivery systems are sized to handle the increased volumes of fuel when using ethanol. Moreover, ethanol is a higher octane fuel and hence sparks timing is advanced in order to achieve the better engine performance and emissions characteristics. Moreover, since the ethanol has a lower calorific value fuel compared to gasoline, a higher fuel flow rate is required to meet the power demand.

The most common commercially available FFV in the market is the ethanol flexible-fuel vehicle, with more than 17 million automobiles and light duty trucks on the roads around the world by early 2009, and concentrated in four markets, Brazil (8.2 million), the United States (almost 8 million), Canada (600,000), and Europe, led by Sweden (311,122). North American and European FFVs are optimized to run on a maximum blend of 15% gasoline with 85% anhydrous ethanol (called E85 fuel). The limit to the alcohol content is set to avoid the cold starting problems (www.wikipedia.com).

The possibility of using various blends of gasoline–ethanol in the fuel tank of FFV vehicles requires means to determine the concentration of ethanol in the fuel line. Thus, a fuel sensor is provided in the fuel system to measure the concentration of ethanol in the blend and optimize the engine management system accordingly.

The Brazilian flex technology avoided the need for an additional dedicated sensor to monitor the ethanol–gasoline mix as earlier FFVs were too expensive. The Brazilians used a lambda probe to measure the oxygen concentration in the exhaust as an indirect means to determine the AFR of the blend being combusted. The results were communicated to the engine control unit (ECU) to optimize the engine management. The software developed and commercialized by the Brazilian engineers was called Software Fuel Sensor (www.wikipedia.com).

5.7 Ethanol Materials Compatibility

Ethanol blended fuels are used in on-road as well as off-road vehicles in two-or four-stroke engines. In most of the four-stroke engines, fuel is introduced in the intake manifold. Thus, ethanol fuel comes in contact with the fuel tank, the fuel delivery system, and the engine cylinder. Soft metals such as zinc, brass, or aluminum, which are commonly found in conventional fuel storage and dispensing systems, are not compatible with ethanol, especially at the higher concentration found in E85 motor fuel. Most automotive manufacturers prefer aluminum for the fuel delivery system in order to reduce the total weight of the vehicle. Earlier older vehicles use lead coated steel tanks for fuel storage. Ethanol blends will react with such materials and partially dissolve them in the fuel. The dissolved materials clog the fuel filter and injectors and thus affect performance and drivability. Aluminum can be safely used if it is hard anodized or nickel plated. Hence, current automobile storage tanks are made of polymer compounds that are resistant to ethanol blends. Moreover, stainless steel and bronze material can be used for fuel delivery systems (Davis 2006). Copper, brass, and bronze also exhibit very little corrosion with the alcohols.

Nonmetallic materials such as natural rubber, polyurethane, adhesives (used in older fiberglass piping), certain elastomers, and polymers used in flex piping. Bushings, gaskets, meters, filters, and materials made of cork also degrade when in contact with ethanol. Nonmetallic materials that are resistant to ethanol blends include thermosetting reinforced fiber glass, thermoplastic piping, neoprene rubber, polypropylene, vitol, and teflon.

The corrosiveness of alcohols may be influenced by the presence of water, carbon dioxide, and oxygen. The carbon dioxide and oxygen are more soluble in methanol and ethanol than water, thus increases the risk of corrosion. Ethanol blends well with gasoline, but it also is completely miscible (mixable) in water. When water infiltrates a tank through sump covers and loose fittings at the top of the tank, the ethanol in the ethanol–gasoline blend will absorb the water. Water mixes easier with ethanol than gasoline and separates from the gasoline. Because of phase separation, the blends in the tank are no longer a homogeneous blend of ethanol and gasoline, but in two layers—gasoline on top and an ethanol layer at bottom. This phase separation causes problems in the fuel lines and in engine performance.

Gasoline distribution systems reach an equilibrium state where varnish, gums, sludge, and water are deposited throughout the system. Ethanol blending will disturb this equilibrium and can lead to severe handling and performance problems. For this reason, it is essential that

- Gasoline should not be mixed with ethanol–gasoline blends in the distribution network.
- The distribution system should be cleaned and dried before allowing ethanol–gasoline blends.
- Ethanol–gasoline and water contact must be avoided completely (CONCAWE 1995).

5.8 Ethanol Opportunities and Challenges

Ethanol is produced from petroleum products as well as biomass sources. Production of ethanol from renewable sources like sugar cane and next generation biomass sources like cellulosic materials reduces the exhaust emissions remarkably. Ethanol is nontoxic, less flammable than gasoline, and less severe when spills or releases of vapor occur. It is safer than gasoline to store, transport, and refuel. Because ethanol is water soluble and biodegradable, land and water spills are usually harmless, dispersing and decomposing quickly. Ethanol emits lower exhaust emissions in comparison to that of gasoline except in unregulated emissions like unburned ethanol and aldehydes. Higher blends of ethanol with gasoline need some modification in the engine fuel systems. In the fuel system noncorrosive materials are to be used in order to avoid corrosion. Use of ethanol reduces the energy insecurity and improves the environment also.

Unmodified ethanol cannot be used as the fuel for a conventional, unmodified CI engine because the ethanol will not ignite by compression. Either the fuel or the engine must be modified and success has been demonstrated with both approaches. Millions of conventional CI engines already exist. Also, because petroleum diesel fuel is likely to be the predominant CCI engine fuel for some years, engine manufacturers are unlikely to build many CI engines optimized to run on ethanol. These considerations suggest that ethanol is more likely to gain acceptance as a CI engine fuel by modifying the ethanol, either by use of additives or by blending the ethanol with petroleum diesel to achieve E-diesel.

There are technical barriers to the commercialization of E-diesel. More detailed data are needed on the efficacy of the emulsifiers as a function of temperature and moisture content. Also, more studies need to be conducted to analyze the effect of E-diesel on the durability of CI engines, including data on E-diesel effect on the elastomers in the fuel system. Higher volatility of E-diesel could cause cavitation and consequent erosion of metal parts in the fuel system. Because E-diesel is more volatile than petroleum diesel, safety in storage and handling is a concern. Fire safety experts and insurance underwriters need to be consulted to determine whether safety codes for storage and handling of gasoline are sufficient for E-diesel. Finally, because E-diesel does not meet the requirements of ASTM D975, the U.S. diesel fuel standard, a new ASTM standard may be needed to provide minimum requirements, including E-diesel properties that are not covered by ASTM D975 (McCormick and Parish 2001).

Engine manufacturers were initially reluctant to extend warranties to engines for which biodiesel was used as a fuel. Only after an ASTM standard was developed for biodiesel were the manufacturers willing to provide such engine warranties. This healthy progress on ethanol in engines will displace gasoline and diesel and improve the environmental and energy security also.

References

Abdel-Rahman, A. A., and M. M. Osman. 1997. Experimental investigation on varying the compression ratio of SI engine working under different ethanol–gasoline fuel blends. *International Journal of Energy Research* 21:31–40.

Ajav, E. A., B. Singh, and T. K. Bhattacharya. 1998. Performance of a stationary diesel engine using vapourized ethanol as a supplementary fuel. *Biomass and Bioenergy* 15:493–502.

Al-Hasan, M. 2003. Effect of ethanol–unleaded gasoline blends on engine performance and exhaust emission. *Energy Conversion and Management* 44:1547–61.

Barnes, K. D., D. B. Kittelson, and T. E. Murphy. 1975. Effects of alcohols as supplemental fuel for turbocharged diesel engines. *SAE* 750469. Warrendale, PA.

Boruff, P. A., A. W. Schwab, C. E. Goering, and E. H. Pryde. 1982. Evaluation of diesel fuel-ethanol micro emulsions. *ASABE Transactions* 25:47–53.

Bullis, K. 2008. *Cellulosic plant opens.* Technology Review, MIT online web site, Cambridge, MA.

Cheung, C. S., Y. Di, and Z. Huang. 2008. Experimental investigation of regulated and unregulated emissions from a diesel engine fueled with ultralow-sulfur diesel fuel blended with ethanol and dodecanol. *Atmospheric Environment* 42:8843–51.

CONCAWE. 1995. *Alternative fuels in the automotive market.* CONCAWE Report No. 2/95.

Corkwell, K. C., and M. M. Jackson. 2002. Lubricity and injector pump wear issues with E-diesel fuel blends. *SAE* 2002-01-2849. Warrendale, PA.

Davis, G. W. 2006. Using E85 in vehicles. *Alcoholic fuels*, ed. Shelly Minteer. Boca Raton, FL: CRC Press.

EPA. 2001. Increased use of ethanol in gasoline and potential ground water impacts. *Environmental assessment of the use of ethanol as a fuel oxygenate: Subsurface fate and transport of gasoline containing ethanol.* UCRL-AR-145380. Livermore, CA: Lawrence Livermore National Laboratory.

EPA 420-F-00-035. 2002. EPA fact sheet. www.epa.gov.

Ferfecki, F. J., and S. C. Sorenson. 1983. Performance of ethanol blends in gasoline engines. *ASAE* 0001-2351/83/2601-0038.

Fernando, S., and M. Hanna. 2005. Phase behavior of the ethanol-biodiesel-diesel micro-emulsion system. *ASABE* stainless steel and bronze material can be used for fuel delivery systems. *Transactions of ASAE* 48; 903–08.

Fikret, Y., and B. Yuksel. 2004. The use of ethanol–gasoline blends as a fuel in an SI engine. *Renewable Energy* 29:1181–91.

Frame, E., and R. L. McCormick. 2005. *Elastomer compatibility testing of renewable diesel fuels.* NREL/TP 540-38834 report of the National Renewable Energy Laboratory. Golden, CO.

Goering, C. E., D. R. Griffith, W. D. Wigger, and L. D. Savage. 1987. Evaluating a dual-fueled diesel tractor. Paper presented at the 1st annual national corn utilization conference, June 11-12, St. Louis, MO.

Goering, C. E., and D. R. Wood. 1982. Over fueling a diesel engine with carbureted ethanol. *ASABE Transactions* 25:576–80.

Goering, C. E., R. T. Parcell, and C. P. Ritter. 1998. Hydrated ethanol as a fuel for a DI, CI engine. *ASABE Transactions* 41:1255–60.

Goering, C. E., T. J. Crowell, L. D. Savage, D. R. Griffith, and M. W. Jarrett. 1992. Compression-ignition, flexible-fuel engine. *ASABE Transactions* 35:423–28.

Hansen, A. C., A. P. Vosloo, P. W. L. Lyne, and P. Meiring. 1982. Farm scale application of an ethanol–diesel blend. *Agricultural Engineering in South Africa* 16:50–53.

Hansen, A. C., M. Mendoza, Q. Zhang, and J. F. Reid. 2000. *Evaluation of oxydiesel as a fuel for direct-injection compression–ignition engines.* Final Report for Illinois Department of Commerce and Community Affairs. Contract IDCCA 96-32434.

Hansen, A. C., Q. Zhang, and P. W. L. Lyne. 2005. Ethanol–diesel fuel blends—A review. *Bioresource Technology* 96:277–85.

Hansen, A. C., R. H. Hornbaker, Q. Zhang, and P. W. L. Lyne. 2001. On-farm evaluation of diesel fuel oxygenated with ethanol. *ASAE* 01-6173. St. Joseph, MI.

Hardenburg, H. O., and A. J. Shaefer. 1981. The use of ethanol as a fuel for compression-ignition engines. *SAE* 811211.

He, B. Q., J. X. Wang, J. M. Hao, X. G. Yan, and J. H. Xiao. 2003a. A study on emission characteristics of an EFI engine with ethanol blended gasoline fuels. *Atmospheric Environment* 37:949–57.

He, B. Q., S. J. Shuai, J. X. Wang, and H. He. 2003b. The effect of ethanol blended diesel fuels on emissions from a diesel engine. *Atmospheric Environment* 37:4965–71.

Jia, L.-W., M.-Q. Shen, J. Wang, and M.-Q. Lin. 2005. Influence of ethanol–gasoline blended fuel on emission characteristics from a four-stroke motorcycle engine. *Journal of Hazardous Materials* A123:29–34.

Kass, M. D., J. F. Thomas, J. M. Storey, N. Domingo, J. Wade, and G. Kenreck. 2001. Emissions from a 5.9 liter diesel engine fueled with ethanol and diesel blends. *SAE* 2001-01-2018 (SP-1632).

Kroeger, C. A. 1986. A neat methanol direct injection combustion system for heavy duty applications. *SAE* 861169. Warrendale, PA.

Kwanchareon, P., A. Luengnaruemitchai, and S. Jai-In. 2007. Solubility of a diesel-biodiesel–ethanol blend, its fuel properties and its emission characteristics from diesel engine. *Fuel* 86:1053–61.

Letcher, T. M. 1980. Ternary liquid-liquid phase diagrams for diesel fuel blends. *South African Journal of Science* 76 (1): 130–32.

Letcher, T. M. 1983. Diesel blends for diesel engines. *South African Journal of Science* 79 (1): 4–7.

Licht, F. O. cited in Renewable Fuels Association, Ethanol Industry Outlook 2008 and 2009, pp. 16 and 29 (www.ethanolrfa.org).

Marek, N., and J. Evanoff. 2001. The use of ethanol blended diesel fuel in unmodified, compression ignition engines: An interim case study. In Proceedings of the air and waste management association 94th annual conference and exhibition, June 24. Orlando, FL.

McCormick, R. L., and R. Parish. 2001. *Technical barriers to the use of ethanol in diesel fuel.* Milestone report NREL/MP-540-32674. Golden, CO: National Renewable Energy Laboratory.

Meiring, P., R. S. Allan, A. C. Hansen, and P. W. L. Lyne. 1983. Tractor performance and durability with ethanol–diesel fuel. *ASABE Transactions* 26:59–62.

Merritt, P. M., V. Ulmet, R. L. McCormick, W. E. Mitchell, and K. J. Baumgard. 2005. Regulated and unregulated exhaust emissions comparison for three Tier II non-road diesel engines operating on ethanol–diesel blends. *SAE* 2005-01-2193. Warrendale, PA.

Minteer, S., ed. 2006. *Alcoholic fuels*. Boca Raton, FL: CRC Press.

Naegeli, D. W., P. I. Lacey, M. J. Alger, and D. L. Endicott. Surface corrosion in ethanol fuel pumps. *SAE* paper 971648.

Palmer, F. H. 1986. *Vehicle performance of gasoline containing oxygenates.* International conference on petroleum based fuels and automotive applications, London.

Rakopoulos, C. D., K. A. Antonopoulos, and D. C. Rakopoulos. 2007. Experimental heat release analysis and emissions of a HSDI engine fueled with ethanol–diesel fuel blends. *Energy* 32:1791–1808.

Rakopoulos, D. C., C. D. Rakopoulos, E. C. Kakaras, and E. G. Giakoumis. 2008. Effects of ethanol–diesel fuel blends on the performance and exhaust emissions of heavy duty DI diesel engine. *Energy Conversion and Management* 49:3155–62.

Schaus, J. E., P. McPartlin, R. L. Cole, R. B. Poola, and R. Sekar. 2000. *Effect of ethanol fuel additive on diesel emissions*. Report by Argonne National Laboratory for Illinois Department of Commerce and Community Affairs and US Department of Energy.

Shi, X., Y. Yu, H. He, S. Shuai, J. Wang, and R. Li. 2005. Emission characteristics using methyl soyate–ethanol–diesel fuel blends on a diesel engine. *Fuel* 84:1543–49.

Shropshire, G. J., and C. E. Goering. 1982. Ethanol injection into a diesel engine. *ASABE Transactions* 25:570–75.

Speidel, H. K., and I. Ahmed. 1999. *Biodegradability characteristics of current and newly-developed alternative fuels. SAE* 1999-01-3518, in *SAE* SP-1482.

Spreen, K. 1999. *Evaluation of oxygenated diesel fuels.* Final report for Pure Energy Corporation prepared at Southwest Research Institute, San Antonio, TX.

Sullivan, N. W., and L. L. Bashford. 1981. Pre-turbocharger alcohol fumigation in a diesel tractor. *ASAE* 81-3581, St. Joseph, MI.

U.S. Department of Energy. 1999. Office of Energy Efficiency and Renewable Energy, Alternative Fuels Data Center, http://www.eere.energy.gov/afdc/altfuel/fuel_properties.html.

Wrage, K. E., and C. E. Goering. 1980. Technical feasibility of ethanol. *ASABE Transactions* 23:1338–43.

6

Dimethyl Ether

Spencer C. Sorenson

CONTENTS

6.1 Introduction

Dimethyl ether, also called DME, is one of the newer candidates for replacement of conventional fuels in the transport, energy generation, and domestic cooking and heating markets. It has been used for many years as an aerosol propellant. DME in the atmosphere decomposes to CO_2 and water in the time frame of a day or so and, therefore, does not react with ozone in the upper atmosphere as fluorocarbons previously used for aerosol propellants did. It is also attractive in this respect due to its vapor pressure, very low toxicity, and high activity as a solvent.

Since its presentation as an attractive alternative diesel fuel in 1995 (Fleisch et al. 1995; Hansen et al. 1995, Kapus and Ofner 1995; Sorenson and

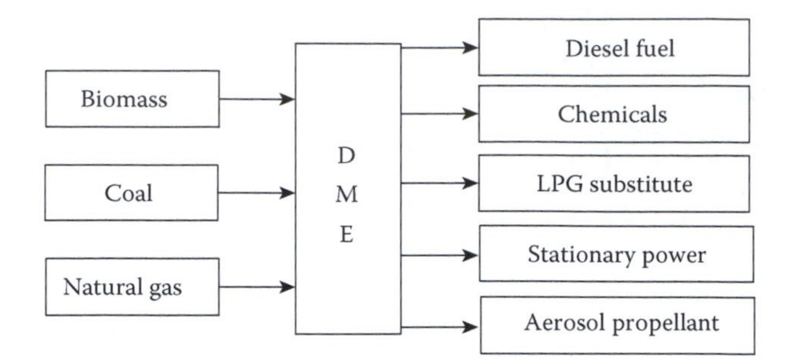

FIGURE 6.1
A schematic view of the flexibility of DME as a multisource–multipurpose fuel/chemical.

Mikkelsen 1995), interest in DME as an alternative to petroleum-based fuels has grown and resulted in several demonstration projects, establishment of international organizations related to DME, and practical implementation of DME in the domestic market in China. In the following, the development of interest in DME will be presented, along with its properties, as well as the advantages and challenges of DME in the fuel area.

A large amount of research and development work has been performed in Japan since 1995. It is reported in the DME Handbook (JDF 2007), published by the Japan DME Forum. All aspects of DME use as a fuel are discussed in this work, and it is recommended to the reader in need of more detailed information than that which can be presented here.

In the following, a discussion of many aspects of DME use will be considered. Briefly, DME can be looked at as a multisource–multiuse chemical, where fuel is a major application. This can be seen in Figure 6.1, which illustrates the raw materials applicable to DME production, and the many uses to which it can be put. Although they will not be discussed further here, the chemical applications are important. DME can be used to produce acetic acid and is an intermediary in the production of propylene. It is interesting to note that much of the DME capacity currently in the process of being constructed is in connection with the production of propylene (Liebner 2006)

6.2 Properties of DME

DME is the simplest of all ethers, organic compounds that consist of an oxygen molecule bonded to two organic radicals. The chemical formula of DME is CH_3-O-CH_3. DME is a colorless gas, with a slight odor at room temperature and pressure. As such, it is necessary to keep DME in closed, low pressure containers for normal use and distribution. This does have

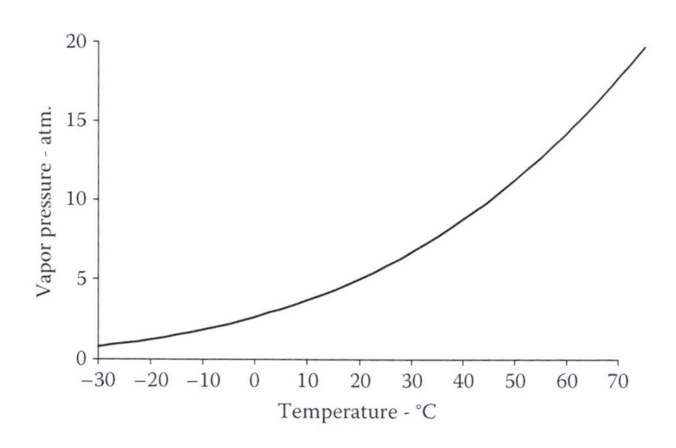

FIGURE 6.2
The vapor pressure of DME as a function of temperature.

the advantage of preventing fuel venting to the atmosphere, though DME is not very reactive in the urban air pollution environment (Bowman and Seinfeld 1995). The vapor pressure of DME is shown as a function of temperature and pressure in Figure 6.2, and lies between those of propane and butane, the major components of liquefied petroleum gas (LPG). The vapor pressure of DME can be calculated by the following equation (JDF 2007):

$$\ln(p_v) = 14.2457 - \frac{2141.93}{T - 25.678},$$

for the vapor pressure, p_v, in atmospheres and the temperature, T, in K. The handling of DME then, requires technology similar to that of the LPG industry.

Other properties of DME are given in Table 6.1. The overall chemical formula is the same as that of ethanol, so the stoichiometric fuel air ratio is the same and heating values of DME and ethanol are close to each other. While the heating value of DME per kg fuel is about 33% lower than that of diesel fuel, the heating value per kg of air at a stoichiometric mixture of fuel and air is about 10% higher than that of diesel fuel (and most other liquid hydrocarbon fuels). Since air capacity is a limiting factor in combustion devices with air as the oxidizer, one would not expect a lower power output with DME as compared to other fuels. For combustion devices with heterogeneous combustion, operation at stoichiometric conditions is not encountered and power levels for DME and hydrocarbon powered equipment can be expected to be similar. Limitations in devices with heterogeneous combustion are typically due to combustion limits such as excess smoke or carbon monoxide, or thermal loading on mechanical parts.

TABLE 6.1

Properties of DME

Property	DME	Diesel
Chemical formula	CH_3OCH_3	$CH_{1.8}$
Molecular weight	46	200–300
Oxygen content: mass %	34.8	0
Stoichiometric air fuel ratio: kg/kg	9.1	≈ 14.8
Liquid density: g/ml @ 15°C	0.668	≈ 0.84
Lower heating value: kJ/kg-fuel	28,800	≈ 42,500
Lower heating value: kJ/liter	≈ 15,400[a]	≈ 37,500
Lower heating value: kJ/kg-air at stoichiometric fuel air ratio	3165	≈ 2871
Boiling Point: °C	–24.9	Range 200–380
Lower heating value: kJ/kg	28,800	≈ 42,500
Viscosity: kg/m-s @ 25°C kg/m-s	0.125	2–4
Vapor pressure @ 25°C: bar	5.1	«1.0
Critical pressure: atm	52	≈ 25
Critical temperature: °C	127	250
Ignition temperature: °C	235	≈ 250
Explosive limits: vol. % in air	3.4–17	≈ 0.5–7

[a] Assuming a maximum liquid volume of 80% in a tank to allow for expansion with temperature increase.

The mass percentage of oxygen in DME is 34.8, an important characteristic for spray combustion, as found in diesel engines and turbines, and DME does not form smoke in heterogeneous combustion. This smokeless combustion is one of the major advantages of DME as an alternative fuel for engines and turbines. The lack of smoke during DME heterogeneous combustion may give a slight advantage compared to hydrocarbon fuel, which produces a highly radiant flame during heterogeneous combustion. In a diesel engine, for example, about 30–40% of the heat transfer during combustion is due to thermal radiation (Ebersole, Myers, and Uyehara 1963). This radiation is essentially blackbody radiation at the flame temperature, which is on the order of 2300 K. Optical studies of DME combustion in a diesel engine have shown the flame to be nonluminous (Oguma et al. 2003; Wakai et al. 1998). A lower radiant heat transfer would result in a slightly higher efficiency for DME and equivalent combustion rates to diesel fuel. Some evidence has been developed to indicate the lower radiation heat loss of DME in a diesel engine (Egnell 2000), though it is indirect and the actual heat transfer was not measured.

The density of liquid DME is about 20% lower than that of diesel oil. This, combined with the lower heating value, and the need for a vapor space in pressurized DME containers, results in a fuel tank size for about twice the volume of that of diesel oil on a vehicle having the same range, as shown by the volumetric heating value in Table 6.1. The viscosity of DME is very low, and special attention needs to be paid to leakage. In addition to low viscosity, DME has low lubricity (Sivebaek and Sorenson 2000) and is prone to accelerated wear in diesel engine pumping systems. The ignition temperature is much lower than that of gasoline, and comparable to that of good diesel fuels. Consequently, DME is exceptionally well suited to combustion in a diesel engine. The high vapor pressure aids evaporation, the fuel-bound oxygen reduces smoke, and the low ignition temperature provides rapid ignition and smooth combustion. It is these properties that first aroused attention to DME as an alternative fuel (Fleisch et al. 1995). Later in the development of DME as a fuel, additional applications have come into focus.

In connection with use in a diesel engine, a property of interest is the compressibility of DME, which is more compressible than diesel fuel. It has been shown that this increases the magnitude of pressure pulsations in injections, which must be taken into account to prevent improper fuel injection (Sorenson, Glensvig, and Abata 1998). A second element is that due to higher compressibility than diesel fuel, DME will increase the amount of energy needed to raise the pressure, and thereby increases fuel heating in an injection system. This is important in the modern common rail injection system, where the fuel is pumped to a high pressure in a recirculating system, and special fuel cooling arrangements may be needed. DME's high vapor pressure reduces the required injection pressures, alleviating this point somewhat, but fuel cooling may still be needed. Figure 6.3 shows the compressibility in terms of the bulk modulus of DME as a function of temperature and pressure.

The compressibility is given in terms of the bulk modulus,

$$K = -\frac{dp}{dv \cdot \rho},$$

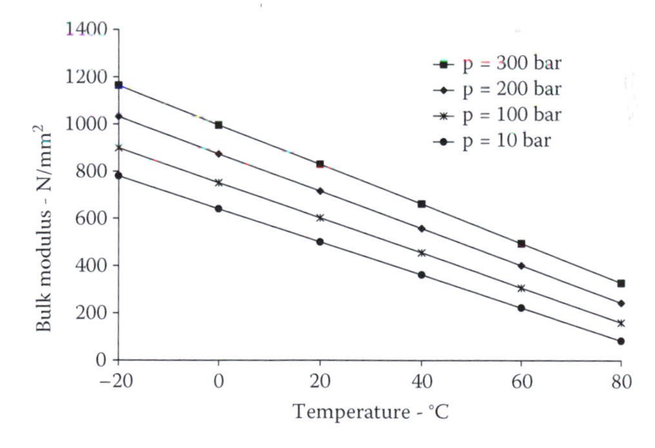

FIGURE 6.3
The bulk modulus of DME as a function of temperature and pressure.

where dp is a change in pressure, dv is a change in volume, and ρ is the density of the fluid.

The bulk modulus can be calculated by,

$$K = A(T) + B(T) \cdot p,$$

where $A = -6.9167T + 628.33$ and $B = -0.0048T + 1.2278$, for T in °C, p in bar, and bulk modulus in N/mm^2. A high value of the bulk modulus means that the fluid is less compressible.

A final aspect of compressibility is that it is related to liquid sonic velocity and hence the rate of propagation of pressure pulsations in fuel systems. The sonic velocity in a fuel is given by

$$V_s = -\sqrt{\frac{K}{\rho}},$$

where V_s is the sonic velocity of the fuel (Obert 1972). For petroleum oil with a bulk modulus of 1448 MPa (N/mm^2) and a density of 840 kg/m^3, the sonic velocity is 1313 m/s, while for DME with a bulk modulus of 400 MPa and a density of 668 kg/m^3, the sonic velocity is 774 m/s. This can be important in the design of fuel injection systems in high speed diesel engines, where pressure pulsations affect injection pressures and thereby the combustion process.

Another important property of DME that is of interest in fuel applications is its high activity as a solvent. For many conventional polymer sealing compounds, DME can extract the plasticizer compounds, resulting swelling, cracking, or embrittlement, and thus a loss in sealing capacity. Two materials have been shown to be effective seals: Kalrez® and Xflour®, polymers containing Fluorine, the former being quite expensive. In some applications, the polymer EPDM has been found effective, though this apparently varies with production method. For static sealing, Teflon® is inert to DME. DME is not known to interact with metal components in fuel systems.

With regard to health and environmental characteristics, DME is currently used as a solvent/propellant in aerosol containers, including cosmetics. As such, considerable effort was expended to examine the health aspects of DME, and it was found not to exhibit any properties that are detrimental to human health. DME in the atmosphere is not prone to form ozone in urban environments, and is less active that most fuel hydrocarbons and organic fuel additives in atmospheric reactions in urban areas (Bowman and Seinfeld 1995). Additional information on properties and calculation methods can be found in the DME handbook (JDF, 2007).

6.3 DME Production

6.3.1 Syngas Production

DME is not found naturally, and must be produced by a chemical process from a feedstock containing carbon and hydrogen. For fuel use, two basic options exist, but they are both based on the use of syngas. Syngas is a mixture of primarily H_2, CO, and some other gasses resulting from the high temperature gasification of organic compounds. The generation of syngas is a key element in the production of DME, and a range of other chemicals such as ammonia, methanol, propylene, and synthetic diesel fuel from carbon containing compounds. For current DME production proposals, natural gas and coal are the most commonly named feedstocks. Biomass from the paper industry is also being investigated for DME production in a more CO_2 neutral fashion (Landälv 2006).

For syngas production from natural gas, there are two types of syngas generators: the steam reformer and the autothermal reactor (Hansen et al. 1995). In the former, steam is mixed with the natural gas and the mixture is passed over a catalyst in a reactor to form the syngas. The catalyst typically is enclosed in tubes, and these are externally heated. The overall reaction occurring with steam reforming is

$$CH_4 + H_2O \rightarrow CO + 3H_2 \quad \Delta H_R = 131.2\ \text{kJ/kMol} - CH_4. \tag{6.1}$$

Steam reforming produces a syngas that is high in hydrogen. This is beneficial in the case of formation of methanol. It is also possible to reform with carbon dioxide:

$$CH_4 + CO_2 \rightarrow 2CO + 2H_2 \quad \Delta H_R = 246.8\ \text{kJ/kMol} - CH_4. \tag{6.2}$$

The syngas produced with this method has equal amounts of hydrogen and carbon monoxide. As shown later, this is the stoichiometric mixture for the overall formation of DME from syngas. The syntheses of methanol and DME are dependent on the relative amounts of hydrogen and carbon monoxide in the syngas, and there are syngas production processes using both of the above reactions.

In the autothermal reactor, which is used for natural gas, there is a partial combustion of the feedstock, with about 35–40% of the stoichiometric amount of an oxidizer. For DME production, oxygen is the most common oxidizer used. The use of air would involve the introduction of a large amount of nitrogen to the gasifier system, which would greatly increase gas flow rates through the systems without giving any production benefit. In the autothermal reactor, the rich combustion occurs in the presence of a suitable catalyst to produce the syngas. The heat of the rich combustion converts the natural gas to syngas, with the added oxygen combining with the carbon to give the CO needed for the syngas. Systems are also known where the steam reformer and autothermal reactor are combined. This is called the combined reforming process. Typically, the steam reformer comes first, followed by the autothermal reactor. Steam can also be added to an autothermal reactor to adjust the product composition.

In this case, the overall reaction for formation of syngas from natural gas can be written

$$2CH_4 + O_2 + H_2O \rightarrow 2CO + 4H_2 + H_2O \quad \Delta H_R = -36.6\ \text{kJ/kMol} - CH_4. \tag{6.3}$$

The composition is well suited to methanol formation.

In the direct process where a ratio of $H_2/CO = 1$ is desired, this can be obtained by recirculating the CO_2 from the products of the DME reactor back to the reformer. Then the reforming reaction can be written

$$2CH_4 + O_2 + CO_2 \rightarrow 3CO + 3H_2 + H_2O \quad \Delta H_R = -16.5\ \text{kJ/kMol} - CH_4. \tag{6.4}$$

This process gives a ratio of $H_2/CO = 1$, which is the stoichiometric mixture for DME formation from syngas.

For coal gasification, the process is more complicated since the fuel is in the solid state and contains many impurities, and various types of gasifiers are used to convert the solid coal to syngas. One type is a fixed bed system, where solid coal particles are found in a bed in the reactor, and hot gasses are fed over them. The volatile portions of the coal are driven off by heating, and may be partially burned by adding an oxidizer to give higher temperature gasses that gasify the remainder of coal, which is predominantly carbon. As the gasification proceeds, slag and ash drop to the bottom of the reactor and are removed. Particles in the gas leaving the gasifier often contain unconverted carbon, and are typically removed from the exit gas in a cyclone or filter, and recycled to the gasifier, where the carbon can be converted to syngas. Fluidized bed gasifiers and entrained flow gasifiers with pulverized coal are also used. Because of slag, ash, and other impurities in coal, catalysts are not used in coal gasification.

Gasification of biomass and waste products is possible but is complicated by variable composition, and a variety of impurities that must be removed. Research is being conducted in Japan relative to the gasification of syngas from waste plastic (JDF 2007).

There is an energy loss associated with the production of syngas. The term used to define this is the cold gas efficiency, which is the ratio of the heating value of the products to that of the reactants. For natural gas as a feedstock, the efficiency is on the order of 80–90%. For coal, it is about 10% lower. For biomass and waste about 15–20% lower (JDF 2007). On an efficiency basis then, natural gas is a preferred raw material, also because it is cleaner and the gasifier construction is simpler and cheaper. However, other issues such as resource availability and environmental sustainability also affect the choice of the feedstock.

6.3.2 DME/Methanol Production

The production of DME and methanol are closely related. Basically to make DME, methanol is formed from syngas, and then DME is formed from the methanol by a dehydration process. These processes can occur in separate reactors or in a common reactor. The former method is called the indirect method, and the latter method the direct method of DME production. Though it is called the direct method, the same reaction steps occur, they just take place in one reactor instead of two. That is to say, DME is not directly formed. Methanol is formed first and the dehydration reaction occurs thereafter in the same reactor.

The following reactions are involved in the production of DME from syngas. The first is the formation of methanol from syngas over a suitable catalyst.

$$CO + 2H_2 \rightarrow CH_3OH \quad \Delta H_R = -89.4\ kJ/kMol - Methanol. \tag{6.5}$$

The water gas reaction also occurs:

$$CO + H_2O \rightarrow CO_2 + H_2 \quad \Delta H_R = -41.7\ kJ/kMol - CO. \tag{6.6}$$

These two reactions form the basis of commercial methanol production, a process that has been used for many years.

In the production of DME, the final reaction is the catalytic dehydration of the methanol to form the DME.

$$2CH_3OH \rightarrow CH_3OCH_3 + H_2O \quad \Delta H_R = -25.5\ kJ/kMol - DME. \tag{6.7}$$

Combining Reactions 6.5, 6.6, and 6.7, the overall process for the conversion of syngas to DME is obtained.

$$3CO + 3H_2 \rightarrow CH_3OCH_3 + CO_2 \quad \Delta H_R = -245.7\ kJ/kMol - DME. \tag{6.8}$$

Combining Reaction 6.4 with Reaction 6.8, the overall formation of DME from natural gas is obtained:

$$2CH_4 + O_2 \rightarrow CH_3OCH_3 + H_2O \quad \Delta H_R = -276.7\ kJ/kMol - DME. \tag{6.9}$$

The same result is obtained by combining Reactions 6.3, 6.5, and 6.7.

Depending on the process catalyst selected and the conditions of the reactions, Reactions 6.5 through 6.7 can occur either in one reactor or Reactions 6.5 and 6.6 can occur in a single reactor, and the DME can be produced by dehydrating the methanol produced in Reaction 6.7 in a separate reactor. These processes are shown schematically in Figures 6.4 and 6.5.

A calculation of the energy efficiency of the direct production of DME from natural gas using the above process, results in an ideal cold gas efficiency (heating value of product relative to heating value of input gas) of 82.8%. Thus, there is an energy loss inherent in the production process, even

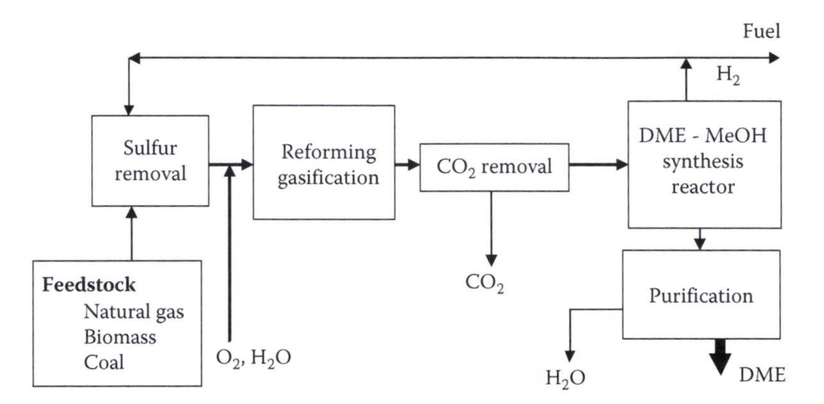

FIGURE 6.4
Schematic diagram of the direct DME production process, where DME is produced directly from syngas in the synthesis reactor.

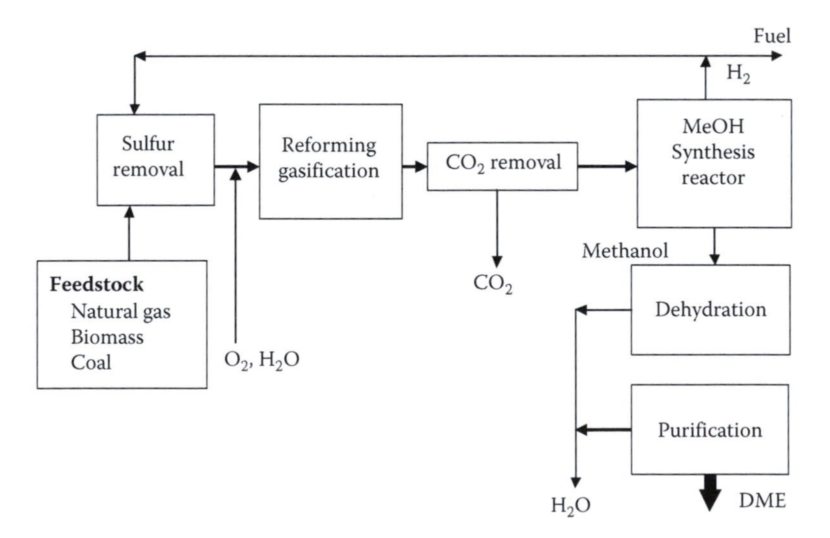

FIGURE 6.5
The two-step process for making DME, involving the separate production of methanol, which is dehydrated to form DME.

under ideal circumstances. When the efficiencies of the various processes involved in a production plant and the recovery of process energy are taken into account, an actual overall process efficiency of about 71% is achievable using natural gas as a feedstock in large scale DME production (Hansen et al. 1995). A cold gas efficiency of 68% has been achieved in testing on a pilot plant scale of 100 tons/day. Scaled up to full size, this gives an efficiency of about 72% (JDF 2007). The scale is on the order of 5000–7000 tons of DME per day through a single reactor. A limiting factor in the maximum size of a DME or methanol production string is the size of the autothermal reactor. The unit has a high thermal loading, and durability is determined by its size. Similarly, technical limitations on the gasifier with coal feedstock result in similar optimum capacities for production of DME. Production facilities at one location for producing more DME would most likely make use of multiple product streams, each with its separate autothermal reactor or gasifier.

The effectiveness of the conversion of syngas to DME or to methanol depends on the temperature and pressure of the reactor and the composition of the syngas. To show these effects, results from chemical equilibrium calculations are shown. In an actual reactor, chemical reaction rates and catalyst composition are important, and reactions may not reach chemical equilibrium. Nonetheless, chemical equilibrium calculations are useful in showing trends and tendencies. Figure 6.6 shows the equilibrium conversion of syngas to methanol as a function of reaction temperature and pressure using Reactions 6.5 and 6.6, and the conversion of syngas to DME using Reactions 6.5, 6.6, and 6.7. The conversions occur at the stoichiometric hydrogen carbon ratios in accordance with Reactions 6.5 and 6.8.

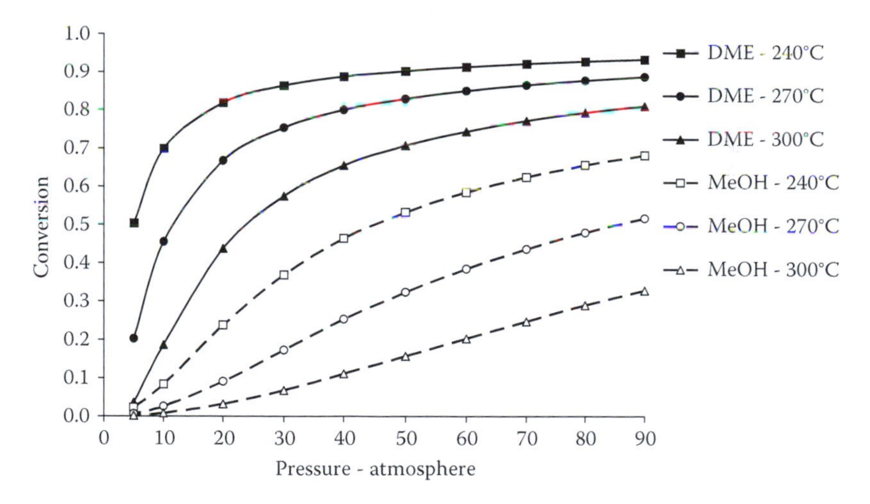

FIGURE 6.6
Equilibrium conversion of syngas to methanol and DME as a function of reaction pressure and temperature. $H_2/CO = 1.0$ for the DME conversion, and 2.0 for the methanol conversion.

It is seen that conversion of both DME and methanol is best at lower temperatures and higher pressures. But the temperatures in actual systems must not be too low, or else reaction rates will be too slow and the products will be far from the equilibrium condition. In practice, the higher the temperatures, the closer the reaction comes to equilibrium, but the lower the equilibrium conversion rate. So a compromise temperature must be found. The conversion of syngas to DME is much higher than that of methanol at lower pressures, which could be considered an advantage to the direct process, as the reactor need not be pressurized as high for making DME. The lower pressure may, though, result in lower reaction rates. A lower reactor pressure would lower construction and investment costs, though the economic analysis later will show that the major cost is that of the feedstock when using natural gas. Temperatures on the order of 260–280°C and pressures between 50 and 80 bar are seen in practice for the achievement of useful reaction rates.

The effectiveness of the conversion of syngas to DME or methanol also depends on the ratio of hydrogen to carbon monoxide in the DME reactor. Figure 6.7 shows the equilibrium conversion efficiency of the syngas to DME as a function of reactor stoichiometry. Three curves are shown, the first is the total amount of carbon and hydrogen converted to DME in the direct method, where Reactions 6.5, 6.6, and 6.7 occur at $H_2/CO = 1.0$. The second curve shows the conversion in the direct method for the case where the water gas Reaction 6.6 is slow and not in equilibrium (JDF 2007), and the third shows the conversion to methanol. In the second case, the reaction is

$$2CO + 4H_2 \rightarrow CH_3OCH_3 + H_2O \quad \Delta H_R = -204.5\ \text{kJ/kMol} - \text{DME}, \quad (6.10)$$

which is a combination of the methanol synthesis, Reaction 6.5 and the methanol dehydration, Reaction 6.7, in the absence of the water gas reaction.

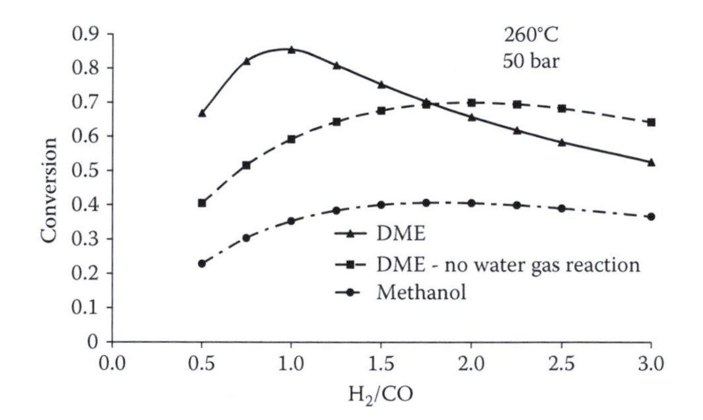

FIGURE 6.7
The equilibrium conversion of carbon and hydrogen to DME and methanol shown as a function of hydrogen to carbon ratio of the syngas. The reaction occurs at a temperature of 260°C and a pressure of 50 atm. DME conversion is also shown where the water gas reaction is considered slow.

This could occur in a case where the catalyst is not sensitive to the water gas reaction.

The maximum overall conversion to DME in the direct process with complete equilibrium occurs at hydrogen to carbon ratio of 1.0, which is the stoichiometric mixture for the conversion of syngas to DME according to Reaction 6.6. For the methanol conversion, the maximum conversion is at a H_2/CO ratio = 2.0, again the stoichiometric value, this time according to Reaction 6.5. If the water gas equilibrium reaction is not active, the conversion of the direct DME process has about the same H_2/CO ratio for maximum conversion as methanol. The highest overall conversion is found for the direct process with a H_2/CO ratio = 1. The direct process at this stoichiometry would fit well with coal, since it has a low amount of hydrogen, while the indirect process with separate methanol production fits better with natural gas, which has a high amount of hydrogen. As seen above, the H_2/CO ratio can be controlled by adding steam, or by gasifying with CO_2. The importance of the water gas shift, Reaction 6.6, is shown by the lower conversion and the shift in the H_2/CO ratio for maximum conversion when it is not active. For a direct formation of DME with a hydrogen rich syngas, the slow water gas reaction gives a slightly higher conversion than with the water gas reaction at the H_2/CO ratio of about 2.

The equilibrium reaction products are also a function of the ratio of hydrogen to carbon in the syngas. The product distribution for the direct conversion process with a H_2/CO ratio of 1.0 is shown in Figure 6.8. As expected from the conversion effectiveness, the yield of DME is highest at a H_2/CO of 1.0, the stoichiometric condition. When this ratio is less than one, there

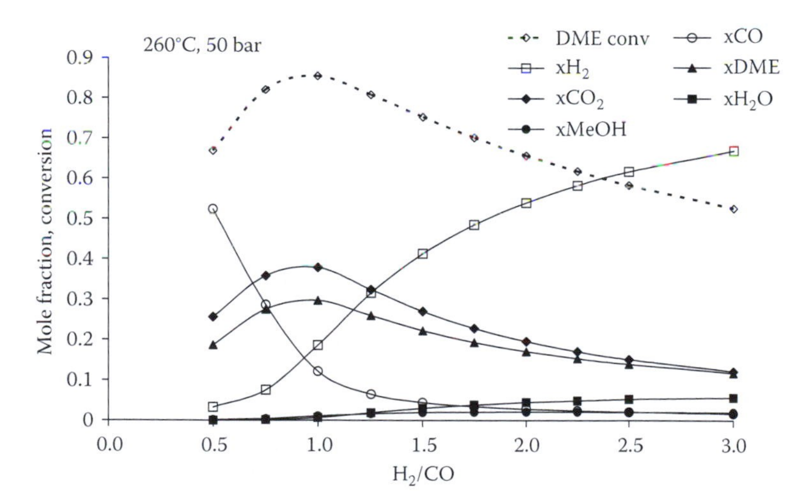

FIGURE 6.8
Conversion of syngas to DME and the equilibrium composition of the products of Reactions 6.5, 6.6, and 6.7 shown as a function of hydrogen to carbon ratio of the syngas. The temperature is 260°C and the pressure 50 atm.

is an excess of carbon, and when it is greater than one, there is an excess of CO. The amounts of water and methanol produced are small, and there is a tendency for the amount of water produced to increase at large H_2/CO ratios. The amount of other products than DME indicates the need to process the gas to remove the DME product and unwanted components, and to recycle unreacted syngas components. In order to maintain H_2/CO ratio at one, the CO_2 can be recycled and added to the reformer, Reaction 6.4.

The most efficient and economic process was stated to be the direct process (Hansen et al. 1995). However, there is an advantage to producing methanol separately and then dehydrating as shown in Figure 6.4, as one can choose the final product mix, and produce the most economically attractive product mix for existing market conditions. It is difficult to determine the economic differences between the two, as comparable units of direct and indirect production of DME at full economic scale are not yet completed at the time of writing.

There is an additional advantage to the use of methanol dehydration to produce DME, especially in the initial stages. This is due to the established methanol market and production system. In recent years, the production of methanol has been shifted to larger plants, which produce methanol in amounts on the order of 5000 tons per day. This gives an economic advantage compared to smaller older plants, and combined with a sluggish methanol market due to the restriction of methyl tert butyl ether (MTBE) in gasoline, has resulted in an excess production capacity for methanol on a worldwide basis. Such existing capacity can be converted to DME production for a very much lower cost than the construction of a new DME plant. A new DME production facility on the scale of 5000 tons/day is estimated to cost in excess of 500 million U.S. dollars (JDF 2007). The addition of a methanol dehydration unit would cost on the order of a 10th of this. So especially in the introductory stage of DME use, conversion of these existing, smaller methanol plants could significantly reduce initial production investment costs, and help to solve the "chicken-egg" problem, so common with the introduction of alternative fuels. The first DME used in the marketplace for home heating and cooking in China has been produced from excess methanol capacity.

DME produced for aerosol cans has been produced by methanol dehydration, and was previously considered to be too expensive to be used as a fuel. This high price has been due to two factors. The first is the low production volume. Total current annual world consumption of DME is on the order of 150,000 tons per year (JDF 2007), much smaller than the proposed value of around a million tons per year, even if it all was concentrated in one place, which is not the case. The second is the requirement for very high purity of DME, since aerosol sprays come in close contact with people and quality must be carefully controlled. DME fuel quality can be much lower than this, which gives a significant price reduction combined with larger scale production.

The major impurities expected in fuel grade DME are methanol and water as seen from Figure 6.8. Figure 6.9 shows the region of acceptable ignition in

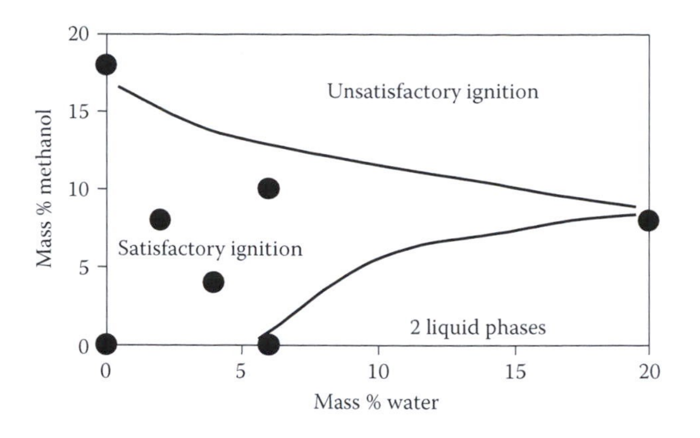

FIGURE 6.9
Regions of good ignition characteristics for DME/water/methanol blends.

a diesel engine operating on mixtures of DME, methanol, and water. It can be seen that up to 4–5% of each of these impurities will operate successfully in diesel engines. Unpublished studies by the author have shown that there is minimal impact of water and methanol in these amounts on the composition of exhaust gas emissions. Similar results have been obtained with gas turbines and industrial burners (JDF 2007)

The source of the synthesis gas is a compound containing carbon. At the present time, the most attractive source of carbon is natural gas. For economic reasons, stranded gas, which is gas that is not easily transportable to other markets, is the most attractive alternative, as it can be purchased at a much lower price than gas used for the home heating market. Production of DME from this source can be seen as a form of gas-to-liquid (GTL) process, as the DME produced is in liquid form at moderate pressure, and can be transported using technology very similar to that for LPG. The term GTL is most often applied to the production of synthetic diesel fuel through syngas using the Fischer–Tropsch process, but DME also represents a form of liquefaction of the gas, and of course LNG does also. Coal is used for DME production in China currently.

Another feedstock option currently under development is that using black liquor from the paper industry as a carbon source (Landälv 2006). This is an attractive option, since the global CO_2 emissions are quite low, and the biomass-based gasification feedstock has already been collected, a significant problem with biomass fuels. In addition, the economics of the process are supported by the income of the paper industry, which already pays for the collection of the biomass material and some of the processing of the feedstock. Currently, the black liquor is burned in a boiler to produce process heat, but this can be replaced with low grade forest waste. The boilers must be replaced periodically, and this leads naturally to the change to DME producing equipment. Particularly in the northern paper producing areas, there is a significant amount of this resource available, and a significant portion of the heavy duty transport markets could be supplied by DME from this source. On a worldwide basis, the annual energy in black liquor in the world was stated to be equivalent to 30 million tons of gasoline.

6.3.3 DME Production Economics

The economics of any alternative fuel is a critical issue. Since many alternative fuels must be made from other raw materials, there is often an economic disadvantage due to the processing needed. This means that the price of the raw material is important, but not the sole factor in determining price.

It is very difficult to quote accurate economic factors related to alternative fuels. This is due to several considerations. First of all, the price of energy resources is quite variable, as shown by the excursions of oil prices in the last decade, where variations of nearly an order of magnitude in crude oil price have been observed. Secondly, proposed alternative fuels may not have been produced in the quantities required for large scale usage yet, giving uncertainties due to estimation of efficiencies and prices of processes not previously conducted on that scale. This uncertainty is small in the case of DME by the indirect method, since the methanol production technology and dehydration of methanol have been in practical use for a long time. Currently, much of the world's methanol is made in so-called mega methanol plants, which are of a similar size (natural gas consumption) to those proposed for fuel use. Dehydration units have been smaller due to the demand for aerosol DME. Direct production methods have so far been tested in smaller pilot scale projects. The JFE direct process in Japan has been successfully tested on a unit of 100 tons per day (Ohno 2006). Thus, though many of the technologies in production are well understood and have been in practice for some time, only methanol production on the scale involved for fuel DME has been practiced recently. Thirdly, prices stated in the literature may be colored by the interests of the presenter. It is not unheard of for an advocate of an alternative fuel to present the most optimistic scenarios, and an opponent to present the most pessimistic.

The factors involved in determining the cost of producing and delivering DME are many. The more important factors are listed below:

- Raw material (feedstock costs)
- Plant investment
- Plant operating expenses: utilities, staff, local taxes
- Shipping expense: long distance for imported DME produced in other countries, short distance for local and regional distribution
- Storage facilities: large scale for imported DME produced in other countries, small scale for local distribution

The most complete estimate of the costs of DME is reported in the DME Handbook (Ozawa, 2007 in JDF 2007). The results for different fuel prices were presented for the time up to 2006, but even in the short time between that and this publication, the oil price has doubled that level, and returned to 2006 levels. The world has also been experiencing a financial and economic crisis that has not stabilized at the time this is being written. In addition, the exchange rate of the dollar relative to many major currencies has also changed. Compared to the euro, for example, the rate was 1.15 $/€ in the January 2002 compared to 0.64 $/€ in 2008, nearly a factor of two. The dollar exchange rate is important, since it is the currency of international oil trading. These factors affect all costs involved in the production and transportation of DME, as well as the process of obtaining capital for building production facilities.

As a starting point, Japanese estimates of DME economics from the year 2002 will be used (JDF 2007). This can give a first indication of at least the relative importance of the factors that go into determining the price of DME. It might be considered as a sort of worst case comparison for DME, since the price of oil was very low at this time. If DME can be competitive at low oil prices, it should also be competitive at high oil prices. In general, it can be assumed that if oil prices increase, other prices will increase as well. The amount of increase relative to the oil price is very uncertain.

In the calculation of DME economics, the price of energy is normally given in U.S. dollars per million Btu (MMBtu). The standard energy content of a barrel of petroleum is considered to be about 5.8 MMBtu or 6.1 GJ, based on the higher heating value of the fuel. Then the price of oil in $/MMBtu can be obtained by dividing the oil price in $/bbl by 5.8. That is, an oil price of $50/bbl corresponds to a price of about 8.6 $/MMBtu.

In 2002 in Japan, the CIF (Cost plus Insurance and Freight) of DME was estimated to be between $3.5/MMBtu and $5/MMBtu based on a natural gas price for producing DME of $1.0/MMBtu to $1.5/MMBtu. Transport costs were estimated at between 0.5 and 1$/MMBtu, depending on location, which ranged from Indonesia to the Middle East. A similar production price was estimated for coal, based on a coal price of $0.5/MMBtu. At the time of that report, the price of crude oil varied between about $3/MMBtu and $5/MMBtu ($18/bbl–$30/bbl). The oil price was based on actual deliveries to Japan while the DME price was estimated. Note that the gas feedstock price for manufacturing the DME is lower than that of crude oil. This emphasizes the idea of using gas that is not readily available for large commercial or domestic consumption, nor is it of a suitable volume for the investment in liquified natural gas (LNG) facilities. This "stranded gas" has a lower price than pipeline gas.

As a comparison to other fuels, for the 2002 time frame, LNG in Japan cost between $4/MMBtu and $4.5/MMBtu, and LPG between $5/MMBtu and $6/MMBtu. Compared to other fuels, then, DME can be seen to be competitive. In a scenario with increasing oil prices, the critical point in the comparison of prices of different fuels would be the increase in price of stranded gas, and it is most difficult to try and estimate this figure, as the result depends on negotiations between buyer and seller in a long-term contract, which would probably be related in some way to oil price.

In a newer study for Asian conditions, it was estimated that for the production of DME from coal in China, the delivered price of DME would be about $10.5/MMBtu (Fryer 2006). Between 2004 and 2006, the oil price went from about $45/bbl to about $70/bbl ($7.8/MMBtu–$12/MMBtu) and the delivered oil price from about $440/ton to $690/ton ($9.4/MMBtu–$15/MMBtu). It was concluded that DME from coal would be competitive with diesel with a crude oil price of $45/bbl or more. The oil price has been at or above this level since 2005. The relation between the coal-based price and the natural gas-based price was not discussed.

The Japanese study also showed the relative importance of different factors in the production of DME. Two cases were considered, the first being production of DME from natural gas and the second production from coal. The natural gas was assumed to be stranded gas. Transportation costs were estimated to Japan. Sites considered for the Japanese situation were Indonesia, Western Australia, and the Middle East. Transport costs are to be expected for Japan, but need to be considered for other parts of the world as well, at least in the case of DME from natural gas. This is due to the need for low price natural gas, and by nature, this gas is located away from markets and pipelines. So the DME would be manufactured at the well and then shipped to potential customers. An approximate summary of the results of the relative production costs from this study is given in Table 6.2.

For the case of natural gas, the feedstock is the major factor, resulting in about 56% of the total production costs, with depreciation and utilities being the other two largest factors. This is for an assumed gas price of $1/MMBtu. Therefore, it is very important to locate production at a place where the gas price is low. For the coal, the feedstock price was considered to be lower, $0.5/MMBtu, though more feedstock is needed because of the lower conversion efficiency of the coal-based process. In this case, the

TABLE 6.2

Relative Costs of Factors Involved in the Production of DME, Based on Energy Prices in 2002

Item	Natural Gas Feedstock (% Total Production Cost)	Coal Feedstock (% Total Production Cost)
Feedstock	56	30
Utilities	11	11
Maintenance	5.2	19.1
Depreciation	22	32
Other	6	8

feedstock price is a much lower fraction of the total cost while maintenance and depreciation are much higher. This is a reflection of the more complicated process involved in making clean syngas from coal, and a higher need for maintenance due to the impurities in the coal and the need to remove them. An additional aspect of the coal process is that there is a substantial amount of heat produced in the gasification process, which has a much lower efficiency (70–80%) than the natural gas. It was proposed that waste heat could be used to generate steam and eventually produce electricity, and an estimate of the effect of selling this electricity was a reduction in the FOB price of DME produced from coal of about 10%. This depends on the plant being located where there is a market for this amount of electricity. The amount of electricity generated was estimated to be on the order of 500 kWh per ton of DME produced for a plant producing 5000 tons of DME per day. This corresponds to an electrical power production on the order of 100 MW.

Since it is likely that DME would be produced at locations remote from the final user, transportation costs need to be considered as well. International DME transport would be made by ship to ports in the receiving country using ships similar to current tankers that ship LPG under low pressure. These tankers are estimated typically to have a capacity on the order of from 45,000 to 65,000 tons of DME. For transport distances on the order of 7000 km, the transport cost was estimated to be on the order of 15–18% of the total DME price.

In the production of DME/methanol from the paper industry, estimates of project economics have been made (Landälv 2006). For a DME production on the order 500 tons/day an investment is required on the order of 135 million €. Using economic conditions valid for Sweden, it was calculated that the payback time on this investment was less than 4 years. For a 2000 ton/day plant the investment was estimated to be 280 million €. In this case, the DME plant was assumed to be a replacement for a worn out black liquor boiler, which would have cost about 110 million €. This savings was applied to the total investment, and the total investment of about 170 million € was calculated to have a payback time of under 4 years. Prices of the fuel generated were stated to be equivalent to gasoline from oil at $30/bbl. In light of the above, this is a very attractive price at current oil price levels.

It is nearly impossible to give a definite price for DME, as there are many variables, and all of the costs involved are strongly related to the price of crude oil, as many energy prices are tied to this. What can be said from the studies presented is that DME produced in large quantities from stranded gas and coal can be quite competitive with other energy sources such as crude oil and LNG, and somewhat cheaper than LPG. It is the author's best estimate that DME would retain its competitive position in the case of rising oil prices, and probably be cheaper than very high priced oil.

6.4 DME Applications

6.4.1 Internal Combustion Engine Fuel

The first diesel engine results presented for DME were in relation to its use as an ignition improver in methanol (Brook et al. 1984; Murayama et al. 1992), which was under consideration as a method to reduce particulate emissions in order to meet the strict particulate emission standards that had been proposed for city busses in the United States for 1991. While methanol can burn without producing soot emissions, its low cetane number makes it difficult to ignite in diesel engines. In order to improve the ignition properties of methanol, nitro-akyl additives were used as ignition improvers. As these are expensive, and produce exhaust particulate matter in diesel engines, DME was thought of as an alternative. However, as Figure 6.9 indicates, so much DME was needed that it would have to be present in larger amounts that the methanol.

At the time of these studies, DME's low self ignition temperature was known, but based on the historically expensive small scale production processes for making high purity DME, it was basically regarded as a rather expensive chemical. Therefore, it was thought most economically sound to use DME in limited quantities as an additive in order to minimize cost, and engine studies were not extended to pure DME operation.

Attention to DME as an alternative fuel by itself ("neat" fuel) first came about as a result of introductory studies in diesel engines (Fleisch et al. 1995; Kapus and Ofner 1995; Sorenson and Mikkelsen 1995). At the same time, it was shown that DME can be produced in large scale at an attractive price (Hansen et al. 1995). Due to its low ignition temperature, DME is well suited to diesel engine operation, and unsuited to operation in a spark ignition engine, where it would knock extensively. Without any modification to the diesel combustion chamber, engines operating on DME were found to operate at the same efficiency as on diesel fuel, while exhibiting smoke-free operation. Because of the lack of smoke, it was possible to operate the engines under conditions where emissions of oxides of nitrogen, NO_x, were quite low. These conditions include retarded injection timing, extended injection duration and high rates of exhaust gas recirculation (EGR). Diesel engines have been operated on DME with EGR amounts of up to 70%, with no smoke emissions.

The reason that DME does not produce smoke in diesel engines is apparently not due to the structure of the fuel itself, but to the presence of the oxygen in the fuel. This is supported by Dec (1997), where it is found that particulate matter in a diesel flame is formed in the products of rich combustion of the fuel inside the burning spray, and not in the primary reaction of the fuel. This is also supported by Figure 6.10, which shows relative carbon formation in diesel engines as a function of fuel oxygen content for a variety of fuels from a series of studies (Beatrice et al. 1996; Cheng and Dibble 1999;

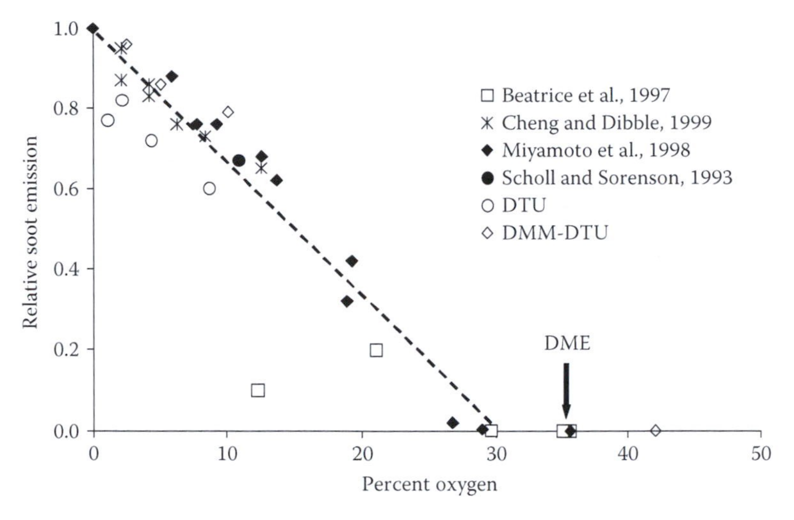

FIGURE 6.10
The smoke formation tendencies of various fuels in diesel engine as a function of fuel oxygen content.

Miyamoto et al. 1998) and unpublished studies of the author. There does not appear to be a specific fuel effect on the reduction of smoke formation for these fuels. The lack of soot formation in a diesel flame has been shown on a detailed level to depend on the atomic C/O ratio of the fuel air mixture (Curran et al. 2001).

The first version of a DME commercial 7 liter diesel engine satisfied emissions standards of up to a decade in advance of existing technology, using EGR and only a passive oxidation catalyst to clean up CO emissions (Fleisch et al. 1995). Thus a DME diesel engine offers significant advantages in that an exhaust gas after treatment system for particulate matter is not necessary. For conventional diesel engines, particulate matter control normally requires a form of filtration, and requires a complicated filter regeneration system, possibly combined with a fuel additive to assist filter regeneration.

Figures 6.11 through 6.13 show the basic situation concerning operation with DME in diesel engines. These data are from the very first engine tests with DME conducted by the author on a small diesel engine, and the trends shown here have been seen in all subsequent engine tests with DME in a wide range of diesel engine types and sizes. In the tests shown, the engine was operated with the standard fuel injection pump and injection nozzle, but with a lower injector opening pressure than the diesel version, 80 bar instead of 200. Since the pump volume was the same for DME and diesel, a larger volume of DME had to be injected due to its lower heating value, and this occurred over a longer period of time. This, and the low injector opening pressure, would be expected to give very high smoke emissions in an engine operating with diesel fuel.

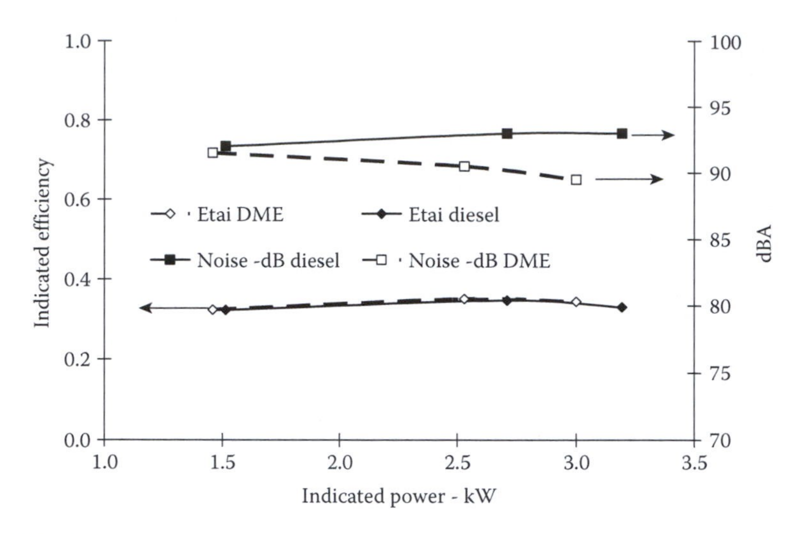

FIGURE 6.11
The indicated thermal efficiency and engine noise of a 0.273 liter single cylinder diesel operating on diesel fuel and DME at a speed of 3000 rpm.

Figure 6.11 shows the indicated engine efficiency, Etai, (i.e., the efficiency of production of work on the top of the piston) and engine noise as a function of engine power. Contrary to expectations due to the longer duration of injection, the DME engine had the same efficiency as the diesel engine and operated with much less noise. The former is thought to be due to a lower heat loss during DME combustion mentioned earlier. The lower noise is another advantage of DME, and is due to good ignitability in a diesel engine. Because it ignites more readily, less DME has been injected at the time of ignition with the diesel engine, and the force of the initial DME combustion is less than that of diesel.

Figure 6.12 shows the nitric oxide (NO_x) and smoke emissions of the DME and diesel engines as a function of the engine power. Here the results are very dramatic. There is no smoke (particulate emissions) from the combustion of DME. The values shown for DME are the detection limit from the test method. Subsequent tests with many other engines verify that DME combustion does not produce soot particles. A DME engine can emit small amounts of particulate matter originating from the lubricating oil, as do all other engines. With modern oil control techniques, these emissions are usually low enough not to need special control. It should be pointed out that when the engine was operated on diesel fuel using the low injector opening pressure, the smoke number was about twice as large as the values shown in Figure 6.12. These values correspond to a very dark diesel exhaust, which is completely unacceptable.

At the same time, the NO_x emissions with DME are about 30% of those with diesel fuel. In diesel fuel emissions research, the NO_x/particulate trade-off

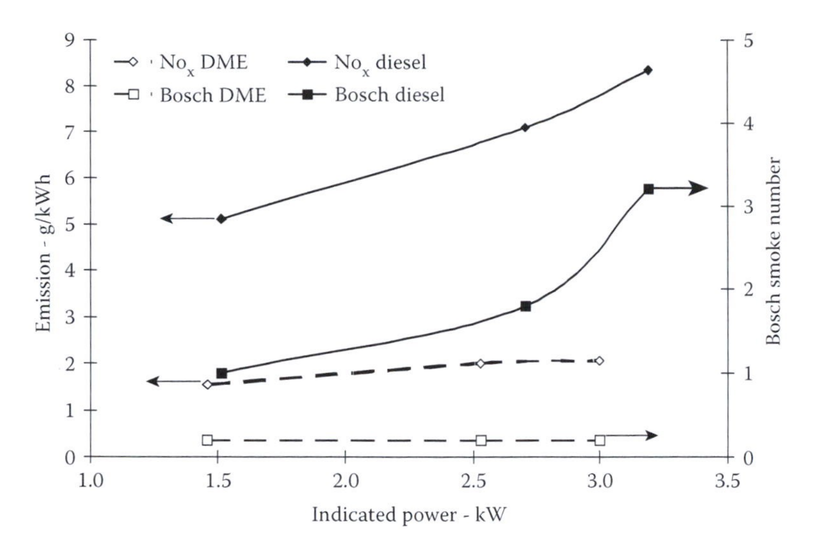

FIGURE 6.12
NO_x emissions and smoke from a 0.273 liter single cylinder diesel operating on diesel fuel and DME at a speed of 3000 rpm.

(Heywood 1998) has been a major hurdle to emissions reductions for many years. Due to the nature of formation of NO_x, which is highest under fuel lean, high temperature conditions, and the nature of formation of soot particles, which is highest under fuel rich, low temperature conditions, efforts to reduce the one emission have normally been at the expense of the other. However, due to its presence in the fuel with DME, oxygen is brought into the combustion zone where the soot is formed in diesel engines (Dec 1997) in an amount adequate to prevent soot formation. Thus with DME, there is no need to consider soot forming processes, and all attention can be focused on the reduction of NO_x emissions instead, eliminating cumbersome and costly particulate trapping systems in the exhaust.

Engines running on DME produce NO_x and the mechanism of formation is the same as that for diesel fuel. The NO_x is formed in a diffusion flame between the products of very rich fuel combustion inside the burning fuel spray and the air outside the flame in the combustion chamber (Dec 1997). It is possible to operate a DME powered diesel engine with high NO_x emissions (Kajitani et al. 1997) by operating with advanced injection timing and other conditions that promote its formation. The soot emissions are eliminated by the DME fuel itself. This enables engine operation on DME under conditions that produce low NO_x emissions. This is the major advantage to using DME in a diesel engine.

The freedom from concern with particulate emissions enables more drastic measures to be used for the reduction of NO_x. The two most applicable techniques used for this are EGR and retarding injection timing. Both of these techniques result in lower temperatures during combustion and, therefore, less formation of NO. Already in the first studies, a reduction in NO_x emissions of about 70% was achieved (Fleisch et al. 1995; Sorenson and Mikkelsen 1995). The limiting factor in the amount of EGR that could be used was the increase in the CO emissions.

Modern engine tests on diesel fuel give NO_x emissions on regulatory emissions testing cycles of about 3 g/kWh. It is normally possible to obtain lower NO_x emissions on DME than with diesel fuel, and at the lowest possible diesel levels, DME engines exhibit a slightly higher thermal efficiency but still do not produce particulate emissions (Gill et al. 2001). Typical results estimated from this study are shown in Table 6.3. In the first three cases the injection timing was retarded to reduce NO_x emissions. The first data column is for diesel fuel, at the condition regarded as the maximum acceptable loss in power, efficiency, and particulate emissions. The next data column shows the emissions from a DME engine at the same NO_x level of 3.5 g/kWh, and indicates the same efficiency, a better power, but with slightly higher CO emissions. It was possible to reduce NO_x emissions with DME to a lower level by retarding injection timing additionally, as shown in the third data column, where the NO_x level is lowered to 2.7 g/kWh. This results in about a 10% loss in power, a loss in efficiency, and increased HC and CO emissions. No particulate emissions were found.

It was possible to obtain even lower emissions from the engine by optimizing the timing and adding 15% EGR to the intake. In this case, shown in the last data column, NO_x emissions are reduced all the way down to about 0.5 g/kWh, with no after treatment of the exhaust, and still no particulate emissions. Power is nearly the same as the 3.5 g/kWh NO_x condition, and the efficiency is only reduced a small amount. This clearly illustrates the excellent combustion properties of DME in a diesel engine.

To meet the lowest future emissions regulations, it may still be necessary to use exhaust gas after treatment for NO_x emissions with DME. But since

TABLE 6.3

Emissions for Diesel Engines for Different EGR Levels and Fuels at a Constant Speed

Fuel	Diesel	DME	DME	DME
NO_x – g/kWh	3.5	3.5	2.7	0.5
EGR – %	0	0	0	15
CO – g/kWh	0.6	1	1.8	8
HC – g/kWh	0.18	0.12	0.2	0.19
PM – g/kWh	0.06	0	0	0
IMEP – bar	10	12	11	11.5
Thermal efficiency	0.403	0.403	0.389	0.398

Source: Modified from Gill, D., Ofner, H., Schwarz, D., Sturman, E., and Wolverton, M. A., *SAE* 2001-01-3629, 2001.

Note: The emissions in the first three columns obtained by retarding injection timing; in the last column, ignition delay and EGR used in combination.

such low engine out emission levels have been shown to be achievable, the amount to be removed in the exhaust should be lower than that for diesel engines, reducing the requirements on this system.

In a diesel engine, the emissions of unburned hydrocarbons/organic matter and carbon monoxide are generally not problems. The values of these emissions are similar in DME and diesel powered engines, as shown in Figure 6.13. Passive oxidation catalysts are a well known and effective technology for diesel engines, and should work well with DME engines.

An interesting example of the advantages of DME as compared to diesel fuel is the development of a small two-stroke cycle diesel engine operating on DME (Hansen et al. 2008). The engine has a bore of only 38 mm, and an engine of this size would require a lot of development to operate with reasonable smoke levels on diesel fuel. However, when operated on DME, there are no smoke emissions, and the engine can be operated with very acceptable emissions and good efficiency. Latest results indicate an efficiency of about 32%, a good value for such a small engine. Such an engine might be used for mopeds or other applications where a small light engine is desired.

From a combustion point of view then, DME is an excellent fuel for diesel engines. However, it does have some properties that require new developments in order to achieve successful long-term operation in commercial vehicles. These are connected with its low viscosity and low lubricity. Viscosity is the resistance of a fluid to shear stress, and is important in fluid film lubrication. Film lubrication occurs when a liquid lubricant is present between two surfaces sliding with respect to each other in a steady motion. Lubricity is related to the ability of a fluid to prevent friction and wear in a situation where two surfaces slide with respect to each other in a reciprocating motion. That is, there are points in the operating cycle where there is no relative motion between the surfaces, and the lubricating film generated by

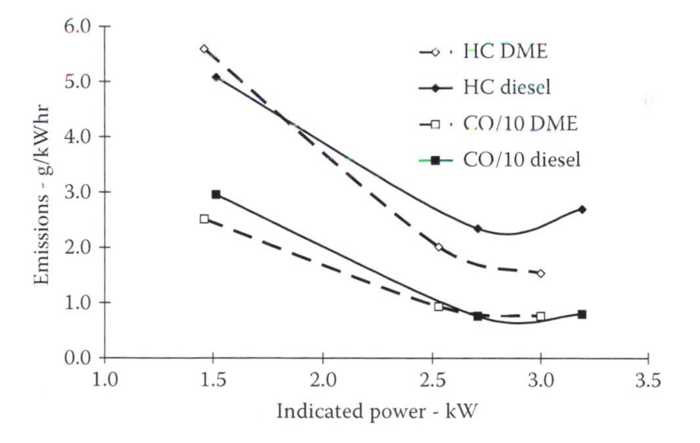

FIGURE 6.13
CO and unburned hydrocarbon emissions of a 0.273 liter single cylinder diesel operating on diesel fuel and DME at a speed of 3000 rpm.

relative motion can break down. This can give rise to contact between the two surfaces, increasing friction, and most importantly causing wear that can lead to the failure of engine parts.

This is important in diesel engine fuel systems. The fuel must be injected into the engine under high pressure (up to 2000 atm in modern diesel engines using diesel fuel, though DME operates satisfactorily with much lower pressures), and this is done by pressing the fuel through a small restriction using a piston in a sealed cylinder of a pump. Tolerances required between the piston and the cylinder are very small (on the order of microns) and wear problems are important. Modern diesel fuel has had aromatic compounds and sulfur reduced in order to combat emissions, with the results that the wear tendencies in injection systems are increased. For modern low sulfur diesel fuel, this has been solved using the addition of "lubricity improvers" to the fuel. These are compounds that apparently attach themselves to surfaces and help prevent the surfaces from contacting each other.

Figure 6.14 shows the results of wear tests in the author's laboratory with DME using different lubricity improving additives. The diesel oil level refers to an acceptable rate of wear on the standard wear test for commercial diesel fuels using the similar high frequency reciprocating rig (HFRR) test procedure. The test uses a reciprocating ball pressed on a plate submersed in fuel and operated for a fixed duration. It is the test used to certify the suitability of diesel fuel for current commercial engines. A modified version of the test has been developed in order to handle DME under its vapor pressure (Sivebeak and Sorenson 2000) and the results shown in Figure 6.14 are for this test. At the left hand side of the curve with no additive, it can be seen that the wear with DME is significantly higher than that of diesel fuel. The test results show that it is possible to improve the lubricity of DME to give a similar result to

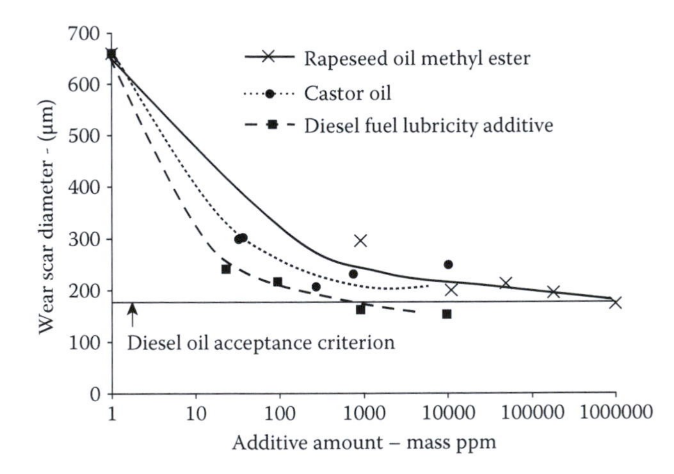

FIGURE 6.14
The wear of a reciprocating ball on plate test with DME with various lubricity improving additives.

that of diesel in the test by adding certain substances. Shown here are a commercial lubricity additive for diesel fuel, castor oil, and rapeseed oil methyl ester (RME). Even though the wear with pure DME is high, it still lubricates somewhat, and has better wear characteristics than no lubricant at all.

It is interesting to note that RME is a lubricity improving agent. Vegetable oil methyl esters are very good diesel fuels (Scholl and Sorenson 1993), come from renewable sources, and should be cheaper than the commercial additives. Thus, a larger amount of RME could be used than other additives. The effects of RME on combustion and emissions in DME are not known. It is likely that there is a limit, since pure vegetable oil and its methyl esters will produce carbon particles when used as pure fuels.

However, experience has shown that even though DME satisfies this test, unacceptable wear can occur in some systems (Nielsen and Sorenson 1999). Thus, since DME is different than diesel fuel, the accepted test procedure does not appear adequate to certify suitability of DME fuel for fuel injection equipment. Currently, there are too few published studies to see how different systems and materials respond to the different additives, and enable a simplified laboratory test for lubricity.

Special fuel injection systems that take DME's special characteristics into consideration are being developed for DME engines (McCandless et al. 2000), but the long-time durability of these systems still needs to be established. Fleet tests of diesel engines converted to DME and running on different kinds of commercial pumps with lubricity additives have accumulated tens of thousands of km, so development is underway (Sato et al. 2008). However, there is still a need to develop new materials for the pumps and new better lubricity additives especially for DME.

6.4.2 Power Generation and Industrial Burners

Since a DME diffusion flame does not produce soot particles, DME is also a natural candidate for use in a gas turbine or other industrial boiler, since these devices also operate with diffusion flames. DME has been successfully tested in gas turbines using injection of both liquid and gaseous forms of DME (JDF 2007). Fuel injection nozzles were modified or specially designed to accommodate the special properties of DME. Tests with gas turbines up to 25 MW indicate that DME can be operated in a gas turbine, with the potential for low NO_x emissions, and emissions levels of CO and unburned fuel comparable to those of conventional turbine fuels such as oil, natural gas, and LPG. As is the case with diesel engine operation, there is no soot emission. At least one manufacturer of gas turbines has approved their operation with DME (Basu and Wainright 2001).

DME has also been shown to be a suitable replacement fuel for other types of industrial boilers. In a retrofit mode, DME has been successfully tested with mixtures of heavy fuel oil and coal. Operation with about 33% DME mixed with coal in an industrial boiler has also been demonstrated. Lower

NO_x concentrations were obtained, but due to the coal, there were still particulate emissions, which were reduced through the use of DME (JDF 2007).

There is a substantial potential for the use of DME in the power generation and industrial sectors. With respect to implementation of a new alternative fuel, this offers a substantial advantage in that the installations can be large, and use significant amounts of fuel at few locations. In contrast to a vehicular application, where there are many consumers, and consumption is spread over a wide geographical area, there are fewer users operating at fixed geographic locations. Thus supply infrastructure is much simpler than for vehicle applications, and with fewer users, it should be simpler to negotiate agreements related to the supply of DME.

6.4.3 Domestic Applications

Since its physical properties are quite similar to those of LPG, it is also to be expected that DME can be used for domestic purposes, such as cooking and domestic heating. This is indeed the case, and in fact, the first commercial use of DME as a fuel has taken place in China for this purpose. Here, excess methanol capacity has been converted to the manufacture of DME by dehydration. The methanol is obtained by coal gasification.

Pure DME burns well in domestic applications (JDF 2007; Sanfilippo 2004). When converting from LPG burners to DME burners, two major considerations must be taken into account. The first is that the stoichiometric air–fuel ratio for DME is 9.0:1 as opposed to about 15.5 for LPG. In order to maintain the same excess air ratio, the fuel nozzle area must be increased when converted to DME. The second consideration is that seals and hoses and other components with plastics and elastomers must be checked for compatibility with DME and changed as needed.

An additional use of DME is as an extender to LPG. Studies have shown that LPG with a DME addition of up to about 20%, can be used without modification in existing LPG equipment (Sanfilippo et al. 2004). When DME is blended into LPG in moderate amounts, studies have shown the material compatibility is not a significant issue. Use of DME as an extender for LPG appears to be a very attractive application for the initial introduction of DME into the market place. There is a large established market, and the DME can be commercially introduced into the market place with very little change in product and infrastructure. This is important, in that a large production serves to verify the credibility of the manufacturing processes, and provides the first real economic basis to evaluate the fuel in practice. The "chicken and egg" situation is known for many fuel alternatives.

6.4.4 Fuel Cells

A final application to be mentioned is that of fuel cells. It has been shown that pure DME can be used to obtain power from a fuel cell (JDF 2007; Kajitani

2006). The performance of the fuel cell was stated to be a half to a quarter of that of a hydrogen fuel cell. Performance is similar to that of a fuel cell operated on methanol. Studies of the direct DME fuel cell have been on a laboratory scale so far.

Another possibility for the use of DME in fuel cells is to use it as a carrier and source of hydrogen for a hydrogen fuel cell. The DME is reformed to produce hydrogen in a similar manner to the formation of syngas with a catalytic reaction at elevated temperature (JDF 2007). Compared to the reforming of methanol, DME reforming requires a higher temperature, but is better than natural gas. The products of reforming are H_2 and CO_2. The reforming process consists of endothermic reactions and thus requires energy to perform. Much of the research to date has been concerned with the development of catalysts to improve the reforming process. Little data are available concerning complete systems using DME to power fuel cells.

6.5 Economic Issues Regarding DME Applications

Since DME is not yet widespread in its use as a fuel, experience with economic implications of the use of DME on the expected applications is lacking. However, enough is known about the properties and potentials of DME to name many of the factors that will have an economic impact, both positive and negative, on the applications where DME can be used. In the following, some of these factors will be mentioned.

6.5.1 Internal Combustion Engines

Diesel engines operating on DME will probably eventually use common rail systems. In this system, one high pressure pump delivers fuel at a selected pressure to all the cylinders, whose injectors are electronically controlled to give a highly flexible fuel injection and combustion process. The elements that are expected to be impacted here are the injection system piping, injectors, high pressure pump, seals, and fuel cooling. The fuel pressure needed for DME is lower than diesel, so the fittings and tubing needed are required to handle lower pressures. The pump will produce lower pressure, but will probably need special treatment or materials for durability. The injectors can be of the same general design as for diesel, though larger nozzle holes are needed. A fuel feed pump between the tank and the engine will be needed as is the case with diesel. A question mark is the seals, as the current options for sealing DME are expensive, and there is a real need for better, cheaper seals for DME in all applications. If fuel cooling is required, this will increase costs.

Since DME does not generate particulate matter in a diesel engine, a particulate filter in the exhaust should not be needed, and this would give substantial savings for the engine and emissions control system. NO_x emissions can be made quite low without exhaust gas after treatment, and it may be possible to satisfy emissions standards with only EGR. In that case, the only exhaust gas treatment necessary would be a passive oxidation catalyst for CO and possibly unburned organic compounds. If needed, the exhaust could be treated with a Selective Catalyst Reduction (SCR), as used on modern diesel engines. Due to the attractive emissions characteristics, the requirements of the SCR system for a DME engine would be less demanding than for a diesel engine.

The engine fuel tank for DME would have to be larger than for diesel, and more expensive due to the pressurization. There are factors acting to both increase and decrease the cost of the engine system relative to diesel. All factors considered, it does not appear that a DME engine system would not be much more expensive than a diesel, and depending on emissions standards, possibly cheaper. Ideally, an existing engine could be converted to run on DME by changing the fuel system, but the sensitivity of the performance to the in-cylinder air flow interaction with the fuel spray makes a retrofit application less attractive than an engine designed to run on DME from the start.

6.5.2 Industrial/Gas Turbine

The areas where special consideration for DME needs to be taken are in the fuel handling system and in particular the fuel injection nozzle. In terms of fuel handling, the question of seals and elastomer compatibility is again relevant. A new fuel injection nozzle would be needed for DME. Studies have shown that this is a reasonably straight forward task (JDF 2007), and it is possible to design a compatible fuel injection nozzle with existing air flow motion in the burner. Thus, retrofitting is a good possibility for changing to DME in these applications. In a new application, DME equipment and that for LNG, LPG, or fuel oil should be of approximately the same cost.

6.5.3 Domestic Use

The economic impact of DME in domestic use depends on whether it is used as an LPG extender or as a neat fuel. In the former case, current results indicate that existing systems can be used without change, so there is no economic effect of using DME in LPG up to about 20% other than fuel cost, which appears to be competitive, possibly cheaper, depending on crude oil prices. For pure DME systems, the nozzles in burners need to be changed to accommodate the different fuel air ratio requirements, and plastics and seals probably need to be changed. Cost is very design dependent, but converting an existing system to DME would require some investment at the user level, though retrofitting is a real possibility here. It appears to be attractive to use neat DME in newly constructed systems, where the replacement/rebuilding question is not relevant. New equipment for neat DME is not expected to cost more than the corresponding equipment for LPG.

6.6 Greenhouse Gas Emissions

When DME burns in a combustion device, it produces CO_2, just like any other carbon containing fuel. There are many aspects involved in the CO_2 production and use of a fuel, and in order to compare fuel and feedstocks, a well-to-wheel approach is needed. A life cycle analysis of many alternate fuels was performed by the European automobile and fuel industry, in conjunction with the European Union (JEC, 2008) using alternative fuels with a light duty vehicle. Two factors are involved, the first the production of the fuel (well to tank) and the second the use of the fuel in the vehicle (tank to wheel). Table 6.4 shows selected results from this report for CO_2 emissions from DME and synthetic diesel fuel for production from natural gas and 3 forms of wood: waste from other production, wood grown for fuel, and black liquor, a waste product for paper production. In all cases, the production of DME produces less greenhouse emission than synthetic diesel fuel, the amount varying from about 4 to 15%. The production of these fuels from biomass in the form of wood produces much lower greenhouse gas emissions than from natural gas, even in the case where there is carbon capture.

The second part of the picture is the use in the application, that is in this case, vehicle emission of greenhouse gasses per driven km. Selected results from (JEC, 2008) are shown in Table 6.5 for different internal combustion/engine fuel combinations. That shows that DME emits fewer grams of CO_2 per km than any other conventional engine. The same holds true for larger vehicles.

Combining all these factors, and taking into consideration that the CO_2 produced from biomass is recycled, a total well to wheel comparison for a diesel powered passenger vehicle according to (JEC, 2008) is shown in Table 6.6, where coal is included as a worst case. It is seen that the source of the carbon for the DME has a huge impact on greenhouse gas emissions. It should be noted that although the greenhouse gas emissions from biomass are very low, there is a problem of availability of resources, and (JEC, 2008) estimates that

TABLE 6.4

Production and Distribution CO_2 and Greenhouse Gas Factors (Well to Tank) for DME and Synthetic Diesel Fuel in g/MJ.

	DME		Synthetic Diesel	
Source	**GHG**	**CO_2**	**GHG**	**CO_2**
Natural Gas-Remote plant	21.1	18.0	22.5	19.1
Natural Gas-Remote plant CO_2 capture	11.1	8.1	13.2	9.7
Wood Waste	4.6	4.3	4.8	4.6
Farmed Wood	6.5	4.1	6.9	4.3
Black Liquor	2.2	2.1	2.4	2.4

Source: Modified from JEC Well-to-Wheels study Version 3. 2008.
Note: GHG = Green House Gas Equivalent.

TABLE 6.5

Co_2 Emissions Factors for a Passenger Car, Tank to Wheel. gCO_2/km.

Spark Ignition Engine	118
Diesel Engine with DPF	106.6
Diesel Engine with neat Biodiesel	111
Diesel engine with synthetic diesel fuel, DPF	103
Diesel Engine with DME	95

Source: Modified from JEC Well-to-Wheels study Version 3. 2008.

TABLE 6.6

CO_2 and Greenhouse Gas Factors (Well to Wheel) for DME and Synthetic Diesel Fuel in g/km for a Conventional Diesel Powered Passenger Vehicle

Source	DME	Synthetic Diesel
Black liquor	5	5
Wood Waste	8	10
Farmed Wood	14	15
Natural Gas-Remote plant	154	171
Coal	338	355

the amount of transportation fuels that could be replace by biomass derived alternative fuels is between 5 and 15% depending on the fuel chosen. For DME it is about 11%, the highest of the carbon based alternative fuels shown. Although the above results are given for a passenger vehicle, other studies give comparable results for other vehicles (Verbeek and Van der Weide, 1997; Semelsberger et al., 2006). For turbines and other burner applications, similar trends should be observed, as it was shown that there are not major differences in the efficiency of device operating on DME or other carbon based fuels. The production and distribution factors from (JEC, 2008) can be used and applied to specific applications.

In recent years, it has been noted that carbon particulate matter may play a significant role in the process of global warming (Taupy, 2007). DME is very advantageous in this regard since its combustion in practical devices does not produce particulate carbon emissions.

6.7 DME Safety Aspects

DME and all other fuels are combustible substances, and caution must be exercised when using them. In the case of DME, since its properties are so similar to those of LPG, similar handling and security measures are appropriate. An examination of the use of LPG-based equipment for DME has shown that existing LPG handling standards in the Netherlands need to be modified slightly to accommodate DME (Elbers and Logtenberg 1999). The most important issues were those of assuring that the sealing material is compatible long-term with DME, and taking measures to maintain a maximum of 86% liquid in the container when filling. A nominal value of 80% is appropriate. This filling amount is needed to prevent high pressures if storage temperatures become too high and the liquid density decreases to the point where there is all liquid in the tank. Above this point the pressures could become very high, compromising the integrity of the container. LPG has a similar restraint, with different allowable values of filling percentages, depending on LPG composition.

Many other aspects of safety have been investigated through the Japan DME Forum (JDF 2007). Briefly, they found that DME is stable in storage, and unlike other ethers, does not form peroxides. Explosion hazards of DME are similar to or better than those of LPG, and a fire emanating from a simulated tank rupture was less dangerous than the equivalent with LPG, since the fireball is smaller and less luminous than that from LPG. The low luminosity greatly reduces the radiation heating of the surroundings. The DME toxicity to living organisms is very low. In general, DME was found to be at least as safe as, and normally safer than, LPG, the latter being a commonly used fuel throughout the world.

It should be mentioned that the DME molecule is polar, and that in fuel handling systems, care should be taken to remove any build up of a static electric charge in DME fuel systems. Also, similar to natural gas, an odorant needs to be added to DME to help users detect leakage. Sulfur compounds (mercapatans) are typically used. DME produced from syngas is sulfur-free.

6.8 Future Scope

DME has been found to be an attractive fuel for combustion in a variety of applications. It is an excellent diesel fuel substitute, producing no smoke and permitting engine operation at very low NO_x emission levels. It can be used in gas turbines, with equivalent power and lower emissions than those of traditional fuels, and is suitable for domestic use for cooking or home heating, either as a substitute for LPG, or as an LPG extender in LPG/DME blends. DME could also be used as a hydrogen carrier for fuel cell systems, and it will function directly in a fuel cell, though not as well as hydrogen. A major advantage of DME is that it produces no smoke during heterogeneous combustion, allowing more aggressive treatment of other exhaust emissions, resulting in low emissions levels with simpler exhaust treatment technology for diesel fuel engines.

There are several options for DME production. DME can be produced from a variety of carbon containing sources: natural gas, coal, and biomass. Thus, it can be viewed as a multisource–multipurpose fuel, as shown in Figure 6.1. This flexibility gives an advantage in terms of market introduction, as a variety of applications are possible for introducing DME into the commercial market. The first commercial fuel applications have taken place in the domestic market in China. In addition, the DME produced from different sources is interchangeable, given realistic fuel standards.

DME appears to be economically competitive with fossil-based fuels when formed from coal, stranded natural gas, and waste products from the paper industry. DME must be handled and transported under a light pressure (<10 bar) in technology equivalent to that of LPG. DME has low viscosity and high activity as a solvent, and care must be taken in the selection of seals, and special DME tolerant fuel pumps need to be further developed for use in diesel engines. In order to obtain a large reduction in greenhouse gas emissions, DME needs to be made from biomass. In this case, a reduction on the order of 80% is attainable.

As with any new technology, the introduction of DME into world markets faces obstacles having to do with uncertainties, economics, and infra structure concerns. Given the benefits available from its use, and its flexibility as a fuel, DME is, in the author's opinion, a very attractive option for future energy systems and has good chances for taking its place in the world's energy market.

References

Basu, A., and J. M. Wainright. 2001. *DME as a power generation fuel: Performance in gas turbines.* Presented at Petrotech-2001 conference, Delhi, India.

Beatrice, C., C. Bertoli, N. Del Giacomo, and M. Lazzaro. 1996. An experimental characterization of the formation of pollutants in DI diesel engines burning oxygenated synthetic fuels. *Application of powertrain and fuel technologies to meet emissions standards*, 261. London: Institution of Mechanical Engineers.

Bowman, F. M., and J. H. Seinfeld. 1995. Atmospheric chemistry of alternative fuels and reformulated gasoline components. *Progress in Energy Combustion Science* 21:387–417.

Brook, D. L., C. J. Rallis, N. W. Lane, and C. Dipolat. 1984. Methanol with dimethyl ether ignition promoter as fuel for compression ignition engines. *Proceedings of the 19th Intersociety Energy Conversion Engineering Conference.* San Francisco, 654–58.

Cheng, A. S., and R. W. Dibble. 1999. Emissions performance of oxygenate-in-diesel blends and Fisher–Tropsch diesel in a compression ignition engine. *SAE* 1999-01-3606.

Curran, H., E. M. Fisher, P.-A. Glaude, N. Marinow, D. W. Layton, W. J. Pitz, P. F. Flynn et al. 2001. Detailed chemical kinetic modeling of diesel combustion with oxygenated fuels. *SAE* 2001-06-0653.

Dec, J. E. 1997. A conceptual model for DI diesel combustion based on laser-sheet imaging. *SAE* 970873.

Ebersole, G. D., P. S. Myers, and O. A. Uyehara. 1963. The radiant and convective components of diesel engine heat transfer. *SAE* 630148.

Egnell, R. 2000. A theoretical study of the potential of NOx reduction by fuel rate shaping in a DI diesel engine. *SAE* 2000-01-2935.

Elbers, S. J., and M. Th. Logtenberg. 1999. *Conversion of LPG distribution guidelines into DME distribution guidelines.* Delft, the Netherlands: Research Center TNO.

Fleisch, T., C. McCarthy, A. Basu, C. Udovich, P. Charbonneau, W. Slodowske, J. McCandless, and S. E. Mikkelsen. 1995. A new clean diesel technology: Demonstration of ULEV emissions on a Navistar diesel engine fueled with dimethyl ether. *SAE* 950061.

Fryer, C. 2006. The renaissance of coal-based chemicals: Acetylene, coal-to-liquids, acetic acid. *Asia Petrochemical Industry Conference*, May. Bangkok.

Gill, D., H. Ofner, D. Schwarz, E. Sturman, and M. A. Wolverton. 2001. The performance of a heavy duty diesel engine with a production feasible DME injection system. *SAE* 2001-01-3629.

Hansen, J. B., B. Voss, F. Joensen, and I.-D.Sigurdardóttir. 1995. Large scale manufacture of dimethyl ether—A new alternative diesel fuel from natural gas. *SAE* 950063.

Hansen, K. R., C. S. Nielsen, S. C. Sorenson, and J. Schramm. 2008. A two-stroke compression ignition engine fuelled by DME. *SAE* 2008-01-1535.

Heywood, J. B. 1998. *Internal combustion engine fundamentals.* New York: McGraw-Hill.

JDF (Japan DME Forum). 2007. *DME handbook.* Minato-ku, Tokyo: Japan DME Forum.

JEC Well-to-Wheels study Version 3. 2008. Online available at http://ies.jrc.ec.europa.eu/WTW.

Kajitani, S. 2006. *A Fundamental Study of Direct type DME Fuel Cell.* Presented at the conference DME2, of the International DME Association, May 17, London.

Kajitani, S., Z. L. Chen, M. Konno, and K. T. Rhee. 1997. Engine performance and exhaust characteristics of direct injection diesel engine operated with DME. *SAE* 972973.

Kapus, P., and H. Ofner. 1995. Development of fuel injection equipment and combustion system for DI diesels operated on dimethyl ether. *SAE* 950062.

Landälv, I. 2006. *Technical and Commercial Impact of Locating a Black Liquor Gasification Plant Producing DME/Methanol at a Swedish Pulp Mill.* Presentation at DME2, International DME Association, May 17, London.

Liebner, W. 2006. *Lurgi's outlook on DME technologies and market.* Presentation at the conference DME-2. London: International DME Association.

McCandless, J. C., H. Teng, and J. B. Schneyer. 2000. Development of a variable-displacement, rail-pressure supply pump for dimethyl ether. *SAE* 2000-01-0687.

Miyamoto, N., H. Ogawa, N. M. Nurun, K. Obata, and T. A. Smokeless. 1998. Low NO_x high thermal efficiency, and low noise diesel combustion with oxygenated agents as main fuel. *SAE* 980506.

Murayama, T. C., J. Guo, and M. Miyano. 1992. A study of compression ignition methanol engine with converted dimethyl ether as an ignition improver. *SAE* 922212.

Nielsen, K., and S. C. Sorenson. 1999. Lubricity additives and wear with DME in diesel injection pumps. *Proceedings of the fall technical conference of the ASME internal combustion division*, October 16–20, Ann Arbor, MI, 145–54.

Obert, E. D. 1972. *Internal combustion engines and air pollution*. New York: Harper and Row.

Oguma, M., S. Goto, M. Konno, Z. Chen, and T. Watanabe. 2003. Chemiluminescence analysis from in-cylinder combustion of a DME-fueled DI diesel engine. *SAE* 2003-01-3192.

Ohno, Y. 2006. *The role of DME in the world: A perspective*. Presentation at the Conference DME-2. London: International DME Association.

Ozawa, K. 2007. Price comparison with other fuels. *DME handbook*, Section 9.2.3. Minato-ku, Tokyo: Japan DME Forum.

Sanfilippo, D., S. Dellagiovanna, M. Marchionna, and D. Romani. 2004. *Dimethyl ether as LPG substitute/make-up*. Presented at DME-1 International DME Association, Paris.

Sato, Y., T. Yanai, A. Kawamura, and H. Oikawa. 2008. *Public road test for practical use of DME vehicles in Japan*. Presented at DME-3/5th Asian DME Conference, Shanghai.

Scholl, K. W., and S. C. Sorenson. 1993. Combustion of soybean oil methyl ester in a direct injection diesel engine. *SAE* 930934.

Semelsberger, T. A., R. L. Borup, and H. L. Greene. 2006. Dimethyl ether (DME) as an alternative fuel. *Journal of Power Sources* 156:497–511.

Sivebaek, I. M., and S. C. Sorenson. 2000. Dimethyl ether (DME)—Assessment of lubricity using the medium frequency pressurised reciprocating rig version 2 (MFPRR2). *SAE* 2000-01-2970.

Sorenson, S. C., M. Glensvig, and D. L. Abata. 1998. Dimethyl ether in diesel fuel injection systems. *SAE* 981159.

Sorenson, S. C., and S. E. Mikkelsen. 1995. Performance and emissions of a 0.273 liter direct injection diesel engine fueled with neat dimethyl ether. *SAE* 950964.

Taupy, J.-A. 2007. *DME a Clean Fuel for the Future.* Paper IPTC 11754, International Petroleum Tehnology Conference, December 4–6, Dubai, U.A.E.

Veerbeek, R. and J. Van der Weide. 1997. Global Assessment of Dimethyl-Ether: Comparison with Other Fuels, *SAE Technical paper* 971607.

Wakai, K., K. Nishida, T. Yoshizaki, and H. Hiroyasu. 1998. Spray and ignition characteristics of dimethyl ether injected by a D. I. diesel injector. *4th international symposium on diagnostics and modeling of combustion in internal combustion engines*, Cambodia, Kyoto, Japan.

7

Liquefied Petroleum Gas

Mohamed Younes El-Saghir Selim

CONTENTS

7.1 Potential of LPG

Liquefied petroleum gas (LPG), a mixture of propane (C_3H_8) and butane (C_4H_{10}) gas, is a popular fuel for internal combustion engines. This popularity comes from many features of the fuel such as its high octane number for spark ignited engines, comparable to gasoline heating value that ensures similar power output. Other features include the possibility of transport and storage of LPG in liquid state because of relatively low pressure of saturated vapor in normal temperature range, better exhaust gas composition, and lower cost per energy unit in comparison with gasoline.

The last feature is a reason of rapidly growing popularity of gas fuel in the world. The above mentioned features of LPG also give the possibility to applying this fuel to compression ignition engines with relatively low-compression ratio as the heating value of stoichiometric mixture of diesel oil and air is similar to the heating value of LPG–air mixture and it gives the possibility to obtain the same power output from the engine fueled with diesel oil and LPG. Also, the high octane number allows compressing of the propane–butane with air in CI engines with low-compression ratio without apprehension of self-ignition appearance and enables nonknocking combustion of the mixture. This is in addition to improved exhaust gas composition and lower cost per energy unit from LPG in comparison with diesel oil.

Literature analysis shows the fueling of IC engines with lean homogeneous mixtures is more and more common. There are systems with spark ignition (SI) of the mixture and systems where the mixture is ignited by pilot diesel oil dose. The first solution seems to be more advantageous because one fuel is used for fueling the engine. But the ignition of lean mixture by spark is more difficult, so such systems require special solutions of ignition. This problem disappears in CI engines. Of course, application of two fuels simultaneously is necessary.

7.2 LPG Production

LPG is formed naturally, interspersed with deposits of petroleum and natural gas. Natural gas contains LPG, water vapor, and other impurities that must be removed before it can be transported in pipelines as a salable product. LPG can be produced from natural gas purification or from crude oil refining. It consists of hydrocarbons that are vapors, rather than liquids, at normal temperatures and pressures, but that turn to liquid at moderate pressures; its main constituent is propane, and it is sometimes referred to by that name.

7.2.1 Natural Gas and Oil Extraction

When natural gas and crude oil are drawn from the earth, a mixture of several different gases and liquids are in fact extracted, with LPG typically accounting for roughly 5% of the whole. Before natural gas and oil can be transported or used, the gases that make up LPG, which are slightly heavier than methane, are separated out.

7.2.2 Crude Oil Refining

The process of refining oil is complex and involves many stages. LPG is produced from oil at several of these stages including atmospheric distillation, reforming, cracking, and others. It is produced because the gases of which it is composed (butane and propane) are trapped inside the crude oil. In order to stabilize the crude oil before pipeline or tanker distribution, these "associated" or natural gases are further processed into LPG.

In crude oil refining, the gases that make up LPG are the first products produced on the way to making the heavier fuels such as diesel, jet fuel, fuel oil, and gasoline. Roughly 3% of a typical barrel of crude oil is refined into LPG although as much as 40% of a barrel could be converted into LPG. Although tied to the production of natural gas and crude oil, LPG has its own distinct advantages and can perform nearly every fuel function of the primary fuels from which it is derived.

7.3 Properties of LPG

While LPG is stored as a liquefied gas under pressure at ambient temperature, it is usually used in the liquid form for internal combustion engines, being directly injected into the engine by the fuel injection system. The term "LPG" covers a range of mixtures of propane and butane stored as liquids under pressure, but it is propane that is mostly used to fuel vehicles.

An LPG-derived refinery can differ from its refinery counterpart. However, whether refinery or natural gas origin, a mixture of liquefied saturated and unsaturated hydrocarbons in the range of C3–C4 is considered boiling. Commercially, LPG is sold to domestic and industrial customers in four grades.

- LPG propane consisting mainly of propane and/or propylene
- LPG butane consisting mainly of n-butane, isobutene, and/or butylenes
- LPG mixture consisting of C3 and C4 hydrocarbons
- High purity propane containing about 95% propane

Table 7.1 also shows the composition of LPG for different countries as butane is mixed with propane at ratios varying from 0% (pure propane used) to 75%. Moreover, LPG contains traces of hydrocarbon impurities such as ethane, ethylene, hexanes, pentanes, and methyl acetylene; sulfur compounds such as hydrogen sulfide, mercaptanes and elemental sulfur and traces of ammonia, nitrogen, and sodium. These sulfur compounds and particularly elemental sulfur in LPG are corrosive toward copper and other metals. Though the present gas compressors and liquid transfer pumps are the nonlubricated type, some compressors used in the system are lubricating oil pumps and there is a chance of leakage of lubricating oil into the LPG. Water is present in LPG naturally when it is produced from natural gas, whereas in refinery LPG, caustic washing generates water. If LPG is to be stored under low-temperature conditions, LPG has to pass through a bed of alumina before it is stored. The correct amount of water dissolved in LPG depends on its

TABLE 7.1

LPG Composition (% by Volume) as Automotive Fuel

Country	Propane	Butane
Austria	50	50
France	35	65
Italy	25	75
Spain	30	70
Sweden	95	5
United Kingdom	100	—
Germany	90	10
Australia	70	30

Source: Saleh, H. E., *Fuel*, 87, 3031–39, 2008. Reprinted with permission from Elsevier Publications.

composition, the temperature, and whether it is stored in a liquid or a vapor state.

LPG and petrol have many similar properties and the good practices appropriate for work on petrol vehicles apply equally to LPG vehicles. The viscosity of liquid LPG (0.15 centi Poise at 15.6°C) is much lower than water (1 centi Poise). The value for LPG vapor at atmospheric pressure of 15.5°C is about 0.08 centi Poise. The flash point of propane is very low, in line with its very low freezing and boiling points. The autoignition temperature of propane is nearly twice as high as that of gasoline providing more safety when exposed to hot surfaces. The stoichiometric air–fuel ratio for propane is only slightly higher than that of gasoline. As liquid, propane is about 68% of the weight of gasoline. Its boiling point is very low, which is typical of lower molecular weight hydrocarbons. Its vapor pressure is much higher than that of gasoline at 100°F. The net calorific value of LPG on mass basis is higher but its net calorific value on volume basis lower than that of gasoline and diesel. The main difference is that LPG vaporizes more rapidly than petrol, so, as LPG is stored in the fuel tank and associated fuel lines at elevated pressure (up to 10 bar g), any leakage will immediately vaporize and disperse. LPG vapor is highly flammable and mixtures in air of between 2 and 10% will readily ignite and explode. The rapid expansion resulting from the conversion of liquid propane into a vapor causes severe cooling of the gas. Anyone exposed to a vapor cloud or in contact with metal surfaces may receive cold burns. LP gases are 1.5–2 times heavier than air. Leaking LPG will sink to ground level rather than rise to the atmosphere whereas natural gas or hydrogen goes up. Table 7.2 shows the physical and chemical properties of LPG as compared to gasoline and diesel fuels.

The octane rating of propane is higher than that of gasoline. Hence, higher compression ratios engines can be used to take advantage of the higher octane number. Though the LPG specifications are specific to the particular

TABLE 7.2
LPG Properties Compared to Gasoline and Diesel Fuels

Property	LPG	Gasoline	Diesel
Research octane number	106–111	80–95	20–30
LHV (MJ/kg)	46.1	44.2	43.25
LHV (MJ/liters)	23.63	31.82	35.9
Relative CO_2 per kJ	0.885	1	1.028
Relative density	0.51	0.74	0.83
Evaporated enthalpy (kJ/kmol stoic. mix)	820	662	330
Stoichiometric A/F ratio (vol. basis)	24.1	60.1	100
Stoichiometric A/F ratio (mass basis)	15.7:1	14.7:1	14.5:1
Flammability limits (φ)	0.55–2.35	0.85–3.55	—
Boiling point (°C)	−42.1	130–150	180–350
Ignition temperature (°C)	400	280	250

Source: Selim, M. Y. E., *Energy Conversion and Management*, 45, 411–25, 2004. Reprinted with permission from Elsevier Publications.

country, the composition of the LPG almost remains the same. The sulfur content in the LPG is much lower than gasoline and has an advantage of low-sulfur emissions and improved catalyst efficiency.

7.4 LPG Modeling Studies

Liquefied petroleum gas (LPG) was first put into use as a SI engine fuel as early as the 1930s when a vehicle was first run on LPG at the eastern coast of the United States. Because of its good performance, it created interest among researchers to introduce it in a larger scale as an alternative/supplement to gasoline. Since the problems of air pollution and the energy crisis were not dominant at that time, not much research was devoted to utilization of LPG as an automotive fuel.

A quick look at the currently available alternatives indicates they may be classified into two main categories, long-term and short-term alternatives. Liquefied petroleum gas, natural gas, alcohol, and many other hydrocarbon fuels are considered among the short-term alternatives as they are finite in nature and are derived from sources that are themselves finite and suffering from overstress and exhaustion. Further examination of the literature reveals that research is concentrated mainly in three basic areas:

1. Searching for alternatives to replace existing fuel (i.e., gasoline)
2. Modifying the existing design with gasoline as fuel
3. Using the modified design with the alternative fuels

As mentioned above, LPG, compressed natural gas (CNG), hydrogen, and many other alternative fuels fall within the first category. As for the second category, the use of the variable stroke length, variable compression ratio, variable valve timing, turbocharger, supercharger, and so on have been tested and used with gasoline. The use of the modified engine design with another fuel constitutes the third category.

Each one aims at introducing improvement in one direction or parameter. To assist these experimental researches, several theoretical studies investigated the performance of LPG as fuel for SI engines. A quasi-dimensional SI engine cycle model has been developed to predict the cycle performance and exhaust emissions of an automotive engine using gasoline and LPG fuels. The model also studies the mechanism of combustion and emission formation of both fuels (i.e., gasoline and LPG). Many simulation models are now available to be used for studying the behavior of LPG as a SI engine fuel. Each one concentrates on certain phenomena.

As for the use of variable stroke length mechanisms concerned, a simulation model for the engine was developed with the different power and efficiency characteristics being computed and presented for different stroke lengths and the corresponding compression ratios. Effect of variable stroke mechanism on LPG engine performance has been theoretically investigated and it has been shown that the engine performance improves with this new design (Ozcan and Yamin 2008)

7.5 Engine Tests

LPG is commonly used in both types of engines. The gasoline engines can be equipped with retrofit devices that enable them to run on LPG as well as gasoline. The gasoline engines can be converted to dedicated LPG fueled engines. In both cases, vehicles have similar characteristics in terms of fuel system and storage systems. In diesel engines dual fuel mode operation, LPG is inducted into the engine as a mixture of LPG + air.

7.5.1 LPG Conversion Systems

Several methods were tested to improve the performance of LPG powered, four stroke SI engines. The addition of coconut oil, gasoline, ethanol, dimethyl ether, and many other supplements to LPG were studied. Each of these supplements showed a certain percentage of improvement. However, the idea of supplementing fuel was not favorable since it is sometimes accompanied with additional accessories and equipments for the additional fuel. LPG conversion systems are classified in three main groups, as follows:

- Mechanically controlled LPG carburetion systems (first generation)
- Electronically controlled LPG carburetion systems (second generation)
- LPG injection systems (third generation)

In countries where vehicles are not required to meet any emission standards, conventional LPG systems are used. A typical conventional LPG system is shown in Figure 7.1 (Karamangil 2007). These mechanical systems are called "first generation" systems. They consist of an electromagnetic LPG shut-off valve, an evaporator/pressure regulator, and a mixing unit. Combined pressure regulator and mixing unit perform the LPG metering of the kit. This is the simplest form of conversion to LPG.

In "second generation" systems, the main metering of the gas flow is still performed by the pressure regulator and mixing unit; however, they are electronically controlled. The input signals of the electronic unit are derived from engine speed and lambda sensor. The electronic unit controls a digital linear actuator (stepper motor), which adjusts the main jet in the dry gas hose between pressure regulator and mixing unit. So, the control system is a closed-loop. Owing to the self-learning properties of the LPG system, it is not necessary to adjust the engine during its lifetime. This system allows adjustment of the control parameters to account for variations in propane/butane composition, supplied by LPG marketing companies, during the year. With

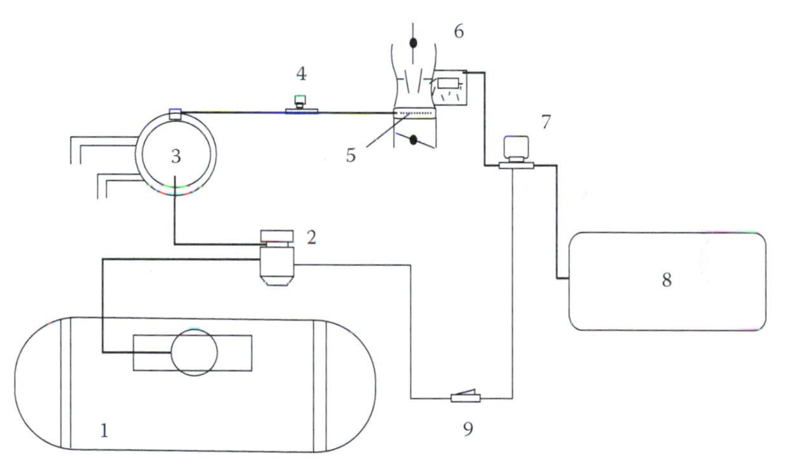

1. LPG tank, 2. LPG solenoid valve, 3. Regulator,
4. Mechanical flow adjustment valve, 5. Mixer,
6. Carburetor, 7. Gasoline solenoid valve,
8. Gasoline tank, 9. LPG/Gasoline selector switch

FIGURE 7.1
Mechanically controlled LPG carburetion system used (first generation). (From Karamangil, M. I., *Energy Policy*, 35, 640–49, 2007. Reprinted with permission from Elsevier Publications.)

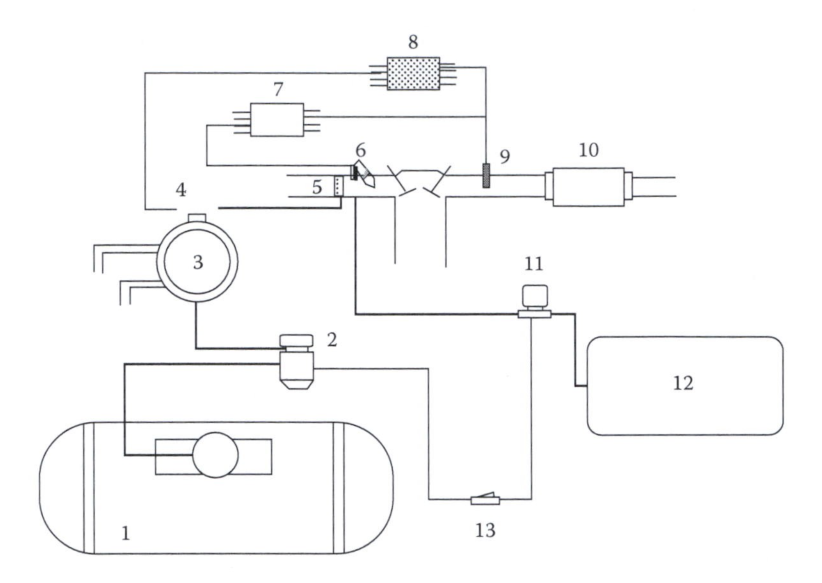

FIGURE 7.2
Microprocessor-controlled LPG system (second generation). (From Karamangil, M. I., *Energy Policy,* 35, 640–49, 2007. Reprinted with permission from Elsevier Publications.)

these electronic control and self-learning features, an LPG-powered vehicle, equipped with second-generation LPG kit, releases less pollutants (Figure 7.2; Karamangil 2007).

Third-generation LPG systems are microprocessor-controlled, self-learning, and without manual adjustments. In these systems, LPG is injected in either liquid or gaseous form. A gas metering unit supplies the same amount of fuel to each cylinder. The LPG is then injected just in front of the inlet valve of each cylinder by the injectors. Signals produced by the sensors and representing performance, emission, and fuel consumption are received and evaluated by the microprocessor to determine the corresponding and correct actuator position, based on the self-learning tables in the memory of the electronic control unit. "Third-generation" systems may be of various configurations, and vehicles equipped with these produce much lower emissions than those equipped with the systems belonging to first two generations. A schematic illustration of the system is given in Figure 7.3 (Karamangil 2007). Closed-loop lambda control system and precise control of fuel measurement make them very suitable for vehicles of recent technology equipped with a three-way catalyst.

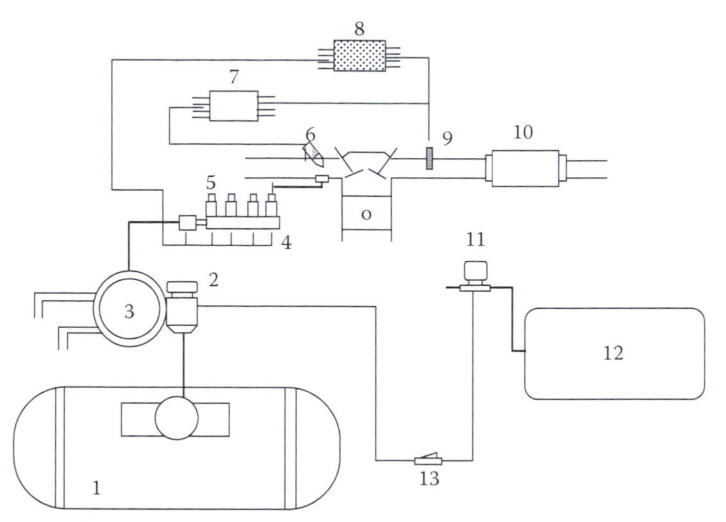

1. LPG tank, 2. LPG solenoid valve, 3. Regulator, 4. Injector rail, 5. LPG injectors. 6. Gasoline injector, 7. Electronic control unit for gasoline, 8. Electronic control unit for LPG, 9. Lambda sensor, 10. Catalytic converter, 11. Gasoline solenoid valve, 12. Gasoline tank, 13. LPG/Gasoline selector switch

FIGURE 7.3
Sequent LPG vapor injection system (third generation). (From Karamangil, M. I., *Energy Policy*, 35, 640–49, 2007. Reprinted with permission from Elsevier Publications.)

7.5.2 LPG in Gasoline Engine Applications

LPG is an environmentally friendly fuel for SI engines that has potential emission advantages over gasoline. LPG is liquefied under pressure and compressed and stored in steel tanks under pressure that varies from 1.03 to 1.24 MPa. It is used for heating, cooking, and can be used as engine fuel. The fuel is liberated from lighter hydrocarbon fraction produced during petroleum refining of crude oil and from heavier components of natural gas. It is also a by-product of oil or gas mining.

Experience shows that LPG has some advantages over gasoline due to the following:

- LPG produces lower exhaust emissions than gasoline
- It reduces engine maintenance
- It offers faster cold starting
- It provides overall lower operational costs

On the other hand, LPG displaces 15–20% greater volume than gasoline. Thus the power output decreases by 5–10%. This reduction can reach up to

30% at very lean conditions. There are many reports about the effects of the LPG on the engine performance and emission characteristics. However, it is apparent that very little information is available about the cyclic variability when using LPG as a fuel. There are some studies about the lean operating performance of LPG-used SI engines. The studies showed that ignition timing strongly affected the overall engine lean misfire limit. The experimental results show that as the fuel/air equivalence ratio leaned, the MBT ignition timing needed to be advanced, where lean mixtures required advancing in ignition timing to provide more time for reaction due to the excess presence of oxygen in the fuel/air mixture. The engine lean misfire limit increases as the engine speed increases, with the same effect of preheating inlet air temperature, and as the engine load increases, the lean misfire limit increases. It has been also shown that the higher laminar flame speed of LPG and good mixing of gaseous fuels with air causes a decrease in cyclic variations, and higher H/C ratio of LPG decreases the engine emissions.

LPG and other gaseous fuels have common properties that provide them some advantages and disadvantages relative to gasoline. Before discussing its usability for engines, its properties are compared with those of other engine fuels in Table 7.1. If the properties of propane are compared with those of gasoline and alcohols, the following benefits and/or shortcomings can be expected when it is used as the SI engine fuel.

Propane has lower density and stoichiometric fuel–air ratio than gasoline, and thus, it could reduce the specific fuel consumption and exhaust emissions. If a propane fueled SI engine operates at the same equivalence ratio as a similar gasoline fueled engine, higher effective power could be expected due to the higher calorific value of propane. However, as will be explained below, this advantage may be balanced by decreasing volumetric efficiency. On the other hand, propane can be used at higher compression ratios due to its higher octane number, and as a consequence of this property, engine performance; that is, engine power and thermal efficiency, would be improved.

The above mentioned properties of propane make it an attractive alternative fuel for SI engines. The most important drawback of this fuel is that it reduces the engine volumetric efficiency and consequently the fresh charge mass, which is mainly due to its rising inlet temperature and its entering the intake system in the gaseous state. This problem can be removed by cooling; that is, offsetting the heat in the inlet manifold.

It has been found that in the case of using LPG in SI engines, the burning rate of fuel is increased, and thus, the combustion duration is decreased. As a consequence of this, the cylinder pressures and temperatures predicted for LPG are higher than those obtained for gasoline. The maximum cylinder pressures and temperatures predicted for LPG are higher. This may cause some damages on engine structural elements. LPG reduces the engine volumetric efficiency and, thus, engine effective power. Furthermore, the decrease in volumetric efficiency also reduces the engine effective efficiency and consequently increases specific fuel consumption. It has been also found

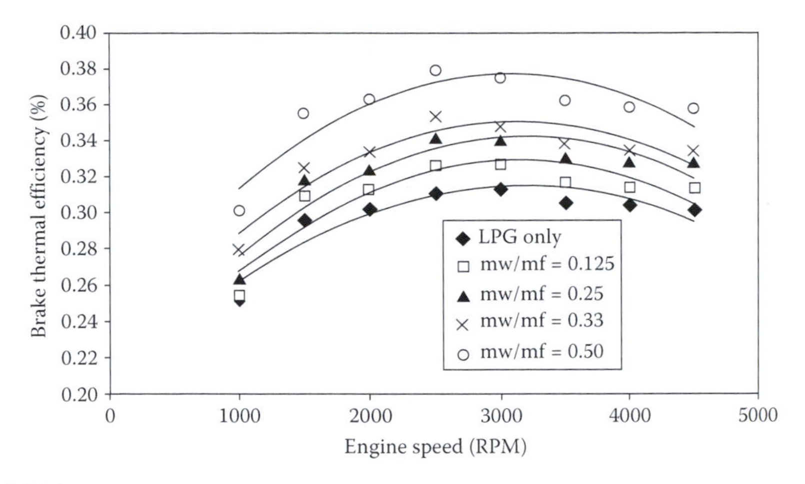

FIGURE 7.4
Brake thermal efficiency variations with engine speed for various water/fuel mass ratios of water injection. (From Ozcan, H. and Soylemez, M. S., *Energy Conversion and Management*, 47, 570–81, 2006. Reprinted with permission from Elsevier Publications.)

that LPG decreases the mole fractions of CO and NO included in the exhaust gases. Furthermore, LPG has negative effects on engine performance, fuel economy, and engine structural elements when it is used at the same fuel–air equivalence ratios as gasoline, however, it has positive effects on obnoxious exhaust emissions such as CO and NO.

Figure 7.4 shows the effect of water addition on engine thermal efficiency for different water to fuel ratios (Ozcan and Soylemez 2006). The results show that as the water to fuel mass ratio increases, the engine thermal efficiency increases due to the decrease in brake specific fuel consumption (BSFC), as shown in Figure 7.5 (Ozcan and Soylemez 2006). For the pure LPG experiments, the average of the thermal efficiency value is 29.3%, whereas it increased to 32% for 0.5 water to fuel mass ratio. As the water percentage in the emulsion increases, the brake thermal efficiency increases. Figure 7.5 shows that the BSFC decreases to a minimum, and then, it begins to increase at high speed as expected. Increasing the water to fuel mass ratio decreases the BSFC due to the increase in brake power with water addition.

7.5.3 LPG in Diesel Engine Applications

The use of LPG as a main fuel in diesel engines that uses the liquid diesel as a pilot fuel is also increasing worldwide. Gaseous fuels, namely LPG and CNG are recognized as clean fuels possessing significant environmental benefits compared to conventional liquid fuels as well as their relatively increased availability at attractive prices. Figure 7.6 shows that

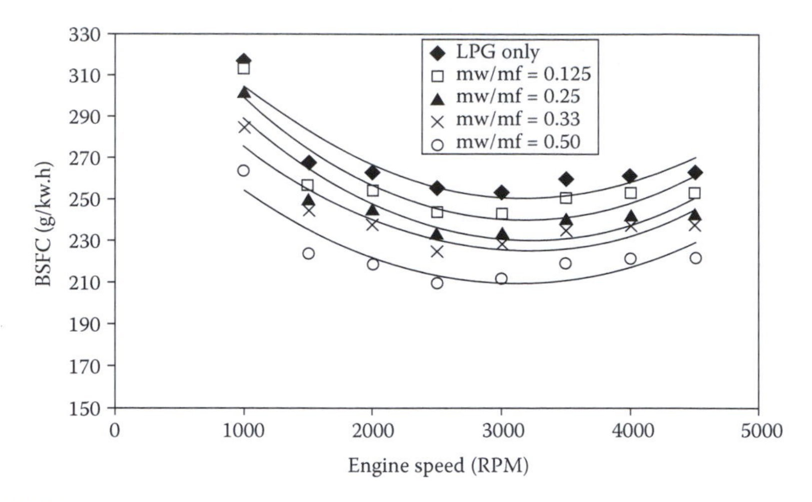

FIGURE 7.5
BSFC variations with engine speed for variations water/fuel mass ratios of water injection. (From Ozcan, H., and Soylemez, M. S., *Energy Conversion and Management*, 47, 570–81, 2006. Reprinted with permission from Elsevier Publications.)

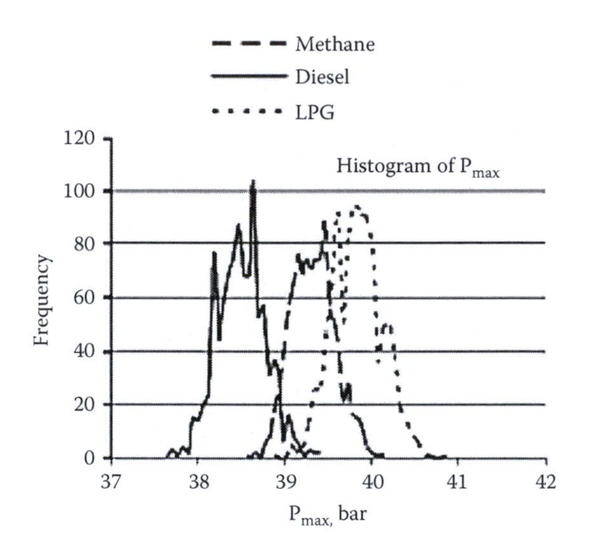

FIGURE 7.6
Histogram of maximum combustion pressure for diesel and dual fuel engine. (From Selim, M. Y. E., *Fuel*, 84, 961–71, 2005. Reprinted with permission from Elsevier Publications.)

using LPG gives slightly higher combustion pressure than using CNG (Selim 2005).

In dual fuel engines, gaseous fuel is usually inducted with the intake air through the inlet manifold. They are mixed and compressed as in a conventional diesel engine. A small amount of liquid diesel fuel, called the pilot, is injected near the end of the compression stroke, which initiates the combustion of the gas–air mixture. Diesel fuel autoignites and creates ignition sources for the surrounding gaseous fuel mixture.

There are other advantages for dual fuel engines using gaseous fuels. Gaseous fuel satisfies the previous requirements as a result of its worldwide usage. It has a high octane number and therefore it is suitable for engines with relatively high compressions ratios. Moreover, it mixes uniformly with air, resulting in efficient combustion and substantial reduction of emissions in the exhaust gas. Further, the exhaust smoke density of dual fuel engines is much lower than for diesel engines. However, some disadvantages of the two fuels currently prevent them from achieving widespread use. Gaseous fuel has a lower energy density relative to gasoline, so vehicles have to carry large gaseous fuel tanks in order to obtain acceptable mileage range. Large tanks occupy a lot of useful space and their additional weight requires significant power. The volume of gaseous fuels displaces some portion of intake air in the engine resulting in a loss of power and torque.

Dual fuel engines also suffer from some disadvantages at part loads as the gaseous fuel concentration is low and the ignition delay period of the pilot fuel increases and some of the homogeneously dispersed gaseous fuel remains unburned and results in poor performance. This is shown in Figure 7.7 as the ignition delay period increases with the decrease in

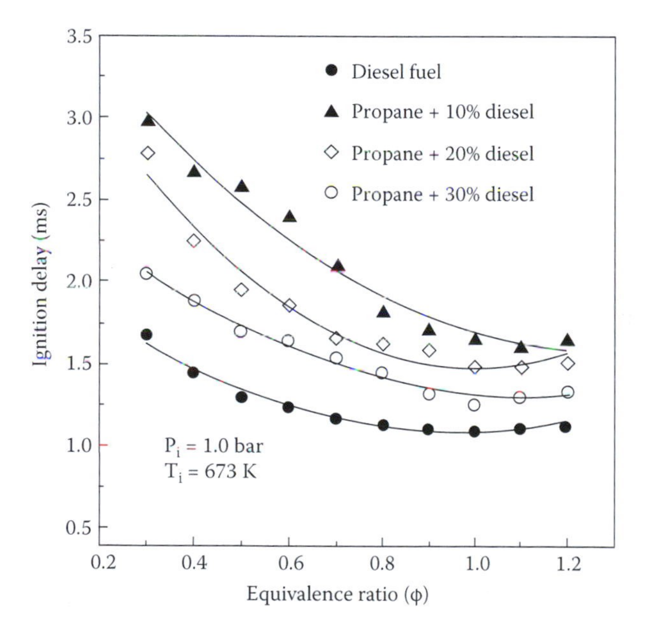

FIGURE 7.7
Effect of equivalence ratio on ignition delay of propane diesel/air mixture. (From Saleh, H. E., and Selim, M. Y. E., *Fuel*, 89, 494–500, 2010. Reprinted with permission from Elsevier Publications.)

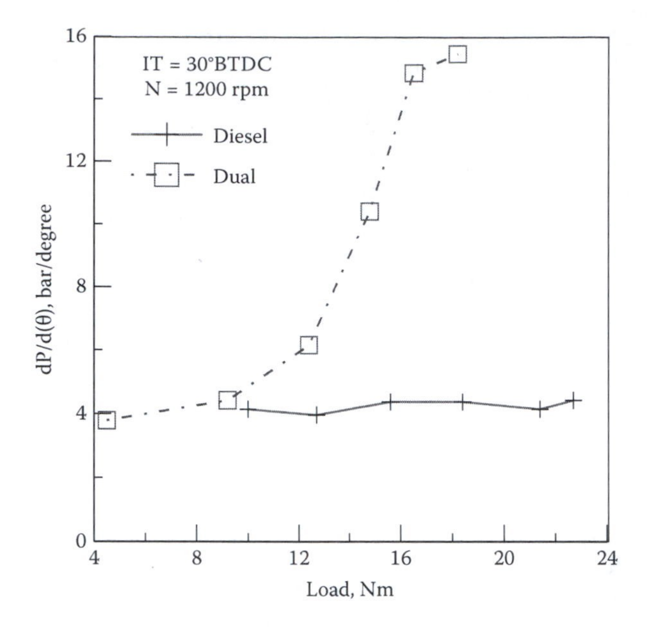

FIGURE 7.8
Effect of engine load on pressure rise rate for diesel and dual fuel engine. (From Selim, M. Y. E., *Renewable Energy*, 22(4), 473–89, 2000. Reprinted with permission from Elsevier Publications.)

the overall equivalence ratio (Saleh and Selim 2010). Figure 7.8 shows the increase in dual fuel combustion noise as compared to base diesel engine (Selim 2000). This combustion noise is well related to the overall engine noise and roughness. A concentrated ignition source is needed for the combustion of the inducted fuel at low loads.

Choice of pilot fuel can help to reduce the engine roughness as illustrated in Figure 7.9 (Selim, Saleh, and Radwan 2008). A pilot fuel that is easy to self-ignite, derived from the Jojoba plant, with reduced ignition delay period, may reduce the maximum pressure rise rate and hence engine noise. Further, injection timing of the pilot fuel, injector opening pressure, and pilot fuel quantity and intake temperature are some of the important variables controlling the performance of dual fuel engines at light loads. Poor combustion of the gaseous fuel at low loads because of dilute fuel–air mixtures results in high carbon monoxide and unburned hydrocarbons emissions. Any measure that lowers the effective lean flammability limit of the charge and promotes flame propagation will improve part load performance. Preheating of the intake charge and increase in pilot diesel quantity resulted in higher thermal efficiency. However, at high loads and high intake temperatures, increased admission of the gaseous fuel can result in uncontrolled reaction rates near the pilot fuel spray and lead to knock. Resorting

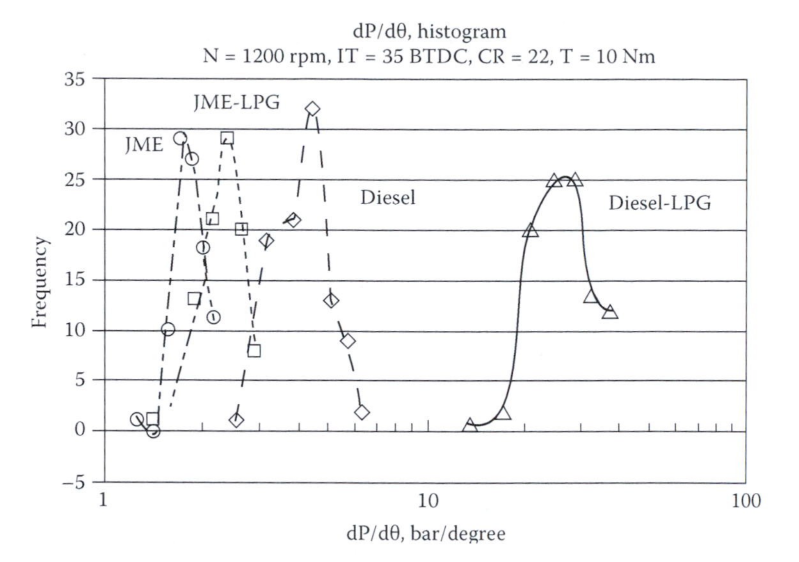

FIGURE 7.9
Histogram of maximum pressure rise rate for diesel (uses diesel or Jojoba methyl ester) and dual fuel engine (uses LPG). (From Selim, M. Y. E., Saleh, H. E., and Radwan, M. S., *Renewable Energy*, 33, 1173–85, 2008. Reprinted with permission from Elsevier Publications.)

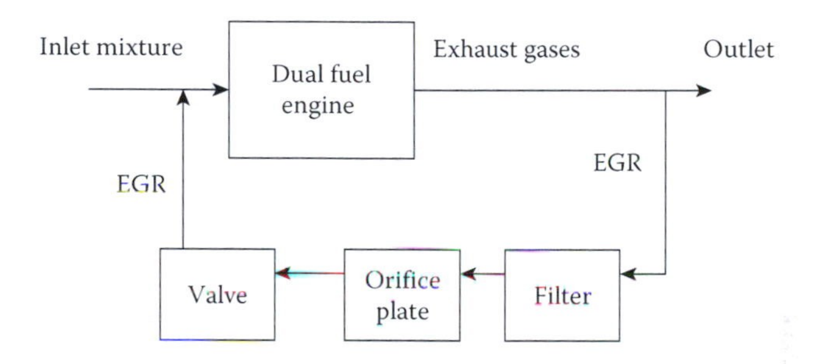

FIGURE 7.10
Schematic diagram of EGR system for dual fuel engine. (From Selim, M. Y. E., Saleh, H. E., and Radwan, M. S., *Renewable Energy*, 33, 1173–85, 2008. Reprinted with permission from Elsevier Publications.)

to hot exhaust gas recirculation (EGR) in dual fuel engines can also help in improving light load performance through increased initial charge temperature. Exhaust gases also seed the charge with active species to promote combustion. EGR is especially suited for engines running with excess air, see Figure 7.10 (Selim 2003).

The potential of different methods of improving the performance of dual fuel engines has to be evaluated to arrive at an optimum combination. It has been shown also that when dual fuel engines utilizing propane, the mixture exhibits longer ignition delay periods when compared to methane–air–diesel mixtures and diesel–air mixtures. For the diesel engine, there has been usually a trade-off between several targets, for example, reducing exhaust emissions, reducing engine noise, reducing fuel consumption, and increasing specific outputs.

As mentioned above, for dual fuel operation the propane is aspirated with the incoming air to the cylinder, so the combustion occurs partially in premixed mode. The diesel pilot flame acts as an ignition source for the propane/air mixture. The effect of energy substituted by pure propane was varied from 0 (conventional diesel engine) to 90% for various engine loads at 1500 rpm with energy conversion efficiency is shown in Figure 7.11 (Saleh 2008).

The results of Figure 7.11 show a slight increase in the thermal conversion efficiency of the dual fuel engine than of normal diesel engine at different engine loads when increasing the propane up to 40%. The thermal conversion efficiency decreases in a normal diesel engine when increasing the percentage of gaseous fuel up to 90%. The main reason for decreasing the thermal conversion efficiency is the increase of the ignition delay period, with the increased percentage of propane and also, the flame propagation in gas–air mixture is much slower at the higher percentage of gas and the lower pilot fuel quantities. It has also been shown that the increase of ignition delay

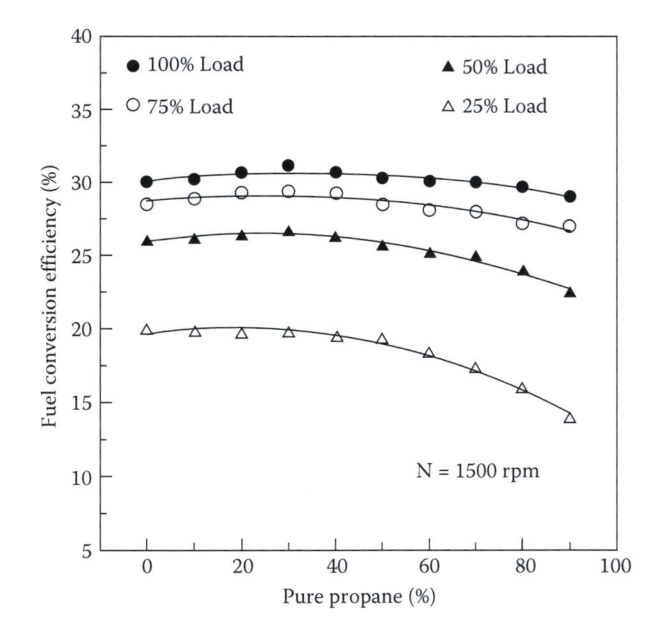

FIGURE 7.11
Fuel conversion efficiency versus percentage of energy substituted by propane gas. (From Saleh, H. E., *Fuel*, 87, 3031–39, 2008. Reprinted with permission from Elsevier Publications.)

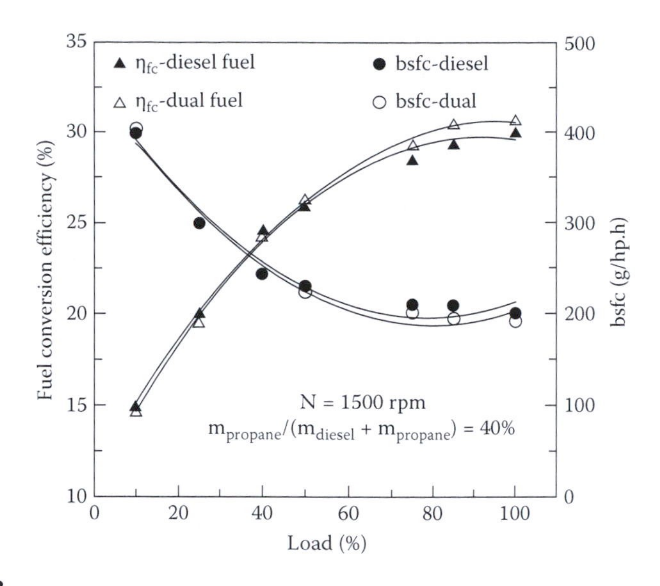

FIGURE 7.12
Fuel conversion efficiency and bsfc as a function of load at N = 1500 rpm. (From Saleh, H. E., *Fuel*, 87, 3031–39, 2008. Reprinted with permission from Elsevier Publications.)

with increase of the LPG mass fraction would be explained by the lowering of cetane number with the increase of LPG mass fraction. It would be noticed that the substitution of 40% of the diesel fuel does not significantly affect the engine performance as shown in Figure 7.12 (Saleh 2008) at overall load conditions. So, the ratio of $m_{propane}/(m_{diesel} + m_{propane}) = 40\%$ substitution of the diesel fuel, is best for maintaining high thermal efficiency comparable to a conventional diesel engine.

The effect of various engine loads on the thermal conversion efficiency and BSFC, of dual fuel combustion (40% of pure propane), comparable to a conventional engine are shown in Figure 7.12. It will be seen that the efficiency increases with load for both dual fuel operation and diesel engine. At high loads, it is found that the dual fuel operation is more efficient than corresponding diesel operation at about 2.3% at full load. This trend may be explained by increased heat release as a result of the overall equivalence ratio increases, and the combustion tends to be more complete, leading to high cylinder pressures and increased power output. The BSFC at full load is decreased about 4.2% as compared to BSFC value of diesel operation as shown in Figure 7.12. At 25% load the thermal conversion efficiency of dual fuel is lower than diesel operation by 2% since the combustion of the gas fuel occurs at very fuel-lean mixtures, burning rate is relatively slower for half load, and slowest for quarter load. This may be due to the incomplete combustion since the flames originating from ignition regions within the pilot envelope cannot propagate fast and far enough within the time available to consume the entire fuel-lean mixture.

The nitrogenous oxides (NO_x) concentration versus engine load for dual fuel mode and diesel engine is illustrated in Figure 7.13 (Saleh 2008). As the load on the engine was increased, NO_x emissions also increased for both dual fuel operation and diesel engine because of an increase in the cylinder combustion temperature and pressure with load. This is considered to be the main reason for the increase in NO_x emissions through the Zeldovich mechanism. It would be expected that the fuels with the highest in-cylinder temperature levels would have the highest NO_x. Since diesel fuel was a higher combustion chamber temperature as indicated from measured exhaust gas temperature, as shown in Figure 7.13, higher NO_x emissions were observed with diesel fuel engine. At full load NO_x emissions reached a value of 575 ppm for diesel engine operation and 450 ppm for dual fuel operation. At 25% load, the NO_x were 243 ppm and 200 ppm for diesel engine and dual fuel operation, respectively. It can be noticed that the difference of NO_x emissions between diesel engine and dual fuel operation is amplified with increase of the engine load, however, the gas fueling fraction increases in dual fuel operation. The main reason is that the propane acts as a diluent in the unburned mixture, increasing the heat capacity of the cylinder charge and reducing effectively the amount of free oxygen that can react with nitrogen to produce NO_x.

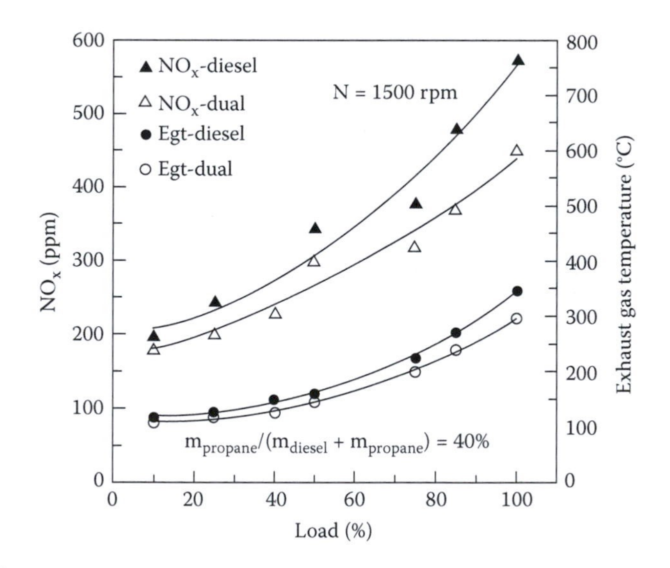

FIGURE 7.13
Variation of NO_x concentration and exhaust gas temperature with engine load. (From Saleh, H. E., *Fuel*, 87, 3031–39, 2008. Reprinted with permission from Elsevier Publications.)

7.6 LPG Material Compatibility

Liquefied petroleum gas (LPG) is commonly used as an alternative fuel for internal combustion engines of vehicles in many countries. LPG is stored and transported based on transportation standards. In order to store LPG in vehicles, the LPG cylinders known as LPG fuel tanks are commonly used and approved by these regulations. The LPG tanks, low-pressure cylinders since their service pressure (SP) is lower than 3.44 MPa (500 psi), can be commercially filled and used in the automobile industry. They are equipped with a refillable two-way hermetic valve, are produced as LPG containers and used in vehicles having water capacities ranging from 35 to 80 liters. There are many studies that determine the burst pressures (BP) and the failure locations of LPG cylinders whose SP and test pressures (TP) are known by the definitions of the standards and rules. The SP is the working (operating) pressure where the cylinders are filled and used in industrial applications. The TP is a given pressure that is applied and released after which the permanent volume expansion of the cylinder must not exceed 10% of the initial measured volume. Finally, the BP is the maximum pressure a cylindrical tank can withstand without bursting. In case of instability of cylindrical shells, analytical formulations are available for ideal shells with perfect geometry and specific boundary conditions. These do not take into account the strain distribution, and nonhomogeneous nonlinear material properties (MPs) and geometrical imperfections including thickness variations in the cylinder shell material. The BP of a vessel was estimated after a single application of internal pressure using mathematical and experimental models for tensile loading.

Guidelines for the design of cylindrical shells can be found in many international codes such as BS-5500 (British Standard) and the ASME codes (Section VIII, Division 1). These rules are restricted mostly to the load carrying capacity under internal pressure. These cylinders are equipped with a valve system and a label welded to the shell body, as shown in Figure 7.14 (Kaptan and Kisioglu 2007).

Cylindrical LPG tanks are usually manufactured within four different groups, which are classified by their water capacities: 35, 45, 60, and 80 liters. Each group of cylinders has a different wall thickness ranging from 2 to 3 mm. In their study, the BP and failure locations of the most commonly used three groups (35, 60, and 80 liters) of cylindrical LPG tanks having 2.5 mm wall thickness are investigated. These tanks consist of three main parts: one cylindrical shell and two torispherical end closures as shown in Figure 7.14. The cylindrical shell is folded up and welded longitudinally and the two torispherical drawn end closures are welded circumferentially at the ends of the shell. Also in the figure, some design parameters are shown such as inner diameter (ID), minimum wall thickness (t), knuckle radius (Rk), crown radius (Rc), and length of the cylindrical shell (L). The LPG tanks are constructed

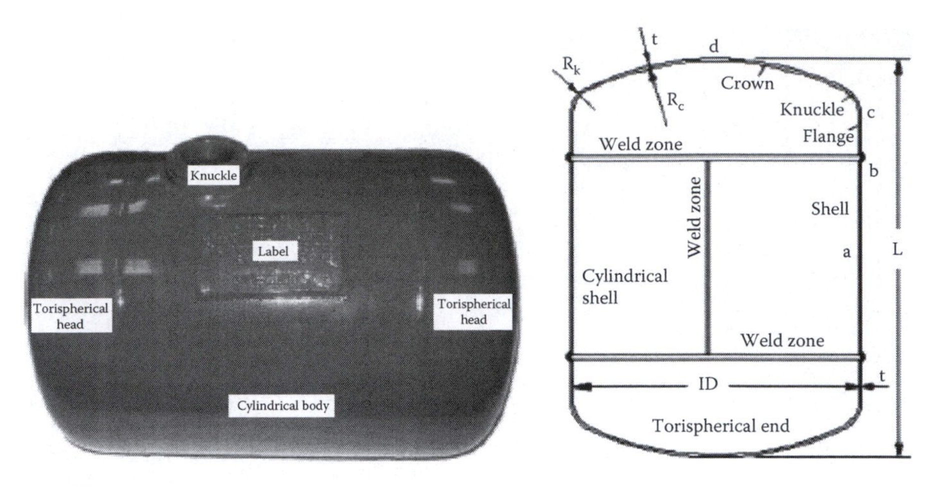

FIGURE 7.14
The vehicle LPG fuel tank and its design parameters. (From Kaptan, A. and Kisioglu, Y., *International Journal of Pressure Vessels and Piping*, 84, 451–59, 2007. Reprinted with permission from Elsevier Publications.)

from Erdemir-6842 steel using the welding process for the shell body and the deep drawing process for the end closures. The Erdemir-6842 is hot rolled steel with 0.18% carbon content and is a ductile material suitable for the cold-forming process used to construct the LPG tanks. The deep drawing process changes the MPs and thickness variations. After completing all manufacturing and welding processes, these tanks have been subjected to a heat treatment process before service.

7.7 LPG Economics

Many gasoline vehicles are converted into propane vehicles. The relatively inexpensive conversion kits include a regulator/vaporizer that changes liquid propane to a gaseous form and an air/fuel mixer that meters and mixes the fuel with filtered intake air before the mixture is drawn into the engine's combustion chambers. Also included in conversion kits is closed-loop feedback circuitry that continually monitors the oxygen content of the exhaust and adjusts the air/fuel ratio as necessary. This device communicates with the vehicle's onboard computer to keep the engine running at optimum efficiency. LPG vehicles additionally require a special fuel tank that is strong enough to withstand the LPG storage pressure of about 130 psi. The gaseous nature of the fuel/air mixture in an LPG vehicle's combustion chambers eliminates the cold-start problems associated with liquid fuels. In contrast to gasoline engines, which produce high emission levels while running cold,

LPG engine emissions remain similar whether the engine is cold or hot. Also, because LPG enters an engine's combustion chambers as a vapor, it does not strip oil from cylinder walls or dilute the oil when the engine is cold. This helps LPG powered engines to have a longer service life and reduced maintenance costs. Also helping in this regard is the fuel's high hydrogen-to-carbon ratio (C_3H_8), which enables propane powered vehicles to have less carbon build-up than gasoline and diesel powered vehicles.

The LPG delivers roughly the same power, acceleration, and cruising speed characteristics as gasoline. It does yield a somewhat reduced driving range, however, because it contains only about 70–75% of the energy content of gasoline. Its high octane rating (around 105) means that an LPG engine's power output and fuel efficiency can be increased beyond what would be possible with a gasoline engine without causing destructive "knocking." Such fine-tuning can help compensate for the fuel's lower energy density. Fleet owners find that propane costs are typically 5–30% less than those of gasoline. The cost of constructing an LPG fueling station is also similar to that of a comparably sized gasoline dispensing system. Fleet owners not wishing to establish fueling stations of their own may avail themselves of over 3000 publicly accessible fueling stations nationwide.

7.8 LPG Safety Aspects

Technology for this fuel is well established, as there are already a significant number of LPG powered vehicles on the road, using "autogas" filling outlets. The major hazards are gross leakage under failure conditions of the main fuel tank (pressure vessel) or piping, fugitive emissions while refueling, and the potential for small continuous leaks from the installation. During refueling, the LPG is passed through a hose to a self-sealing connector that is locked onto the refueling connection on the vehicle. The fuel is transferred until the tank is full, at which point the transfer is stopped. Other hazards are failure to disconnect the transfer hose before driving away, and leakage of the self-sealing coupling allowing gas to escape. Icing due to the rapid boil-off of liquefied gas can also present a hazard if rain-water has entered the coupling. This can freeze self-sealing or self-locking parts of the coupling, and ice formation can cause a major fire hazard. Prevention by good design is required, and the currently proposed standards for use with automotive LPG cover these aspects. Three systems coexist in Europe at present, and adaptors are available at most LPG filling stations to accept nozzles that are standard in other countries. Within the standards, potential for failure of the pressure vessel is already well known, and the tank must be fitted within the main strong section of the passenger cage, so that impacts to the vehicle minimize the risk of striking the tank.

The potential for small continuous leaks is always present, and it is necessary to store vehicles in well-ventilated garages. LPG being a dense gas, the use of low-level vents will allow the heavy vapors to disperse. The potential for explosions of leaked LPG to occur is known, and at least one has occurred in Belgium and resulted in both the loss of the vehicle and of the house above the basement garage where the vehicle was stored overnight. The ignition source was probably the central locking system being remotely operated, fortunately for the driver, who escaped being injured by being far enough away from the vehicle at the time of ignition. Gross spillage of LPG is similar to a spillage of LNG, in that rapid boil-off occurs on contact with the ground, until the ground has cooled to the atmospheric boiling point of the LPG. Leakage of LPG from a fractured pipe would form a large persistent flammable atmosphere, which is likely to be ignited. This would burn back to the fracture, and if the flame were to impinge on the tank, a boiling liquid expanding vapor explosion (BLEVE) would be likely to occur eventually. Even small cylinders are affected unless over-temperature protection is fitted. Normally the propane cylinders engulfed in fire generally did not rupture quickly when fitted with relief valves, but continued to relieve until the top of the cylinder became overheated and the steel weakened. The cylinder would then rupture. However, on cylinders fitted with *brass* outlet valves, the valve itself became weakened by the high temperature and blew out, thus relieving the pressure and avoiding a rupture. Hence it would seem prudent to fit LPG tanks with both an over-pressure relief valve and an over-temperature relief device, such as a fusible plug.

Other potential explosion hazards are hydraulic locking by overfilling, and rupturing the cylinder due to over-pressure. This can be avoided by careful design of the filling arrangement or a suitable relief valve, which will of course have to vent the LPG to a safe place. The potential problems of concentration of nonvolatiles encountered with LNG is less of a problem with LPG, as the fuel is invariably taken off in the liquid form, relying on the storage tank pressure to maintain the fuel as a liquid until it is injected into the engine.

For LPG refueling facilities, there are a number of hazards that are well known, with the most significant hazards being

- Boiling liquid expanding vapor explosion (BLEVE). This results from the loss of pressure following containment (i.e., tank) failure. In this case the escaping liquid expands very rapidly (boils), and, if ignited at or near the course of release, burns at a great rate. Heat radiation is the main hazard associated with a BLEVE.
- Unconfined vapor cloud explosion (UVCE). This occurs when a cloud of LPG, which has leaked from the tank without ignition taking place, is ignited at a later time at a source perhaps some considerable distance from the release point. The exact mechanics and

thermodynamics are not perfectly understood, but the result is LPG burning at a rate roughly equivalent to an explosion.

- Flash fire. This is a lesser form of UVCE, usually involving less gas and hence energy output. In terms of consequences, heat radiation is more significant than overpressure.
- Jet fires. These are caused by the ignition of LPG escaping from a facility due to fracture of pipelines, valves and fittings, hoses, and so on and minor leakage such as flange weep.
- Pool fires. These are caused by leakage and ignition of the liquid phase when it does not immediately drain away. Pool fires are less likely to be a problem with proper plant layout and drainage.

7.9 LPG Merits and Demerits

LPG is made up of two major ingredients, namely propane and butane. The percentage of the two depends upon the season, as a higher percentage of propane is kept in winter and the same for butane in summer. It is a nonrenewable fossil fuel that is prepared in a liquid state under certain conditions.

With more and more people buying vehicles running on LPG, most of the gas stations provide refueling systems for LPG-run cars. LPG turns out to be a lot cheaper and efficient in comparison to petrol and diesel. After petrol and diesel, LPG is the third most extensively used fuel for transportation over the world. The LPG-fitted cars are very popular in countries such as Japan, Italy, Canada, and Austria. However, people making use of LPG cylinders for cooking is not allowed, as the cylinders in many countries are available at fairly low rates compared to the ones available at gas stations.

Today, the LPG kits that are available in the market offer dual fueled or bi-fueled systems. Automatic and manual switching to LPG from petrol or diesel or vice versa is available. Using LPG increases the fuel efficiency of the vehicle as LPG has a high octane value. It causes less corrosion of the engine because less water is vaporized, however, not everybody is aware of the safety risks and conservation issues that surround it. Being a flammable gas, LPG is potentially hazardous. The major disadvantage of using LPG in a vehicle is that because it does not use lead or any other substitute for combustion, it damages the valves, resulting in a decrease of the life of the engine. Moreover, as it is a low-density energy fuel, in comparison to petrol or diesel, LPG is consumed more but because of the subsidized rates available, it proves to be a lot cheaper.

Further, LPG is not recommended for mountains or any kind of rough terrain as it does not provide power and torque to the vehicle, as with other fuels. Using LPG means the vehicle drives 20% less than with other sources of fuel, resulting in more frequent refueling. In contrast to petrol or diesel vehicles, starting is always a problem with LPG driven vehicles under 32°F (cold conditions), because at lower temperatures it has a lower vapor pressure. It is considered to be eco-friendly as it reduces the emission of carbon dioxide by more than 40%. The use of LPG in cars is growing day by day, so in future a gradual increase in its consumption can be seen. Also the lack of many refilling stations is another drawback that needs to be improved in future with the availability of the fuel itself. The other disadvantage is that a small amount of trunk space of the vehicle may be lost in order to make room for the LPG tank.

References

Kaptan, A., and Y. Kisioglu. 2007. Determination of burst pressures and failure locations of vehicle LPG cylinders. *International Journal of Pressure Vessels and Piping* 84:451–57.

Karamangil, M. I. 2007. Development of the auto gas and LPG-powered vehicle sector in Turkey: A statistical case study of the sector for Bursa. *Energy Policy* 35:640–49.

Ozcan, H., and A. A. J. Yamin. 2008. Performance and emission characteristics of LPG powered four stroke SI engine under variable stroke length and compression ratio. *Energy Conversion and Management* 49:1193–1201.

Ozcan, H., and M. S. Soylemez. 2006. Thermal balance of a LPG fuelled, four stroke SI engine with water addition. *Energy Conversion and Management* 47:570–81.

Saleh, H. E. 2008. Effect of variation in LPG composition on emissions and performance in a dual fuel diesel engine. *Fuel* 87:30–39.

Saleh, H. E., and M. Y. E. Selim. 2010. Shock tube investigation of propane–air mixtures with a pilot diesel fuel or cotton methyl ester. *Fuel* 89:494–500.

Selim, M. Y. E. 2000. Pressure-Time characteristics of diesel engine fuelled with natural gas. *Renewable Energy* 22:473–89.

Selim, M. Y. E. 2003. Effect of exhaust gas recirculation on some combustion characteristics of dual fuel engine. *Energy Conversion and Management* 44:707–21.

Selim, M. Y. E. 2004. Sensitivity of dual fuel engine combustion and knocking limits to gaseous fuel composition. *Energy Conversion and Management* 45:411–25.

Selim, M. Y. E. 2005. Effects of engine parameters and gaseous fuel type on the cyclic variability of dual fuel engines. *Fuel* 84:961–71.

Selim, M. Y. E., H. E. Saleh, and M. S. Radwan. 2008. Improving the performance of dual fuel engines running on natural gas/LPG by using pilot fuel derived from jojoba seeds. *Renewable Energy* 6:1173–85.

8

Compressed Natural Gas

Gattamaneni Lakshmi Narayana Rao and Arumugam Sakunthalai Ramadhas

CONTENTS

8.1 Introduction

Natural gas (NG) is a naturally occurring form of fossil energy and therefore nonrenewable. However, NG has some advantages compared to gasoline and diesel from an environmental perspective. Natural gas vehicles represent a cost-competitive, lower-emission alternative to the gasoline-fueled vehicle. Utilization of NG as fuel for internal combustion engines was almost restricted to stationary applications prior to World War II, such as oil field operations, power generation, and water pumping, where no on-site fuel storage was required. During World War II, fuel shortages in Europe resulted in the utilization of several alternative fuels for vehicles, including methane. By the end of the 1960s NG proved that it could contribute to reducing the emission of air pollutants. The country with the most developed program for the utilization of NG as a fuel for transportation is Italy, where almost 300,000 vehicles are fueled by NG. However, the acceptance of NG for vehicles somewhat reduce the fuel dependency. Compressed Natural Gas (CNG) was successfully applied to over 1,000,000 vehicles in many parts of the world such as Argentina, Russia, and Italy and is gaining increasing acceptance particularly for transport vehicles such as taxis, buses, and garbage tippers. Currently more than 1,000,000 vehicles worldwide are powered by NG (Carvalho 1985).

Natural gas occurs as gas under pressure in rocks beneath the earth's surface or often in solution with crude oil as a volatile fraction of petroleum. It does not require elaborate processing or refining as petroleum liquid fuels. It is a natural hydrocarbon energy formed in the earth's crust by millions of years of biological action on organic matter. It occurs along with oil deep in the earth's crust and is recovered from wells under very high pressure.

Natural gas can be used in combined cycle power plants, as the efficiency is much greater than conventional steam cycle plants. Natural gas produces less than half the CO_2 emissions per unit of generated electricity compared to the conventional fuels, which is an important factor as the emission of greenhouse gases, sulfur dioxide, and dust are becoming increasingly unacceptable worldwide. Natural gas is difficult to store or transport because of its physical nature and needs high pressures and/or low temperatures to increase the bulk density, whereas oil is readily stored in large, relatively simple, and cheap tanks and then transported in huge tankers. Natural gas reaches the public for use for transportation through pipelines, liquefied natural gas (LNG), gas to liquids (GTL), converted as electricity, CNG, and in metal hydrates (Thomas and Dawe 2003). Natural gas can be utilized for transport or automotive purposes in two ways: Direct and indirect applications. The applications of NG are depicted in Figure 8.1.

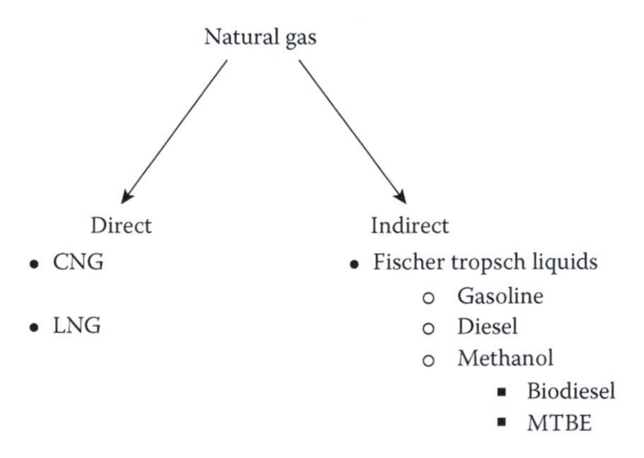

FIGURE 8.1
Applications of natural gas.

8.2 Natural Gas Properties

Natural gas is composed primarily of methane (CH_4; > 80%) but has other HC compounds in small amounts (e.g., propane, butanes, and pentanes). It is a cleaner burning fuel than that of gasoline and diesel as they are more complex mixers of HC and other compounds. However, NG also contains active compounds, such as sulfur, and inert compounds such as nitrogen and CO_2. Excess amounts of higher hydrocarbons must be removed to prevent them from condensing under the high pressures in the gas transmission network. These hydrocarbons are valuable feedstock for gasoline/LPG production. The mixer of minor components varies from plant to plant depending on the source and processing of gas. In order to ensure consistent combustion behavior natural gas specification limits the percentage of propane, butane, and higher hydrocarbons, and the heating value by volume and Wobbe Index.

The high-propane levels in CNG may liquefy at pressures of CNG storage and hence pose problems in fuel systems. Natural gas is in gaseous state and its low boiling point presents problems in storage. Natural gas is stored in cylinders in CNG at high pressure in the range of 15–25 bar or as a liquid at very low temperatures (–170°C and 70–210 bar) usually called LNG. The properties of CNG in comparison with other fuels are shown in Table 8.1.

8.2.1 Octane Number

Natural gas is high-octane fuel (>100 octane number), which is more suitable for spark ignition engines or specially designed gas engines with higher compression ratios. Typical compression ratios for gas engines are in the

TABLE 8.1

Properties of Natural Gas in Comparison with Other Fuels

Property	Natural gas	Gasoline	Diesel
Chemical formula	CH_4 (83–99%) C_2H_5 (1–13%)	C_4 to C_{12}	C_3 to C_{25}
Molecular weight	16.04	100–105	200
Composition by weight, %			
Carbon	75	85–88	87
Hydrogen	25	12–15	30
Density, kg/m³@15.5°C	128	719–779	848
Boiling temperature, °C	−164 to −88	27–225	180–340
RON + MON/2	120+	86–94	–
Auto ignition temperature,°C	482-632	257	316
Flash point, closed cup, °C	−184	−43	60–80
Flammability,%	5.3–15.0	1.4–7.6	1.0–6.0
Lower calorific value, MJ/kg	47.13	43.44	42.78
Stoichiometric air–fuel ratio	17.2	14.7	14.7

range of 10:1–13:1. At higher compression ratios, higher thermal efficiencies can be achieved. Supercharged NG engines provide higher performance and emission characteristics. In order to measure the knock resistance of NG blends, a methane number (MN) scale is used. For this purpose, reference fuel mixtures of methane and hydrogen are used. Pure methane has a MN of 100 and hydrogen has a MN of zero. The relationship between MN and Motor Octane Number (MON) was reported by Kubesh, King, and Liss (1992) and CONCAWE (1995).

$$\text{Motor Octane Number MON} = 0.679 \times \text{MN} + 72.32$$

$$\text{Methane Number MN} = 1.445 \times \text{MON} - 103.42$$

with R^2 in each case greater than 0.95.

Due to its low-cetane number, in general, for satisfactory use in a diesel cycle engine it must be used in conjunction with a high-cetane fuel to provide ignition in dual fuel mode operation. In dual fuel mode, diesel is injected into a premixed charge of air and NG. Natural gas is used in stationary engine applications in dual fuel mode.

8.2.2 Flammability

Natural gas has a very high-autoignition temperature than that of gasoline/diesel. CNG is likely to be safer than gasoline or diesel, since its low density, high-ignition temperature (540°C) and high-flammability limits give the gas a high-dispersal rate and make it less likely to ignite. The minimum spark energy required for methane ignition is much higher than that required for other petroleum products. Hence, NG engines require high-performance ignition systems that can generate more than three times the energy required by gasoline engines (30–40 mJ).

It also has a wider flammability range that allows the engine to operate lean mixtures also. However, there is a marginal reduction in engine performance as the gas enters the engine through the intake manifold—it reduces the engine volumetric efficiency. Moreover, methane combustion is very slow in nature and hence it causes deterioration in performance due to increase in heat losses to the combustion chamber walls. However, this can be overcome by creating turbulence.

8.2.3 Wobbe Index

Wobbe index is actually the correct representation of the heating value of NG arriving, from the gas line, at the orifice where a burner is located. It is the bulk property that accounts for many gas components. Wobbe index or Wobbe number of a gaseous fuel is determined by its composition. It is the ratio of higher calorific value of gas divided by the square root of specific gravity. It can be calculated as

$$\text{Wobbe index } W = \frac{CV}{\rho},$$

where CV = calorific or heating value of gas by volume in MJ/m^3; and ρ = specific gravity. Wobbe index is directly proportional to the heating value of gas that flows sub-sonically through the orifice in response to a given pressure drop. Change in Webbe index will result in nearly proportional changes in rate of energy flow and air–fuel ratio. Webbe index is a good indicator of the change in air–fuel ratio.

Analyzing the effect of variation in Webbe number for gaseous fuels is the same as varying fuels volumetric energy content in gasoline-engine applications. Variation in Webbe number of gas will produce similar variation in air–fuel ratio for gas metering systems on vehicles. A higher Webbe number results in rich mixtures where as a lower Webbe number results in a leaner mixture. The variation in Webbe number may affect the engine performance and emissions. Modern stoichiometric spark ignition engines with closed loop control systems adjust the amount of gas required depending upon the Webbe number.

Pure methane has a Webbe index of 1363 and NG that is supplied to homes in the United States is usually 1310–1390. The Webbe index for ethane and propane is 1739 and 2046, respectively (www.emersonprocess.com). Increasing concentrations of higher hydrocarbons such as ethane and propane increase the Webbe index, while increasing the concentration of inert gases lower it. Hence, the optimization of the two is required to meet specifications (Owen and Coley 1990).

8.2.4 Calorific Value

Natural gas provides mass heating values around 10% higher than conventional fuels. But lower-calorific values as compared to liquid fuel on a volume basis. The calorific value of gas reduces with increasing inert gas compounds. The stoichiometric ratio of gas varies from 14 to 17 depending on the gas. Natural gas contains 75% carbon by weight whereas gasoline/diesel fuels have 86–88%, hence less CO_2 emissions with NG. Moreover, under rich-mixture conditions, CO is lower for methane than conventional fuels. As the combustion reaction is slow with NGs, it reduces the NO_x emissions as well.

8.3 CNG Storage

In the production stage, NG is recovered and collected in NG and oil fields, then transported by pipeline to NG processing plants. The processing stage separates high-value liquids (e.g., propane, butane) from NG and removes impurities like sulfur and CO_2 resulting in pipeline quality NG. The transportation of NG is through pipelines from the processing plant to local distribution companies. The distribution of NG to refueling stations consists of high-pressure NG from a transmission pipeline being depressurized and delivered (MacLeana and Lave 2003).

Natural gas is used in automobiles; hence it is necessary to store the NG in compact form in order to reduce the weight of the vehicle. There are four types of vehicles in the market: dedicated fuel systems, dual-fuel modes, after-market conversions, and hybrid electric systems. Storage systems for these of vehicles are classified into three types: liquefied gas, compressed gas, and adsorbed gas.

Liquefied gas systems are generally used for storing NG for long hauling vehicles like trains, buses, and trucks. For passenger cars, NG is stored in cylinders in compressed form. Natural gas is compressed to about 200 bar and is stored in cylinders of capacity ranging from 20 to 100 liters. However, regulations require that cylinders to be capable of handling pressures of 600 bar in case of accidental exposure to fire. Cylindrical and conformable shaped storage tanks are used for storing the CNG. Conformable shaped tanks are preferred for on board storage of CNG. However, high-strength pressure vessels require uniform internal stresses. The low-energy density of CNG also causes storage problems onboard vehicles, along with the issue of heavy pressurized storage cylinders necessary to store the gaseous fuel. Cylinders are made of steel, high-strength aluminum, aluminum wound with fiberglass, or resin composites of glass or carbon fiber.

Adsorbent NG is the latest technology to store natural gas. Activated charcoal or metallic oxides are used as storage materials. The advantages of adsorbent technologies include mass storage capacity and use of lower pressures (30–40 bar). This reduces the costs associated with compression and storage. Uniform stresses are a natural consequence of cylindrical or spherical-shaped tanks, but achieving uniform stress levels for conformable-shaped tanks requires specialized designs. Among the various options to store natural gas, storing NG in compressed gas cylinders is preferred by the automotive industry because of its space optimization in vehicles and to use the existing infrastructure.

8.4 CNG Distribution

There are two types of distribution systems: service stations where distribution is centralized and domestic compressors where consumers can refuel vehicles in their own garage.

Fuel refueling systems consist of compressor, storage vessel, and dispenser. Compressors are used to boost the gas pressure in the distribution line to about 330 bar. Unlike gasoline stations, NG must be stored above ground. Most of the CNG refueling systems include dryers that remove water vapor, foreign matter, and hydrogen sulfide from NG before it is compressed. The water vapor can condense in the vehicle fuel system, causing corrosion, especially if hydrogen sulfide is present. The dispenser shape is similar to that of a gasoline/diesel filling pump but the internal parts are different and suitable for gasoline-fuel applications. CNG dispensers are made from aluminum and stainless steel. CNG dispensing nozzles have a finite lifetime typically characterized by the total number of refuelings completed. Nozzles should be replaced at the end of the lifetime, before failure. The CNG dispensing hoses are usually stainless steel wire wrap. The inner tube can be Teflon or nonporous. Like nozzles, the hoses also have finite lifetimes and should be replaced before its failure. CNG gas refueling systems can be divided into two categories: slow fill and fast fill.

Fast-fill stations allow vehicles to pull up and refuel in a short period of time. In order to fuel vehicles quickly, gas must be drawn from prepressurized storage vessels. Cascade type storage vessels are used to store the high-pressure gas. Cascade storage divided into several compartments can be independently connected to the refueling pump. The pressure of cascade is higher than the maximum storage pressure of the cylinders on the vehicle. As a general rule of thumb, only 40% of the stored gas in a three bank cascade arrangement is available for refueling. This means that a 30,000 cubic foot storage cascade will deliver about 12,000 cubic feet of NG quickly. This equates to about 96 equivalent gallons of gasoline. A priority sequential panel is used to direct compressor discharge to the high, then medium, then low, bank. When filling, a vehicle will draw first from the low bank, then the medium bank, and top off from the high bank. The compressor will replenish the cascade by filling the high bank first, then the medium bank, and finally the low bank (http://www.mckenziecorp.com/cng_storage.htm). Fast-fill systems dispense the fuel at a faster rate and the gas cylinders have no time to lose heat to the environment. At the end of fill, gas in the cylinders becomes slightly warmer than the atmosphere. Hence, it is difficult to complete refueling by the fast-fill method. Moreover, as the cylinder cools, atmosphere pressure drop will occur. Fast-fill CNG car dispensers require three minutes to fill a car and one minute to fill an auto rickshaw with a flow range of 0–15 kg/min. Bus CNG dispensers, with a flow range of 0–75 kg/min, require 6 minutes to fill.

Slow-fill systems are designed to refuel the vehicles even when the cylinders are hot. Slow-fill systems do not require cascade storage. Cylinders are connected in parallel to the compressor. The working pressure of CNG cylinders is about 200 bar (CONCAWE 1995). Figure 8.2 shows the schematic of a refueling system.

CNG systems essentially incorporate two kinds of refueling principles based on

i. Mother-daughter concept
ii. On-line station concept

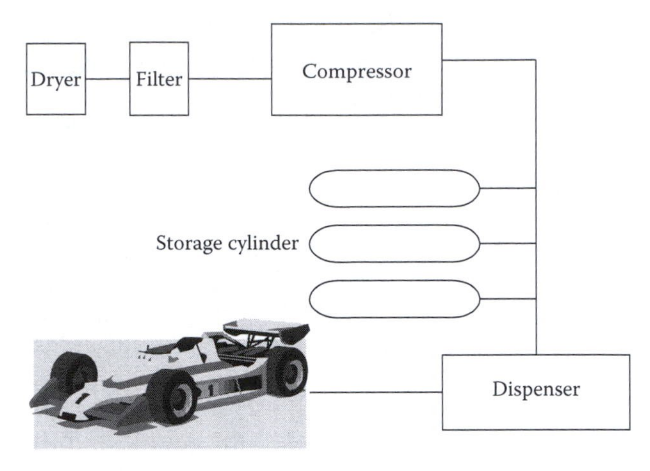

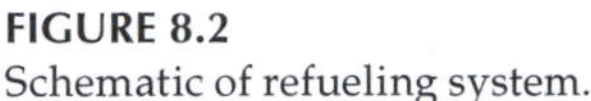

FIGURE 8.2
Schematic of refueling system.

In the mother-daughter concept, the mother station is a compressor station, which caters compressed gas. The CNG thus produced is transported to various refueling stations with the help of light commercial vehicles mounted with a cascade of cylinders. The refueling stations where retail dispensing of CNG takes place are called daughter stations.

In the on-line station concept, the CNG is tapped from the gas pipeline and is stored and distributed in the same way. The transportation through a pipeline reduces the cost of transportation of CNG, but it depends upon the linkage of stations connected through the pipeline network.

8.5 CNG Safety

Compressed natural gas has some excellent physical, chemical, and combustion characteristics that make it a safe engine fuel. The safety aspects of CNG fueled engine operations are generally superior to those of engines fueled with conventional fuels such as gasoline or other alternative fuels. Natural gas or methane is a nontoxic gas that is lighter than air. This means that it will not puddle (like gasoline) or sink to the ground like propane, which is heavier than air. Instead, NG will rise and dissipate in the atmosphere. Karim and Wierzba (1992) reviewed the safety measures associated with operation of engines with various alternative fuels. Based on the observations made by them, the following is summarized.

Natural gas is compressed and stored in metallic cylinders. The safety of the cylinders depends upon its strength and the conductive nature of metallic cylinders. This is unlike the relatively fragile conventional liquid fuel tanks that may have the tendency in a fire of developing hot spots leading to serious safety problems and eventual rupture, and spilling out the contents of the tank with devastating effects. Fuel tanks/cylinders must meet all the safety regulations in the country of use and be tested at regular intervals. All fuel tubes, connectors, pressure switches, indicators, regulators, and so on, and associated equipment must also be of appropriate design and should comply with safety regulations.

Moreover, NG should be free from impurities like sulfur and water and so on, or have acceptable levels of purity. Presence of water may cause tank corrosion problems and reduce the strength of the material. Heating of gas cylinders would increase the pressure inside the cylinder. If the increase in pressure is very high, it may cause an explosion of the cylinder if the pressure relief valve is not functioning. Currently, cylinders are made of special materials that at high pressures will not fragment, instead a crack will develop and release the gas very slowly.

Natural gas has a relatively high flammability limit as compared to the other fuels. For continuous flame propagation, natural gas requires a minimum of 5% by volume as compared to around merely 2% for propane and 1% for gasoline vapor. Hence, it is safer than other conventional fuels. Methane, by virtue of its relatively narrow range of flammability limits, can generate only a small amount of combustible mixture relative to other fuels. The discharge of one unit volume of methane in air would generate a maximum volume of combustible mixture that is around 40% of that following the release of an equivalent volume of propane and at 20% the volume of the mixture formed following the release of a similar amount of gasoline vapor.

The rate of discharge of a gaseous fuel following an accident will depend on the nature of the leak, and the associated area of discharge. The discharge from the high-pressure cylinder into the atmosphere will be at rates that will be proportional to the volume of the area of discharge and inversely proportional to the density of the gas. However, the explosion depends on how the combustion mixtures formed, and at what rate, and the ignition source availability. The dispersion rate of methane is much higher than gasoline. Hence, if there is any leakage of CNG from cylinders, gas will be discharged at a very high velocity jet into the surrounding area, aiding greatly in the rapid dispersion of fuel (Table 8.2).

CNG has a higher autoignition temperature than gasoline. The minimum energy required for ignition of a fuel–air mixture within the flammability range is relatively high for methane gas mixtures. Hence, NG is much more difficult to ignite/fire than other fuels. Quenching of methane–air flames by cold surfaces such as through metallic meshes is much easier than in the case of flames involving LPG or hydrogen. Thus, flame traps are more successful in suppressing methane fires than those involving hydrogen or propane (Karim and Wierzba 1992).

Astbury (2008) reviewed the safety of vehicles using various alternative fuels and was summarized as follows. He reported that leakage of NG into

TABLE 8.2
Ranking Order of Hazards for Fuels

Property	Ranking Order of Hazard
Lower flammability	Diesel > Petrol > Natural gas
Autoignition temperature	Diesel > Natural gas
Energy content by volume	Diesel > Petrol > Natural gas
Ignition energy	Almost equal
Heat release rate	Diesel > Petrol > Natural gas

the interior of the car could result in an explosion being ignited by the remote central locking system being operated when the car is to be used, or by the switch for the interior light being operated when the door is opened. CNG powered vehicles stored in domestic garages present a risk of explosion. Small leaks of CNG will tend to rise toward the roof-space of a garage, as the density of CNG is only 60% of that of air. Thus a small leak will accumulate by the fluorescence light fitting, and this may present a source of ignition. For domestic CNG powered vehicles, it would be prudent either to park the vehicle outside at all times, or to make sure that the garage is ventilated at both low and high levels to ensure adequate natural ventilation.

Inspection and maintenance of the cylinders would be difficult, particularly if cylinders were exchanged at different sites or were owned by different companies. Leakage due to poor sealing of the connection is high, as it is probable that minor leakage would not be noticed or would be ignored by the average nontechnical user of the system (Astbury 2008).

8.6 Engine Tests

Natural gas is not pure methane or a homogeneous mixture, but varies in composition by location, season, and through transmission network. It does not have a narrow range of specification like gasoline or diesel. The fuel composition affects the vehicular emissions. Concentration of methane and Webbe number describes the characteristics of NG.

CNG can be used in engines as sole fuel or bi-fuel operation. Bi-fuel engines are Otto cycle (spark ignited) that run on *either* NG *or* gasoline. The gasoline vehicles can be converted for CNG/gasoline operation; that is, the engine uses either CNG or gasoline for its operation. Dedicated NG engines are Otto cycle (spark ignited); that is, operated only on NG. They tend to be optimized; that is, they have a compression ratio designed to take advantage of the 130 octane of NG, and have been designed to take into consideration the combustion characteristics of the fuel so that the engine is very low polluting.

8.6.1 Natural Gas in SI Engines

Aslam et al. (2006) converted gasoline engines into bi-fuel engines, converted to a computer integrated bi-fueling system from a gasoline engine, and was operated separately either with gasoline or CNG using an electronically controlled solenoid actuated valve system. CNG was stored under a maximum pressure of 200 bars.

Before entering into the carburetor, CNG passes through the three-stage conversion kit. The conversion kit supplies CNG to the engine carburetor at approximately atmospheric pressure (~0.8 bar). Shown in Figures 8.3 and 8.4, the CNG operation increases NO_x emissions by 33% and reduces the HC emissions by 50%. Moreover, CNG produced much less concentration of CO (80%) and CO_2 (20%) emissions compared to gasoline.

Natural gas injection systems introduce the fuel under pressure that provides accurate metering of quantity of fuel injected. Injection systems may be single point or multi point. Single point fuel injection systems are most commonly used in passenger cars. Electronic control unit (ECU) of the engine controls the quantity of fuel injected and the metered quantity of fuel is injected into the engine. In multi point injection systems, fuel is injected into each cylinder by a separate injector.

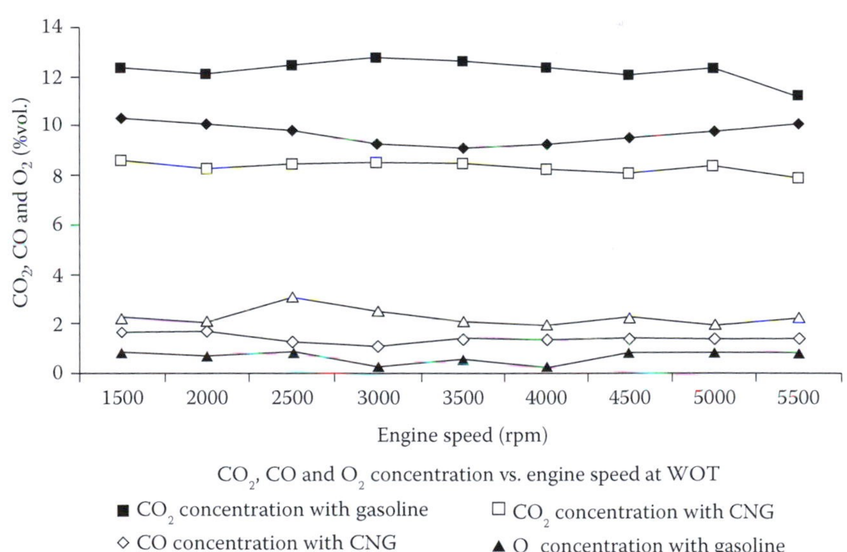

FIGURE 8.3
CO_2, CO, and O_2 concentration versus engine speed at WOT. (From Aslam, M. U., Masjuki, H. H., Kalam, M. A., Abdesselam, H., Mahlia, T. M. I., and Amalina. M. A., *Fuel*, 85, 717–24, 2006. Reprinted with permission from Elsevier Publications.)

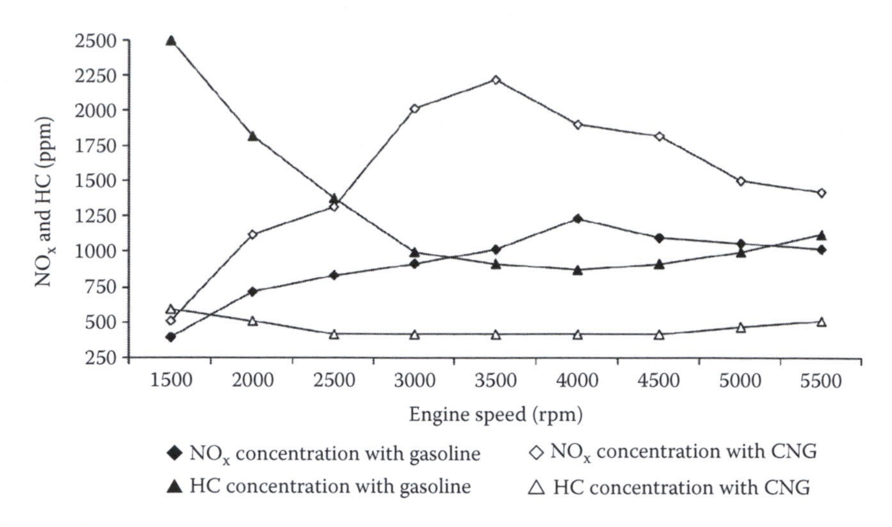

FIGURE 8.4
NO_x and HC concentration versus engine speed at WOT. (From Aslam, M. U., Masjuki, H. H., Kalam, M. A., Abdesselam, H., Mahlia, T. M. I., and Amalina. M. A., *Fuel*, 85, 717–24, 2006. Reprinted with permission from Elsevier Publications.)

8.6.2 Natural Gas in CI Engines

Dual fuel NG engines are based upon diesel technology. Diesel engines can be converted as dual-fuel engines, in that diesel is injected through the injectors and NG is carbureted along with fresh intake air and enters into the engine through the intake manifold. These engines also can operate on 100% diesel fuel. When the engine runs at idle, it tends to operate on 100% diesel. As the vehicle begins to move to full load performance, an increasing amount of NG replaces the diesel fuel to 80% or more. Dual-fuel NG replaces 80–90% diesel that reduces about 25% CO_2.

Dual-fuel engines are throttle controlled using a fumigation system that adds NG to the engine as higher speed is required. The computer controlled dual-fuel engine ensures that the optimal ratio of NG and diesel fuel is delivered to the engine depending upon load and performance requirements (www.unece.org).

The Dual-Fuel™ ECU can be interfaced with Original Equipment Manufacturers' (OEMs) ECU. This enables the Dual-Fuel™ ECU to control the engine ECU, ensuring optimum conditions for dual-fuel combustion. It has been reported that a contemporary integrated system will deliver at least 70% overall gas substitution with CO_2 savings of 26 tonnes yearly. Clean Air Power's integrated C12 product is already certified to Euro 4 standards (http://www.cleanairpower.com/dueltechnology.php).

Natural gas is injected into the combustion chamber shortly before the end of the compression stroke, a diesel pilot that precedes the NG injection provides the ignition source. Varying the fuel composition has a significant impact on emissions. The influence of fuel composition varies strongly with combustion timing. With N_2 dilution, the increase at early timings has been attributed to reaction zone impingement, while the relatively small reduction at late timings is a result of the emissions being primarily volatiles. Car makers are making dual fuel engine CNG–gasoline for passenger cars.

The effect of fuel composition on engine exhaust emission is shown in Figure 8.5. The increase in adiabatic flame temperature with the addition of ethane, propane, or hydrogen to the fuel results in higher NO_x emissions. It has been reported that a 1% change in adiabatic flame temperature results in a 5% increase in NO_x. Changes in fuel composition and adiabatic flame temperature influence the mixture fraction and these affect reaction zone temperatures, temperature post reaction gases, and the time before these gases mix with a cooler charge affect the NO_x emissions. NO_x emissions depend strongly on combustion timing, with, in general, smaller increases in NO_x with more advanced timings. The reduction in HC emissions with ethane and propane are largest at the latest combustion timings suggesting that these fuels may be helping to delay the onset of bulk quenching. At early combustion timings, CO emissions are increased, while at late timings, they are reduced for virtually all the fuel blends. The reduction in CO emissions at the latest timing is consistent with the delay in the onset of

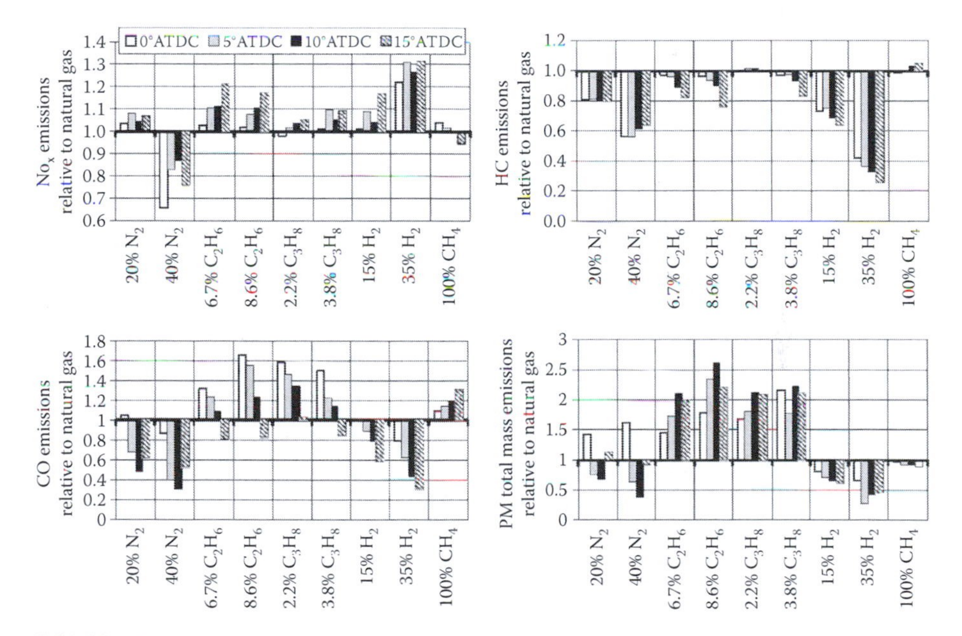

FIGURE 8.5
Effect of fuel composition on emissions, relative to equivalent natural-gas-fueled timing condition (L–R: 50% IHR at 0°, 5°, 10°, and 15° ATDC). (From McTaggart-Cowan, G. P., Rogak, S. N., Munshi, S. R., Hill, P. G., and Bushe, W. K., *Fuel*, 89, 752–59, 2010. Reprinted with permission from Elsevier Publications.)

TABLE 8.3

Comparison of Tail-Pipe Emissions

	Typical Bi-Fuel Vehicles Grams Per Kilometer		
Emission Type	**CNG**	**Gasoline (Benzene)**	**Reduction %**
Carbon monoxide	99	3.34	–86%
Nonmethane hydrocarbons	0.04	0.23	–83%
Nitrogen oxides	0.09	0.12	–25%
Carbon dioxide	246	311	–21%

bulk quenching. The increase in PM with ethane and propane is primarily due to increased black carbon emissions. As total PM mass increases with the heavier hydrocarbons, this result indicates that ethane and propane are increasing both BC and volatile PM mass emissions.

The emission reduction potential of CNG vehicles is depicted in Table 8.3.

8.7 CNG Advantages

1. Environmental: CNG vehicles produce far less of all regulated pollutants compared to gasoline or diesel vehicles, including NO_x and particulate matter. Natural gas has a low C/H ratio and hence lower CO and HC emissions. CNG vehicles produce far less unregulated air toxics and greenhouse gases. Due to proper combustion of gas–air mixtures, reduced unburned HC emissions will reduce the environmental pollution of visible photochemical smoke.
2. Energy security: NG usage reduces the consumption of gasoline and diesel fuel.
3. Operating cost: NG is cheaper at the pump than gasoline and diesel fuel.
4. Distribution efficiency/safety: Natural gas has a higher ignition temperature than gasoline or diesel. Natural gas is lighter than air and hence disperses quickly in the event of leakage of fuel. This is the safest and most efficient energy distribution system. The explosive limit of NG–air mixtures is higher than diesel–air mixtures. For continuous flame propagation, NG requires a minimum of 5% by volume as compared to around merely 2% for propane and 1% for gasoline vapor. Hence, it is safer than other conventional fuels.
5. Flexibility: CNG vehicles can be produced as dedicated and bi-fuel versions. Dedicated vehicles are most appropriate where vehicles tend to operate in an area where NG fueling is available. Bi-fuel vehicles have both NG and gasoline storage tanks on board, and

can operate on either fuel at the flip of a switch. Bi-fuel vehicles are most appropriate where the driver may need to travel to areas not currently served by NG stations. Compared to other fuels it is also economical and environmentally friendly.

6. Transition to hydrogen: Since hydrogen is a gas, hydrogen-powered vehicles will require changes in a number of areas, including buildving codes and standards, mechanic/inspector/user training. NGVs require many of the same changes. Therefore, a growing NGV market today is smoothing the path for a hydrogen vehicle market tomorrow.

8.8 Challenges for CNG

1. Fueling: NG refueling process is slow. Low-volumetric efficiency of engine and low-energy density result in low-engine performance.
2. Vehicle cost: NG vehicles cost more than comparable gasoline or diesel models because of its low-production volumes and the greater cost of fuel storage tanks.
3. Driving range: Compared to a volumetric gallon of gasoline or diesel fuel, there is less energy in an energy gallon equivalent of NG (both CNG and LNG). Therefore, the driving range of vehicles operating on NG is less. Natural gas vehicles need pressurized fuel storage tanks. The present storage capacity has a range of about 150 km.
4. On-board NG fuel tanks are larger than comparable gasoline or diesel fuel tanks. There are inconsistent NG fuel properties around the world.

References

Aslam, M. U., H. H. Masjuki, M. A. Kalam, H. Abdesselam, T. M. I. Mahlia, and M. A. Amalina. 2006. An experimental investigation of CNG as an alternative fuel for a retrofitted gasoline vehicle. *Fuel* 85:717–24.

Astbury, G. R. 2008. A review of the properties and hazards of some alternative fuels. *Process Safety and Environment Protection* 86:397–414.

Carvalho, A. D. 1985. Natural gas and other alternative fuels for transportation purposes. *Energy* 10: 187–215.

CONCAWE. 1995. *Alternative fuels in the automotive market.* CONCAWE Report No. 2/95.

Karim, G. A., and I. Wierzba. 1992. Safety measures associated with the operation of engines on various alternative fuels. *Reliability Engineering and System Safety* 37:93–98.

Kubesh, J., S. R. King, and W. E. Liss. 1992. Effect of gas composition of octane number of natural gas fuels. *SAE* 922359.

MacLeana, H. L., and L. B. Lave. 2003. Evaluating automobile fuel/propulsion system technologies. *Progress in Energy and Combustion Science* 29:1–69.

McTaggart-Cowan, G. P., S. N. Rogak, S. R. Munshi, P. G. Hill, and W. K. Bushe. 2010. The influence of fuel composition on a heavy-duty, natural-gas direct-injection engine. *Fuel* 89:752–59.

Owen, K., and T. Coley. 1990. *Automotive fuels reference book.* Troy, MI: Society of Automotive Engineers.

Thomas, S., and A. R. Dawe. 2003. Review of ways to transport natural gas energy from countries which do not need the gas for domestic use. *Energy* 28:1461–77.

9

Hydrogen

Fernando Ortenzi, Giovanni Pede, and Arumugam Sakunthalai Ramadhas

CONTENTS

9.1 Introduction

Hydrogen is a very promising alternative fuel and has been receiving more attention all over the world in recent years. Research and development activities were carried out out of curiosity, mainly with an objective to evaluate the suitability of hydrogen as an engine fuel. During an energy crisis and World Wars I and II, hydrogen is used as fuel in a dual fuel mode or as a sole fuel. The abundant availability of petroleum products suppresses the commercialization of hydrogen as fuel for vehicles. Currently the search for alternative fuels has picked up all over the world and hence, the research and demonstration of alternative fueled vehicles have begun. For the past few decades many renewable clean-burning alternative fuels have been studied.

Hydrogen is the ideal candidate as an energy carrier for both mobile and stationary applications while averting adverse effects on the environment and reducing dependence on imported oil for countries without natural resources. Biodiesel, alcohols (both ethanol and methanol), compressed natural gas (CNG), liquefied petroleum gas (LPG), biogas, producer gas, and hydrogen were investigated as alternative fuels for both spark ignition (SI) and compression ignition (CI) engines.

Hydrogen is sometimes called the fuel of the future and certainly has a number of inherent attractive features when considered as a fuel for road vehicles. Hydrogen is by far the most abundant element in the universe (90% on the basis of number of atoms), and in the earth's crust it is one of the most abundant elements. On Earth, hydrogen is almost exclusively found in chemical compounds, as opposed to free molecular hydrogen that is virtually unseen in nature. A consequence of the latter is that hydrogen is not a source of energy, but rather a convenient tool for handling energy. In many respects an illustrative parallel may be drawn to electricity, which is also not a source of energy, but instead an appropriate intermediate in energy transport and conversion.

The production of industrial hydrogen is currently based mainly on fossil fuels, but to some extent electricity is also used. If considered as an alternative fuel, hydrogen should not be produced from fossil fuels, since this would not lead overall to decreased emissions of greenhouse gases (GHGs). Hydrogen is a very attractive transportation fuel in two important ways. It is the least polluting fuel (as it does not contain carbon) that can be used in an internal combustion engine (ICE), and it is potentially available anywhere there is water and a clean source of power. Hydrogen-fueled internal combustion engines (H_2-ICEs) operate as clean and efficient power plants for automobiles. Hydrogen produces water only when it is combusted in the ICE and makes it a very environmentally clean fuel. Hydrogen combustion does not produce any of the major pollutants such as CO, HC, SO_x, smoke, lead, or other toxic metals except NO_x. Sulfuric acid deposition, benzene and other carcinogenic compounds, ozone and other oxidants, are intrinsically absent in a well-designed neat hydrogen engine.

A special reason for the technological interest in hydrogen is that hydrogen works very well within fuel cells. Most fuel cells are basically powered by hydrogen, even though the primary fuel is not always pure hydrogen. Using hydrogen in a fuel cell leads to optimized energy efficiency (for the conversion of chemical to mechanical energy) compared with use of hydrogen in an ICE. Conversion efficiencies approaching 70% may be possible (significantly depending though on operation mode and conditions) and this is at least two times better than the conversion efficiency observed for ICEs. In this chapter the hydrogen production, fuel properties, how to use gaseous fuel in vehicles, storage, safety, and challenges associated with hydrogen economy are discussed.

9.2 Hydrogen: Energy Carrier

9.2.1 Need for Energy Carriers

The available energy sources on earth are: nuclear reactions; sun radiation (directly or indirectly considering that wind and water receive energy for their movement by sun radiation); and fossil fuels. They are considered in this document as *primary energy sources*. Hydrogen is not among them.

Potential energy from primary sources need to be converted into other forms to be of use. The most important energy forms for their final use by mankind are: thermal energy (used for heating); mechanical energy (for moving objects); and electric energy (to feed purely electric appliances—computers, consumer electronics, etc.).

This conversion can in principle be made in two ways: concentrated in a few power plants or in several small devices located near the corresponding final utilization devices. It is obvious that for the latter option to take place, the final energy form needs to be easily transportable, which is not often the case. For technical reasons, potential energy from nuclear reactions cannot be converted in small distributed systems, it needs large power plants. At present, indeed, energy from nuclear reactions is converted in large power plants first into heat. The heat is then converted into mechanical movement, then into electricity that is fed into the electric transmission and distribution network. Electricity is a very good means of transferring energy from concentrated plants to their place of final use.

As far as potential energy from fossil fuels is concerned, it can be converted into usable energy either in a concentrated (power plants) or distributed (ICEs for vehicles, boilers for ambient heating, etc.) manner. The concentrated conversion has the advantage that the environmental pollution created during the process can be controlled much better in large than in small plants. Also conversion efficiency can be much higher, just to make a single example, fuel energy can be converted into electricity by modern combined-cycle power plants with efficiencies that reach 58% while the best automotive ICEs barely reach 40%. Moreover, concentrated conversion is made in places located far from densely populated urban areas and therefore the emissions produced can be distributed over very large areas before reaching large masses of persons, thus far reducing any public health impact from the emissions. It can be concluded that if an *energy carrier* is available that is able to transport energy into places of final utilization, and able to generate energy in the final form with negligible pollution, it can bring large environmental advantages.

9.2.2 Role of Electricity

Today by far, the most widely used energy carrier is electricity. Electricity is a very flexible means of transferring energy from concentrated conversion points into the areas of final utilization. However, electricity has the disadvantage of requiring wiring to be conveyed. This is exploited in railways using electrified lines and catenary for electricity distribution to trains. A way to exploit electricity in mobile vehicles not having wiring connecting them to earth is to store it onboard. This can be done converting chemical potential energy into onboard-installed reservoirs that are the well known "batteries." Indeed, this is not very effective because of two reasons:

- Energy per unit of volume that can be stored in batteries and mass is much lower than that of a fossil fuel (more than an order of magnitude).

- "Refilling time," which for a battery is recharging time, is much larger than the filling time of a traditional fuel tank (more than an order of magnitude).

One can thus conclude that although electricity is a very useful and flexible energy carrier, other carriers that would overcome the need of wires to distribute and/or can be stored in a more effective way, would have a large impact on the future way of life for mankind.

9.2.3 Role of Hydrogen

When hydrogen is put in contact with oxygen, it reacts with it, releasing large quantities of energy. If this contact is direct, the released energy takes the form of heat. If it is made by first ionizing hydrogen and using different paths for ions and electrons, it takes the form of electricity (thermodynamics of the two reactions are different, the differences in the amounts of potentially available energy are not very large however, and are disregarded in this reasoning). As a consequence of this, given the wide availability of oxygen in the air, it can be said that hydrogen carries energy with itself. On the other hand, hydrogen is not available free in nature, it is embedded in a lot of natural substances such as water, carbohydrates, and hydrocarbons. To draw hydrogen from these substances, energy must be spent. If, for instance, hydrogen is taken out of water using electrolysis, the theoretical energy needed for this operation is the same theoretically made available during hydrogen recombination with oxygen. These twin reactions that allow creation of hydrogen from natural substances (e.g., water) with energy expense and recreation of water by hydrogen reaction with oxygen, makes it possible to use hydrogen as an energy carrier.

9.3 Hydrogen Production

Hydrogen can be produced from renewable as well as nonrenewable fuels sources. Renewable fuel sources includes water, biomass, and from conventional fuels. Fuel processing technologies convert a hydrogen containing material such as gasoline, ammonia, or methanol into a hydrogen rich stream. Fuel processing of methane is the most common hydrogen production method in commercial use today. Figure 9.1 shows the various hydrogen production technologies (Holladay et al. 2009).

9.3.1 Hydrocarbon Reforming

There are three primary techniques used to produce hydrogen from hydrocarbon fuels: steam reforming, partial oxidation (POX), and auto thermal

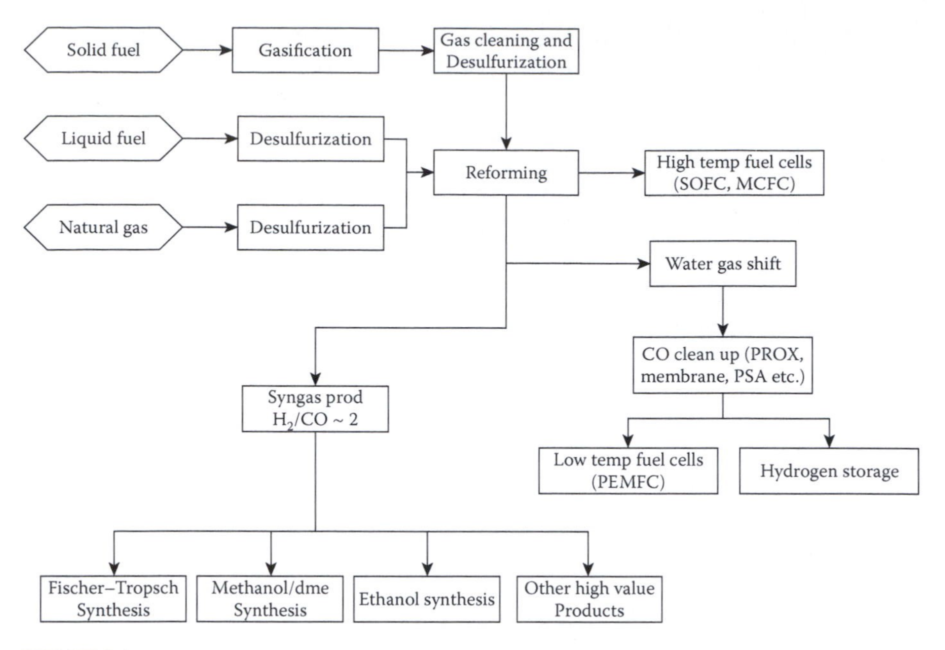

FIGURE 9.1
Fuel processing of gaseous, liquid, and solid fuels for hydrogen production. (From Holladay, J. D., Hu, J., King, D. L., and Wang, Y., *Catalysis Today*, 139, 244–60, 2009. Reprinted with permission from Elsevier Publications.)

reforming (ATR). The reforming process produces a gas stream composed primarily of hydrogen, carbon monoxide, and carbon dioxide. Steam reforming does not require oxygen, has a lower operating temperature than POX and ATR, and produces reformate with a high H_2/CO ratio (3:1). Partial oxidation converts hydrocarbons to hydrogen by partially oxidizing (combusting) the hydrocarbon with oxygen. Auto thermal reforming uses the POX to provide the heat and steam reforming to increase the hydrogen production resulting in a thermally neutral process. Since POX is exothermic and ATR incorporates POX, these processes do not need an external heat source for the reactor. However, these require either an expensive and complex oxygen separation unit in order to feed pure oxygen to the reactor or the product gas is diluted with nitrogen.

9.3.2 Pyrolysis

Pyrolysis converts the hydrocarbon into hydrogen and carbon without air/water/oxygen. Pyrolysis can be done with any organic material. If air or water is present then significant CO_2 and CO emissions will be produced. Among the advantages of this process are fuel flexibility, relative simplicity and compactness, clean carbon by-product, and reduction in CO_2 and CO emissions. The chemical reaction can be written as,

$$C_nH_m \rightarrow nC + \frac{1}{2}mH_2.$$

9.3.3 Plasma Reforming

In plasma reforming the overall reforming reactions are the same as conventional reforming, however, energy and free radicals used for the reforming reaction are provided by plasma typically generated with electricity or heat. When water or steam is injected with the fuel, H, OH, and O radicals in addition to electrons are formed, thus creating conditions for both reductive and oxidative reactions to occur. This process operates at lower temperatures than traditional reforming and is high sulfur tolerant.

9.3.4 Biohydrogen

Biohydrogen production is explained in Chapter 13.

9.3.5 Hydrogen from Water

Electrolysis of water can produce very high purity hydrogen with high efficiency. Electrical current passes through two electrodes to separate water into hydrogen and oxygen. Commercial low temperature electrolyzers have system efficiencies of 56–73%. Solid oxide electrolysis cells (SOEC) electrolysers are more efficient. The SOEC technology has challenges with corrosion, seals, thermal cycling, and chrome migration. Electrolyzers are not only capable of producing high purity hydrogen, but recently, high-pressure units.

$$H_2O \rightarrow H_2 + \frac{1}{2}O_2.$$

9.4 Hydrogen Properties

9.4.1 Physical Properties

Hydrogen is a clean fuel. Hydrogen is the lightest element and has one proton and one electron. Hydrogen with atomic weight 1.00797 and atomic number 1 is the first element in the periodic table. Three isotopes of hydrogen are hydrogen, deuterium, and tritium, respectively. Hydrogen is colorless, odorless, tasteless, and is about 14 times lighter than air. It diffuses in air at a faster rate than any other gases. The physical properties of hydrogen are given in Table 9.1 (Saxena et al. 2008).

A stream of hydrogen from a leak is invisible in daylight. Hydrogen has a very low density of about 0.09 kg/m^3 20°C. The liquid density of hydrogen is 70.03 kg/m^3. Higher cylinder capacity is required to store sufficient amounts of hydrogen for the operation of a vehicle. Moreover, as the energy density reduces, the power output also decreases. On cooling, hydrogen condenses to liquid at –253°C and to solid at –259°C. Hydrogen in gaseous form has a heat capacity of 14.4 kJ/kgK (Pant and Gupta 2009).

Hydrogen engines will not face any starting problems as it has a very low boiling point. Even in very severe winter temperatures a hydrogen engine

TABLE 9.1

Physical Properties of Hydrogen

Property	p-Hydrogen	n-Hydrogen
Triple Point		
Temperature (K)	13.803	13.957
Pressure(kPa)	7.04	7.2
Density solid (kg/m^3)	86.48	86.71
Density liquid (kg/m^3)	77.03	77.21
Density vapor (kg/m^3)	0.126	0.130
Boiling point (K)	20.268	20.39
Heat of vaporization (J/mol K)	898.3	899.1
Liquid Phase		
Density (kg/m^3)	70.78	70.96
Cp (J/mol/ K)	19.70	19.7
Cv (J/mol/K)	11.60	11.6
Enthalphy (J/mol)	–516.6	548.3
Entropy (J/molK)	16.08	34.92
Viscosity (mPa s)	13.2×10^{-3}	13.3×10^{-3}
Velocity of sound (m/s)	1089	1101
Thermal conductivity (w/mK)	98.92×10^{-3}	100×10^{-3}
Compressibility factor	0.01712	0.01698
Gaseous Phase		
Density (kg/m^3)	1.338	1.331
Cp (J/mol/ K)	24.49	24.6
Cv (J/mol/K)	13.10	13.2
Enthalphy (J/mol)	381.61	1447.4
Entropy (J/molK)	60.41	78.94
Viscosity (mPa s)	1.13×10^{-3}	1.11×10^{-3}
Velcoity of sound (m/s)	355	357
Thermal conductivity (w/mK)	60.49×10^{-3}	16.5×10^{-3}
Compressibility factor	0.906	0.906

Source: From Saxena, R. C., Seal, D., Kumar, S., and Goyal, H. B., *Renewable and Sustainable Energy Reviews*, 12, 1909–27, 2008. Reprinted with permission from Elsevier Publications.

will start easily. Expansion ratio is the ratio of the volume at which a gas or liquid is stored to the volume of the gas or liquid at atmospheric pressure and temperatures. Hydrogen's expansion ratio is 1:848; that is, gaseous hydrogen occupies a volume of 848 times more than at a liquid state. When hydrogen is stored as a high-pressure gas at 250 bar and atmospheric temperature, its expansion ratio to atmosphere is 1:240. This necessitates that a large volume of hydrogen is required to be carried for adequate running of a vehicle.

9.4.2 Chemical Properties

Hydrogen atoms are chemically very reactive. When a small amount of ignition energy in the form of spark is provided to the hydrogen–air mixtures, the molecules react with air in the atmosphere very actively and release significant amounts of heat and water vapor. At room temperatures these reactions are very slow, but is accelerated by catalysts such as platinum and spark. Very high temperatures (>5000 K) are needed to dissociate hydrogen molecules into atomic hydrogen completely. Hydrogen is considered an energy carrier to store and transmit energy from primary energy sources.

9.4.3 Fuel Properties

Hydrogen is a suitable gaseous fuel for SI and CI. The fuel properties of hydrogen are given in Table 9.2 (Saxena et al. 2008). The self ignition temperature of hydrogen is very high and best suited for SI engines. This temperature plays an important role in storage pressure of hydrogen because as pressure of the hydrogen inside the cylinder increases, temperature will also increase. The higher self ignition temperature of hydrogen allows the use of larger compression ratios without causing premature ignition. It is well known that as the compression ratio of the engine increases its thermal efficiency will also increase. Hydrogen can be ignited at its low ignition energy of 0.02 mJ as compared to 0.24 mJ for gasoline and 0.28 mJ for methane at stoichiometric.

9.4.3.1 Minimum Ignition Energy

Minimum energy required for ignition is the order of magnitude less than that required for gasoline. The minimum ignition energy is a function of equivalence ratio. At the equivalence ratio nearer to 1, minimum ignition energy for hydrogen–air mixtures is very low. As very little energy is required for the combustion of hydrogen, any hydrogen–air mixture can be ignited due to wide limits of flammability of hydrogen. The hot spots in the combustion chamber may cause premature ignition in the combustion chamber and flash back also.

TABLE 9.2

Fuel Properties of Hydrogen

Property	Hydrogen
Density at STP (kg/m^3)	0.084
Heat of vaporization (J/g)	445.6
Lower heating value (kJ/g)	119.93
Higher heating value (kJ/g)	141.8
Thermal conductivity at STP (mW/cmK)	1.897
Diffusion coefficient at STP (cm2/s)	0.61
Flammability limits in air (v%)	4.0–75
Detonability limits in air (v%)	18.3–59
Limiting oxygen index (v%)	5.0
Stoichiometry composition in air (v%)	29.53
Minimum energy of ignition in air (MJ)	0.02
Autoignition temperature (K)	858
Flame temperature (K)	2318
Maximum burning velocity in air STP (km/s)	3.46
Detonation velocity in air STP (km/s)	1.48–2.15
Energy of explosion mass related g TNT (g)	24.0
Energy of explosion volume related g TNT (m^3) STP	2.02

Source: From Saxena, R. C., Seal, D., Kumar, S., and Goyal, H. B., *Renewable and Sustainable Energy Reviews*, 12, 1909–27, 2008. Reprinted with permission from Elsevier Publications.

9.4.3.2 Flame Speed

The flame velocity of hydrogen–air mixtures is a function of equivalence ratio. Hydrogen has a very high flame speed of 1.85 m/s as compared to 0.42 for gasoline and 0.38 for diesel.

9.4.3.3 Flame Temperature

Flame temperature of hydrogen–air mixtures is about 2207°C as compared to 1917°C for methane and 2307°C for gasoline (Pant et al. 2009). But the rapid combustion allows very little heat loss to surroundings and hence high, instantaneous, local temperatures are produced that leads to formation of high nitric oxides.

9.4.3.4 Diffusivity

Diffusivity of hydrogen in air is much higher than gasoline or diesel. The higher diffusivity helps in formation of homogeneous mixtures of hydrogen and air. Moreover, if there is any leakage of hydrogen from the cylinder, it will disperse quickly in the atmosphere and hence form lean mixtures and reduces the risk of sudden firing. The hydrogen molecules are smaller than other gases and hence it can diffuse through many materials considered air tight or impermeable to other gases. Leakage of liquid hydrogen evaporates quickly as its boiling point is very low (–253°C).

In the case of leakage of hydrogen, buoyancy and diffusion effects in the air are often overshadowed by the presence of air currents from a slight ambient wind and very slow vehicle motion. In general, these serve to disperse the leaked hydrogen more quickly.

9.4.3.5 **Flammability**

Flammability limits are the range that the engine can operate over the wide range of air–fuel mixtures. Hydrogen–air mixtures are flammable in the broad range, –4–75% concentration, whereas gasoline flammability range is less than 10. In the hydrogen–air mixture, 4% hydrogen can provide combustion mixtures whereas the stoichiometric air–fuel ratio is 29.5% (Norbeck et al. 1996). Hydrogen provides stable operation even in dilute condition also. In terms of equivalence ratio the flammability range for hydrogen is 0.1–7.1 and for gasoline is 0.7–4.0. Hydrogen engines can run on more lean mixtures than gasoline engines. Lean fuel engine operation reduces the fuel consumption as well as combustion chamber temperature. However, lean operations reduce the power output of the engine due to the reduction in volumetric efficiency of the engine.

9.4.3.6 **Quenching Distance**

Combustion flames are typically extinguished from a certain distance from the cylinder wall due to heat losses that is called quenching distance. The quenching distance for hydrogen (0.64 mm) is lesser as compared to gasoline (2 mm). So, hydrogen flames travel very close to the wall before they are extinguished. The smaller quenching distance will increase the tendency of backfire and pass nearly to the intake valve.

The ignition and flammability properties of hydrogen in comparison with methane and gasoline is given in Table 9.3.

9.4.3.7 **Air–Fuel Ratio**

Air–fuel ratio for gasoline is nearly 15:1 whereas for hydrogen 34:1. As the gaseous hydrogen fuel enters through the intake system, it occupies more combustion chamber volume than liquid fuel and hence reduces the amount of air entering into it (Figure 9.2). At stoichiometric conditions, hydrogen displaces about 30% of the combustion chamber, compared to about 1–2% for gasoline. This reduces the volumetric efficiency of the engine as well as power developed. As hydrogen has a wide flammability range, the hydrogen engine can work in the A/F range of 34:1–180:1.

TABLE 9.3

Ignition and Flammability Properties of Hydrogen in Comparison with Other Fuels

Property	Hydrogen	Methane	Propane	Gasoline
Minimum ignition energy (mJ)	0.02	0.28	0.25	0.25
Ignition temperature (K)	858	810	783	530
Adiabatic flame temperature (K)	2384	2227	2268	2270
Limits of flammability (% in air)	4.1–75	4.3–15	2.2–9.5	1.5–7.6
Maximum laminar flame velocity (cm/s)	270	38	40	30
Diffusivity (cm^2/s)	0.63	0.20	—	0.08
Minimum quenching distance at 1 atm (cm)	0.06	0.25	0.19	—
Normalized flame emissivity (200 K and 1 atm)	1.00	1.7	1.7	1.7

Source: From Das, L. M., *International Journal of Hydrogen Energy*, 21, 703–15, 1996. Reprinted with permission from the International Association of Hydrogen Energy and Elsevier Publications.

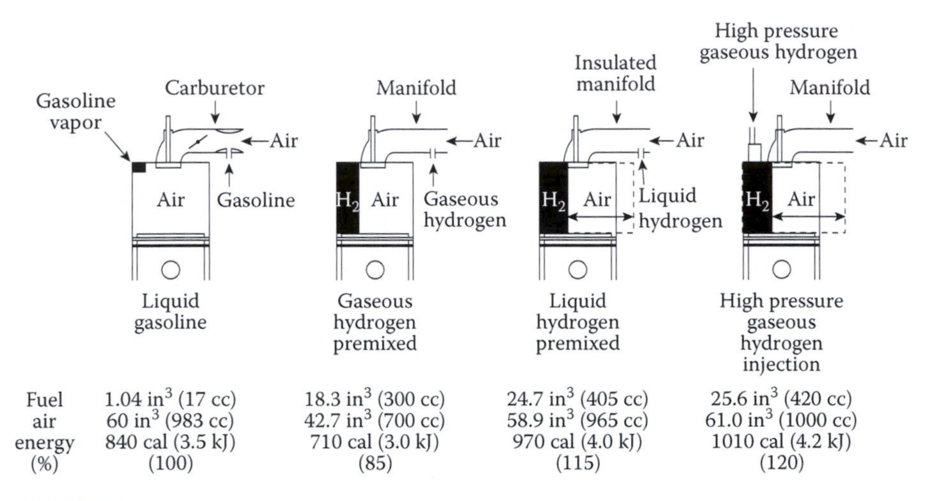

FIGURE 9.2
Combustion chamber volumetric and energy comparison for gasoline and hydrogen fueled engines (www.eere.energy.gov).

9.4.3.8 Hydrogen Embrittlement

Owing to hydrogen embrittlement, the mechanical properties of metallic and nonmetallic materials of hydrogen systems may degrade and fail resulting in leaks. Hydrogen embrittlement depends upon surrounding temperature and pressure, concentration and exposure time to hydrogen, strength and quality of material, physical and mechanical properties, surface conditions, and so on. Control of hydrogen embritlement can be achieved by oxide coating,

TABLE 9.4

Comparison of Energy Content of Various Fuels

Fuel	Chemical Formula	State	Energy (MJ/kg)	Energy (gJ/m³)
Gasoline	$C_{5-10}H_{12-22}$	Liquid	47.4	34.85
LPG	$C_{3-4}H_{8-10}$	Liquid	48.8	24.4
LNG	CH_4	Liquid	~50.0	~230.0
Methanol	CH_3OH	Liquid	22.3	18.10
Ethanol	C_2H_5OH	Liquid	29.9	23.60
Liquid hydrogen	H_2	Liquid	141.9	10.10
Hydrogen	H_2	Gaseous	141.9	0.013
Natural gas	CH_4	Gaseous	~50.0	0.040

Source: From Das, L. M., *International Journal of Hydrogen Energy*, 21, 789–800, 1996. Reprinted with permission from the International Association of Hydrogen Energy and Elsevier Publications.

removing stress concentration, additives to hydrogen, selection of alloy materials, and so on. The internal and environmental hydrogen embrittlement maximizes in the temperature range of between –73 and 27°C whereas hydrogen reaction embrittlement occurs at temperatures above room temperature (Rigas and Sklavounos 2009).

9.4.3.9 Calorific Value

Calorific value of hydrogen in comparison with other fuels is depicted in Table 9.4. Hydrogen contains about 2.75 times the energy as compared to that of gasoline on a mass basis. However, on volume basis it is of low energy content.

9.5 Well-to-Wheels Considerations

The term "Life Cycle Assessment" is very familiar to people in the environmental field. It is used to assess the total environmental performance of a product all along its lifetime, often referred to as from cradle to grave. Also other terms, such as life cycle analysis and eco-balance are used. When talking about fuels, the proper term in use is "Well-to-Wheel Analysis." In order to be able to examine the complete fuel-cycle of a traffic fuel, the chain is often divided into the following five stages:

1. Feedstock production
2. Feedstock transportation

3. Fuel production
4. Fuel distribution
5. Vehicle use

These stages can be divided further into a Well-to-Tank (WTT) and Tank-to-Wheel (TTW) portion of the Well-to-Wheel analysis. The WTT portion of the whole analysis considers the fuel from resource recovery to the delivery to the vehicle tank; that is, the feedstock production + transportation and the fuel production + distribution. This kind of division enables one to consider the analysis in smaller sections when needed and also helps the researchers to concentrate on the key areas. Well-to-Wheel analysis of fuels helps to evaluate different fuel consumption parameters during fuel's lifetime.

9.5.1 JRC/EUCAR/CONCAWE Report

EUCAR (European Council for Automotive Research and Development), CONCAWE (the oil companies' European association for environment, health, and safety in refining and distribution), and JRC (the Joint Research Centre of the EU Commission) have performed a joint evaluation of the Well-to-Wheels energy use and GHG emissions for a wide range of potential future fuel and power train options CONCAWE. The specific objectives of the study were to

- Establish, in a transparent and objective manner, a consensual well-to-wheels energy use and GHG emissions assessment of a wide range of automotive fuels and power trains relevant to Europe in 2010 and beyond
- Consider the viability of each fuel pathway and estimate the associated macro-economic costs
- Have the outcome accepted as a reference by all relevant stakeholders

The considered WTT matrix was given in Table 9.5. Results of interest for this study, with reference to different pathways of hydrogen production, are reported in Figure 9.3.

Key-point, from the same reference, was the following: "In the short-term, natural gas is the only viable and cheapest source of large scale hydrogen. WTW GHG emissions savings can only be achieved if hydrogen is used in fuel cell vehicles albeit at high costs. Hydrogen ICE vehicles will be available in the near-term at a lower cost than fuel cells. Their use would increase GHG emissions as long as hydrogen is produced from natural gas. Hydrogen from nonfossil sources (biomass, wind, nuclear) offers low overall GHG emissions. Renewable sources have a limited potential for the foreseeable future and

TABLE 9.5
Well-to-Tank Matrix

Fuel	Gasoline, Diesel, Naptha (2010 Quality)	CNG	Hydrogen (Comp, Liquid)	F-T Diesel	DME	Ethanol	FAME	Methanol	Electricity
Crude oil	X								
Coal			X					X	X
Natural gas		X	X	X	X			X	X
Pipe									
Remote		X	X	X	X			X	X
Biomass									
Woody biomass			X	X	X	X		X	
Farmed wood			X	X	X	X		X	X
Sugar beet						X			
Wheat						X			
Rapeseed							X		
Sunflower							X		
Wind									X
Nuclear									X
Electricity			x						

Source: www.ies.jrc.ec.europa.eu/uploads/media/ WTW_Report_220104.pdf

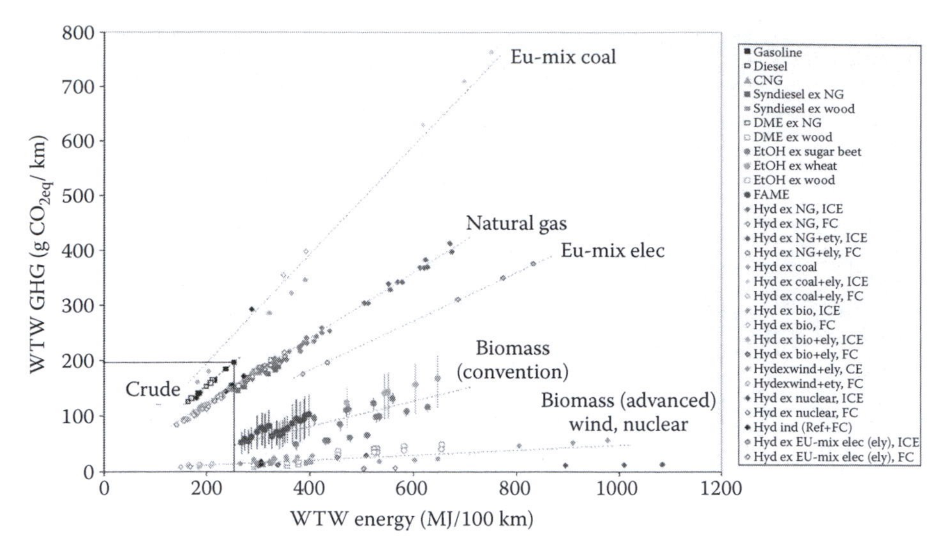

FIGURE 9.3
WTW energy and GHG emissions for all pathways and power train combination. (From ies.jrc.ec.europa.eu/uploads/media/WTW_Report_220104.pdf)

are at present expensive. More efficient use of renewable may be achieved through direct use as electricity."

9.6 Hydrogen Economics

Apart from the hardware and infrastructure costs, fuel costs are, of course, of vital importance when network operators are to choose between different propulsion technologies. For test and demonstration purposes present fuel costs are not that important—but future operation costs (where fuel costs are one of the important parameters) have to be roughly at the same level as the case for traditional technologies if hydrogen powered vehicles are to get a breakthrough. At present hydrogen fuel is more expensive than diesel but that is unlikely to remain so forever. When discussing the future supply and the corresponding likely future prices on diesel and other fossil fuels the relevant question to ask is not: "For how many years have we got available oil, gas, and coal resources?" but "When can the supply no longer keep up with the demand?"

According to a report from 2006 by Ludwig-Bolkow-Systemtechnik for the European Hydrogen Association this will happen in just a few years time (http://www.lbst.de). When the supply of fossil fuels no longer can keep up with demand the prices will rise and renewable energy sources will become increasingly competitive. The oil production profile is shown at the bottom of Figure 9.4. According to this scenario, oil supply will peak before 2010, and with the current demand of 4000 MTOE forecasted to rise to almost 6000 MTOE by 2030 it is clear that oil and diesel prices will rise significantly in the coming years. Figure 9.4 also illustrates the huge gap in total energy supply that is created by declining oil and gas production. At best, coal might flatten the total decline for a few years (Zittel 2005). This means that in the entire energy sector, the transport sector especially, with a high degree of oil dependency, will have to substitute the quantity of declining oil with an alternative fuel. The means of transport that make this transition fastest will no doubt gain a competitive advantage toward the later adopters.

There are a number of different studies trying to project future hydrogen prices. A publication from the OECD/International Energy Agency from 2005 (Prospects for Hydrogen and Fuel Cells) estimates that in the mid to long-term, various centralized production options can produce hydrogen at less than USD 10–15/GJ H_2 (1.2–1.8 USD/kg) corresponding to a price of 0.27–0.39 Euro per liter diesel equivalent, values very close to the results of other studies (Figure 9.5).

An IEE study in progress estimates that hydrogen can be produced via electrolyses using off peak electricity at a price of roughly 0.6 Euro per liter diesel equivalent. The current prices on industrial surplus hydrogen, is roughly 0.5 Euro per liter diesel equivalent. The stated prices are not including taxes, compression/fuel station costs, or interest payments on the electrolyser (www.europarl.europa.eu).

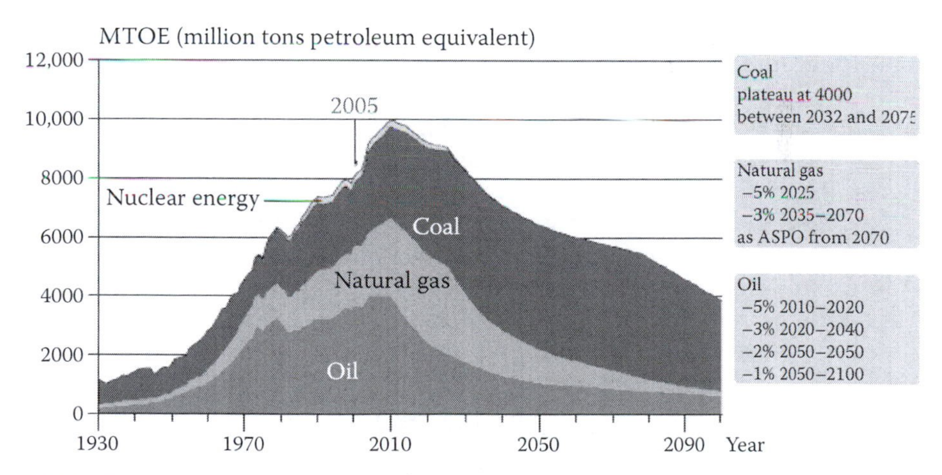

FIGURE 9.4
Oil production profile. (From BP Statistical Review of World Energy Outlook, International Energy Agency, 2004.)

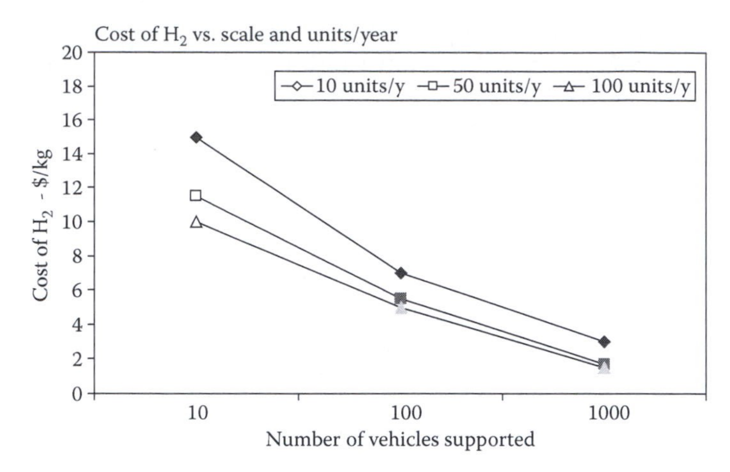

FIGURE 9.5
Hydrogen costs projections according to (GE Energy & Environmental Research, 2002.)

9.7 Hydrogen Safety

The safety use of hydrogen was elaborately analyzed by Carcassi and Grasso (2004) and the summary is as follows: We analyze the safety risks (unintended hazards due to natural or technological causes) rather than security related risks (intended hazards, i.e., malicious acts) and in particular the risks due the use of hydrogen. Technologies are a key element in the world we are living in. Public authorities, society and community, besides making technologies possible and favoring them, should be responsible for their safety. Every technological innovation should guarantee a level of safety and reliability that prevents potential hazards for human beings. It follows that, in every technology, rules must regulate technical aspects while the choices on responsibility and resulting risks management must be taken at a political level (Carcassi and Grasso 2004).

Therefore, an in-depth knowledge of the relation between the technical and legal points of view is indispensable to manage any potential hazard; this knowledge is possible only after a comprehensive agreement on the basic philosophy, whose acceptance depends on the position taken up from society in relation to the technologies and the related hazards. At the same time, in view of the technology importance in life and work, conformation of the limits imposed by technological processes need to be defined at a political level.

Hydrogen related technology is a potential future energy option with the potential to limit the increasing global climate problem and the greenhouse effect. However, increasing safety requirements have to be fulfilled to minimize industrial risks and to harmonize the safety culture also in the field of transportation. A large body of literature, related to hydrogen safety, is now available, as well as many sites on the internet that contain hydrogen safety information. The starting point of a safety assessment is the knowledge of the hydrogen hazard. The preliminary hazard lists posed by the properties of gaseous hydrogen are as follows (www.inl.gov/hydrogenfuels/projects/docs/h2safetyreport.pdf):

1. Physical properties
   - Lighter than air, odorless and colorless gas
   - Low viscosity (leaks easily)
   - Highly diffusive
2. Pressure
   - High-pressure storage can result in pressure rupture, flying debris
   - High-pressure gas jet impingement on body can cut bare skin
   - Oxygen displacement in confined spaces
3. Chemical
   - Flammable, with nonluminous flame, no toxic combustion products
   - Explosive, 4–74% by volume can deflagrate (typically only a modest over pressure, ~a few psi in open areas) can also detonate (high over pressure shock wave, ~several atmospheres)
   - Low ignition energy, 0.02–1 mJ spark to ignite a deflagration
   - Modest autoignition temperature, 574°C
4. Temperature
   - Could be stored at room temperature
5. Materials issues
   - Embrittlement of metal and plastics
6. Toxicological
   - Asphyxiation in confined spaces
   - No other toxic concern effects

The hazards related with liquid hydrogen because of its properties are given below.

1. Physical properties
   - Odorless, colorless, cannot easily be odorized since odorants will freeze out at cryogenic temperatures
   - Boiloff gas quickly warms and then is lighter than air

- Boiloff gas is highly diffusive
- Flow-induced static charge generation
- Boiloff vent rate from storage tanks/fuel tanks is typical to maintain cold temperature in tank
- Liquifies quickly, easily boils by heat transfer into the 20 K liquid
- Rapid phase transition from liquid to gas can cause pressure explosions
- Liquid quickly contaminates itself by condensing gases from air contact

2. Pressure
   - Stored under modest pressure to suppress boiling
3. Chemical
   - Evolved gas is cold, otherwise the same concerns as gaseous hydrogen
4. Temperature
   - Cryogenic burns, especially eyes
   - Lung damage by cold vapor inhalation
   - Possible hypothermia working near these systems
   - Condensation of air near LH_2 systems if insulation allows heat leak paths; can lead to oxygen rich zones near systems
5. Materials issues
   - Mechanical stresses generated by thermal contraction
   - Embrittlement of metal
   - Mild steels susceptible to cracking at cryogenic temperatures
   - Materials have low specific heats at cryogenic temperatures, easy heat transfer
6. Toxicological issues
   - Asphyxiation in confined spaces
   - Frostbite from acute exposure
   - Hypothermia possible from long exposure
   - No other toxic concerns

From the list, the hazard due high-pressure, for gaseous storage, and rapid phase transition from liquid to gas, in cases of liquid storage, seems the most dangerous scenario for the transport sector due to the fact that these hazards develop into the ultimate scenario of fire and explosions. These issues are of particular importance for the risk related to hydrogen

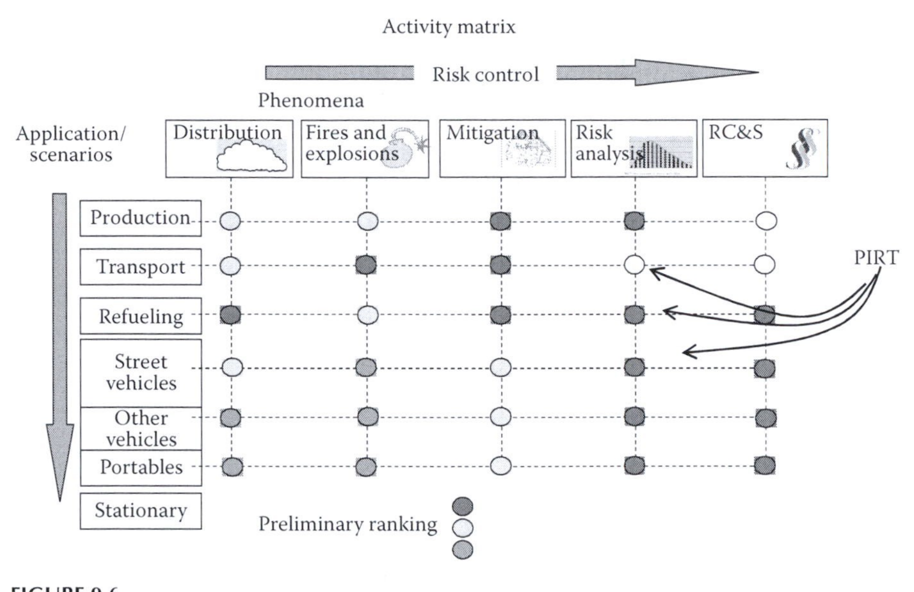

FIGURE 9.6
HYSAFE activity matrix.

storage when vessel rupture, in general and especially in the tunnels, and mitigation measures and tests are highly studied. Worldwide projects, related to Hydrogen Safety Issues such as: HYSAFE, IEA Task 19–Hydrogen Safety, are studying these hazards (Carcassi and Grasso 2004). As an example it is interesting to show how the HYSAFE project deals with safety issues: Figure 9.6 shows the activities. The studies are organized as a matrix where the rows are the risk.

9.8 Hydrogen in Fuel Cells

Hydrogen can be combined with oxygen in systems that first ionize it in such a way that hydrogen ions and electrons follow different paths and this results in generation of electricity. This is the principle of operation of hydrogen fuel cells, and it is shown in a simplified way in the Figure 9.7. There exist other kinds of fuel cells that have a similar operation principle to the one shown in Figure 9.7 but use ions different from the hydrogen ions (H^+) as electricity carriers in combination with electrons. This difference, however, is beyond the scope of this executive summary. The fuel cells of interest to the present executive summary are those that generate electricity by combining hydrogen and oxygen to form water. A typical fuel cell schematic is shown in Figure 9.7.

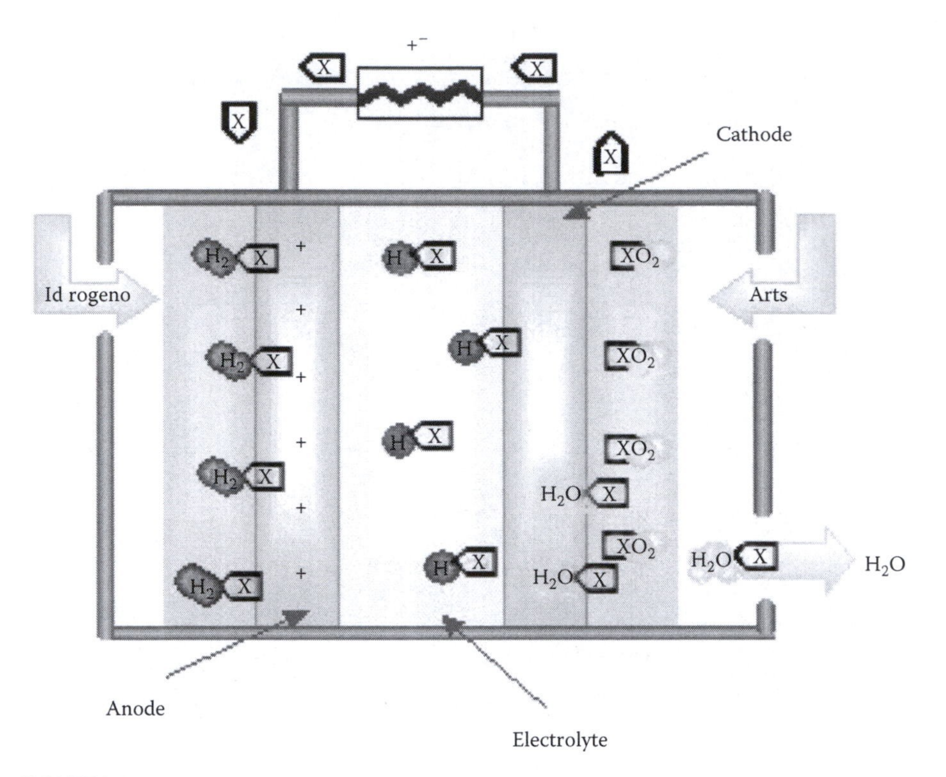

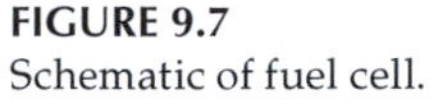
FIGURE 9.7
Schematic of fuel cell.

There are some possible choices:

- Alkaline Fuel Cells (AFC): They have been used widely in space in the last few decades. Their terrestrial utilization is limited by their ease of being contaminated: the very CO_2 content of the air is a poison for the cell. Their limitations have brought about a progressive abandonment of their use and will not be considered further in this document.
- Polymer Electrolyte Fuel Cells (PEFC): These are the most widely used in transportation because they operate at lower temperatures (60–80°C); they can be started-up and shut-down very rapidly; they are compact and efficient; and their cost has steadily decreased during recent years, halving each 1–2 years.
- Phosphoric Acid Fuel Cells (PAFC): These operate at around 200°C and have a technology that is sufficiently mature for stationary utilizations; units with power outputs of a few hundred kW are feasible and increasingly common.

- Hydrogen Molten Carbonate Fuel Cells (MCFC): They operate at a temperature of around 650°C and are currently being considered mainly for stationary applications.
- Hydrogen Solid Oxide Fuel Cells (SOFC): They are the highest in temperature (900–1000°C). As with the MCFC, they are currently being considered mainly for stationary applications.

Fuel-cell based systems for vehicle propulsion have been considered in recent years mainly for road applications (buses, cars, even scooters); for these the most adequate typology is constituted by PEFCs, because of their lower operating temperature; however, for application onboard trains it may happen that higher temperature options are competitive with the PEFC choice. A single fuel cell is capable of delivering only very small amounts of energy: a typical operating voltage is around 0.6 V, less than half of a common toy battery. However, they are easily stacked to reach higher voltages, and stacks of up to 100 kW of nominal power have been built, although it is often preferred to manufacture smaller stacks (up to 50–60 kW) and combine several stacks to raise the system power. Stacked fuel cells are not systems for generating electricity yet: auxiliary systems to perform several side actions such as hydrogen and air circulation, cooling, and managing membrane humidity are needed, as can be inferred from Figure 9.8, showing a typical stack plus auxiliaries arrangement.

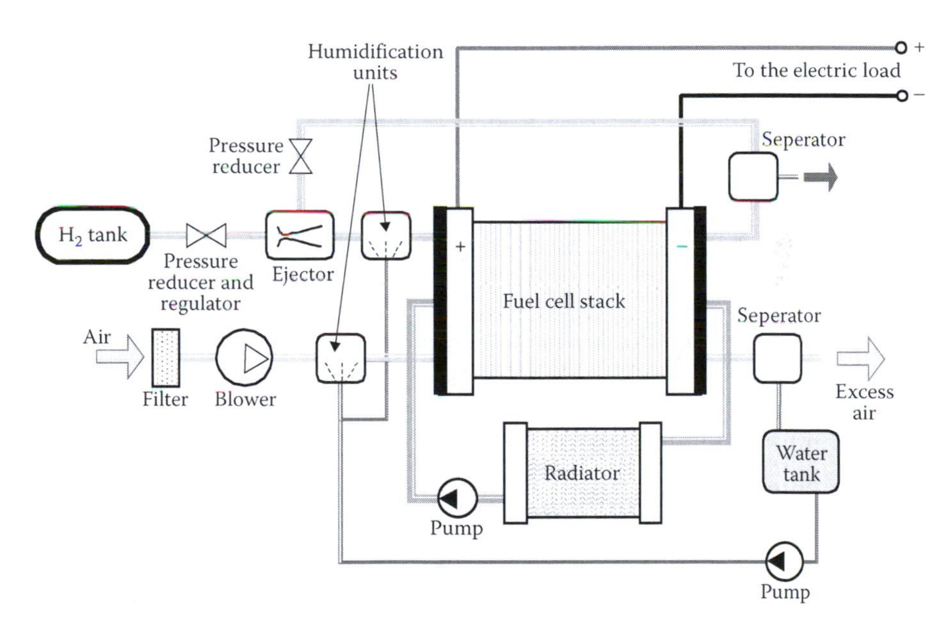

FIGURE 9.8
Fuel-cell based electricity generation system (FCG).

The lifetime of the PEM fuel cell generator is limited by membrane degradation, it was limited to 1500 hours with last year's technology; today it seems possible to reach a few 1000 hours. However the reliability required by railway applications is over 100,000 hours of working time, and for a tramway the stop–start cycle between stations is over 5 million (RAPSDRA 1995; cycling time about 1 minute). Only PAFC reaches 40,000 hours, corresponding perhaps to 1,000,000 km, but at the moment they are used only for stationary applications.

9.9 Fuel Cell Vehicles Hybridization

Fuel cell generators are not well adapted to power variation. Research projects to introduce FC in mobile applications include battery energy storage to deliver the traction power in combination with FC generators. Briefly, a storage unit can be employed together with FCs in order to:

- Reduce the size of the FC, the most expensive component, thus reducing the drivetrain total cost. Projected cost of automotive storage systems is lower, even in case of mass production, than fuel cells projected cost. Thus it is easy to predict that, for a given vehicle performance, a hybrid power train will be cheaper than a pure fuel cell, especially in those applications (urban cycles) where the ratio of peak/average power is higher.
- Reduce power transients, not to reduce emissions as in motor generators, but to simplify the "balance of plant" configuration and to enhance its efficiency: mainly in the case of FC vehicles with fuels other than hydrogen, because the time constant of some subsystems (typically fuel processors) are much longer than required by the driving cycle. In this case an electrical storage system acts like a buffer during peak power, making a vehicle comfortably drivable.
- Reduce start-up power transients that may be important depending on the FC system configuration, for example, direct hydrogen or methanol powered. Indeed, in case of cold start-up, the on board energy storage can be used also to accelerate system heating and to move the vehicle throughout this phase.
- Recover braking energy that moreover is produced and available in the form of electricity (otherwise this energy will be dissipated thermally). The amount of energy saved just with braking energy recovering is in the range of 3.5–20%, enough to justify alone the realization of a hybrid system.

Finally, there is no longer the need for FC optimization during the cycle because the generator efficiency curve matches in a better way to the energy use during urban cycles. Indeed, in comparison to a thermal motor, the peak is more toward the middle–low power and therefore it is better adapted to urban cycles. However, the fuel–cell system could still benefit from some size reduction to prevent excessive operation at light load or on/off operation due to minimum power requirements.

In conclusion, if the power required by the motor is shared between two devices, an economic benefit from cost reduction exists as storage devices are heavier but cheaper than fuel cells, particularly when the ratio of peak power/average power is high. The management (fuel) cost also is reduced when compared with a full-power system since braking energy recovery can also be performed. This is particularly true in those applications (urban transportation) where braking is very frequent and therefore the braking energy recovered is likely larger than the energy loss of the storage.

9.10 Hydrogen in Spark Ignition Engines

Hydrogen can be used in spark-ignition ICEs in a way that is very similar to gasoline use in ordinary engines. The use of ICEs with hydrogen produces very few pollutant emissions. This is due to two factors:

- In theory, nitrogen oxides (NO_x), coming from the oxidation of atmospheric nitrogen are the only undesirable emissions produced by this engine.
- Low explosion limit of hydrogen allows a stable combustion, even under diluted conditions. Even with a mix of a small percentage of hydrogen with air, the temperatures of combustion are low and the speed of NO_x formation is low. The explosion limit of hydrogen also contributes to good engine output with hydrogen under weak loads.

Moreover, H_2 ICEs seem to be efficient. BMW obtained an efficiency of 37.5% with its car. Therefore, the characteristics of hydrogen combustion seem to allow an efficient and environmental friendly propulsion system. This advantage could be obtained with very reasonable costs, at today's prices well below those of fuel-cell based electricity generators. Current research efforts aim at developing hydrogen ICEs with improved power densities and reduced NO_x emissions under higher loads.

Among the most recent experiences we cite:

- The HyFleet project proposes to test 14 hydrogen internal combustion buses in Berlin in cooperation with TOTAL, BVG, and Vattenfall Europe. These buses cost approximately 50% more than a conventional bus. Two types of bus were developed, using different strategies of combustion, an H_2 urban bus equipped with a 150 kW engine that uses petrol and H_2 in stoichiometric proportion, and an H_2 urban bus equipped with new and a more efficient 200 kW lean burn engine with exhaust gas turbo charging and inter-cooling.
- The new BMW 12 cylinder hydrogen engine that is able to run on both hydrogen and petrol.
- A simplified comparison between the use of hydrogen in ICEs and in FCGs can be based on the following items:

  FCGs have much higher efficiencies and pollute much less (they are low noise and their gaseous emissions are null, for low temperature FCs, or very low) at present ICEs have much lower costs and longer useful life.

Hydrogen-IC engines are much cheaper than Hydrogen-FCs, both directly and in terms of fuel cost (with high fuel purity requirements for the H_2-FCs). Furthermore, using IC engines allows bi-fuel operation (engine can run on gasoline and hydrogen). Though FC vehicles consistently achieve the highest fuel efficiency, the H_2-ICE can serve as a bridging technology and might help in the development of the infrastructure needed for hydrogen fuel. Hydrogen can be used in SI engines in three ways: manifold induction, direct injection of hydrogen into the cylinders, and as a supplement to gasoline.

9.10.1 Manifold Induction

Hydrogen can be inducted into the engine through carburetion. The engine was able to operate smoothly within the equivalence range of 0.4–0.8 without frequent symptoms of backfire. This method allows quality governing, ultra-lean operation, high thermal efficiency, and low exhaust emissions in the equivalence range of 0.3–0.5. These carburetted versions of the engine systems, apart from developing low power outputs (as compared to the gasoline-fueled engines) also exhibited severe operational combustion-related problems such as backfire, preignition, combustion knock, and rapid rate of pressure rise (www.imtuoradea.ro).

Das (2002) reported that normal spark plugs sometimes act as a vulnerable potential hot spot leading to premature ignition. Moreover, it causes backfire and might be caused by hot spots in the combustion chamber and residual particulate matter from the oil. Lynch (1983) suggested "Parallel Induction," which is similar to intake port injection and has proved successful in getting

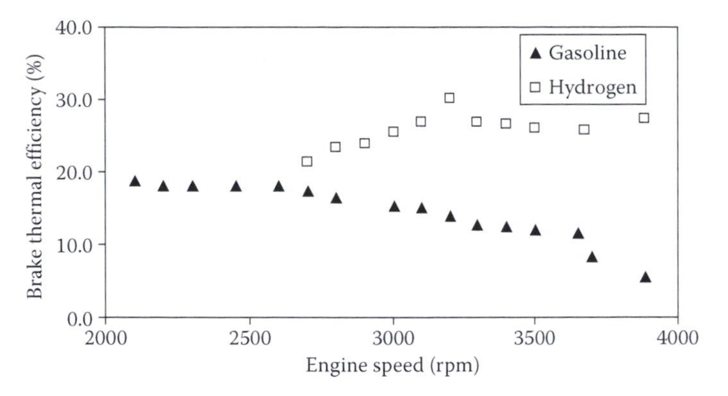

FIGURE 9.9
Brake thermal efficiency in hydrogen fueled SI engine. (From Kahraman, E., Ozcanli, S. C., and Ozerdem, B., *International Journal of Hydrogen Energy*, 32, 2066–72, 2007. Reprinted with permission from the International Association of Hydrogen Energy and Elsevier Publications.)

over the problems associated with backfire. He also reported another method of hydrogen induction technique through a copper tube placed inside the air intake port. A sleeve-type valve-seat mechanism built on the original intake valve is used to control the system. This method of delayed hydrogen admission proved quite effective in suppressing the undesirable combustion phenomena.

Kahraman, Ozcanli, and Ozerdem (2007) analyzed the behavior of hydrogen manifold inducted engines. They reported that hydrogen fuel has a higher brake thermal efficiency and can even operate at lower engine loads with better efficiency (Figure 9.9). It can be noticed that brake thermal efficiency is improved to about 31% with hydrogen-fueled engines compared to gasoline fueled engines. Traces of CO and HC emissions presented in hydrogen-fueled engines are due to the evaporating and burning of lubricating oil film on the cylinder walls. Short time of combustion produces lower exhaust gas temperature for hydrogen. Significant decreases in NO_x emissions are observed with hydrogen operations (Figure 9.10). Carburetor systems do not require the hydrogen supply to be under high-pressure like other fuel delivery systems. Gasoline engines fitted with carburetors can be easily converted to operate on neat hydrogen or blends of gasoline/hydrogen. Carburetor engines are more susceptible to preignition and backfire. Moreover, the accumulation of hydrogen/air mixtures within the intake manifold aggravates the effects of preignition.

9.10.2 Port Injection

The port injection fuel-delivery system injects fuel directly into the intake manifold at each intake port, rather than drawing fuel in at a central point. Air enters at the beginning of the intake stroke to dilute the hot residual

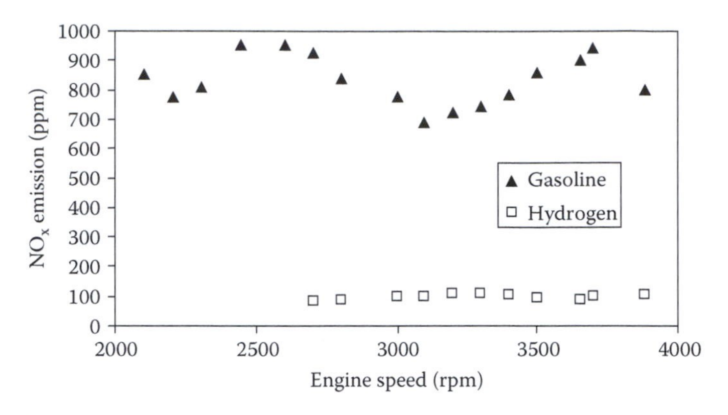

FIGURE 9.10
NO_x emissions in hydrogen fueled SI Engine. (From Kahraman, E., Ozcanli, S. C., and Ozerdem, B., *International Journal of Hydrogen Energy*, 32, 2066–72, 2007. Reprinted with permission from International Association of Hydrogen Energy and Elsevier Publications.)

gases and cool any hot spots in the combustion chamber, and hydrogen is injected into the manifold after the beginning of the intake stroke. Since less gas (hydrogen or air) is in the manifold at any one time, the probability for premature ignition is reduced. This method reduces the risk of backfire also. The inlet supply pressure for port injection is higher than for carbureted systems, but less than for direct injection systems. Currently, electronic fuel injection (EFI) systems meters the hydrogen to each cylinder. Each cylinder is fitted with individual fuel injectors.

Combustion of lean hydrogen–air mixtures with fuel-to-air equivalence ratios of less than 0.5 ($\lambda > 2$) results in extremely low NO_x emissions. Due to the excess air available in the combustion chamber, the combustion temperatures do not exceed the NO_x critical value of approximately 1800 K (Eichlseder et al. 2003).

Exceeding the NO_x critical equivalence ratio results in an exponential increase in oxides of nitrogen emissions, which peaks around a fuel-to-air equivalence ratio of 0.75 ($\lambda \sim 1.3$). At stoichiometric conditions, the NO_x emissions are at around 1/3rd of the peak value (Eichlseder et al. 2003). Figure 9.11 shows the theoretical power density of port fuel injected hydrogen engines in comparision with gasoline engines.

9.10.3 Hydrogen Direct Injection

Hydrogen direct injection (DI) system was developed to achieve near zero emission with an increase in power. Latest model hydrogen engines use direct injection systems for fuel-delivery purposes. Early injection generally refers to any hydrogen DI during the early compression stroke shortly after intake valve closing, whereas late DI refers to strategies with the injection

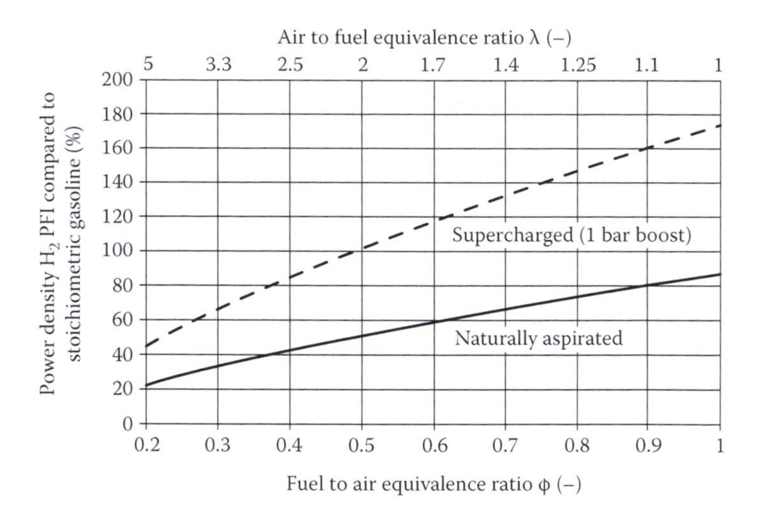

FIGURE 9.11
Theoretical power density of a PFI H_2 engine compared to stoichiometric gasoline operation as a function of equivalence ratio and charging strategy. (From Verhelst, S. and Wallner, T., *Progress in Energy and Combustion Science,* 35, 490–527, 2009. Reprinted with permission from Elsevier Publications.)

late in the compression stroke generally ending just before spark timing. In order to avoid displacement of a fresh charge by hydrogen of low density, the start of injection even for early injection is usually set after intake valve closing. During fuel injection, the inlet valve is closed completely thereby avoiding premature ignition during the suction stroke. The DI fuel system requires higher fuel rail pressure than other fuel systems.

In order to flow hydrogen directly into the combustion chamber, the pressure inside the injection system has to exceed the pressure inside the cylinder; that is, critical injection condition. The amount of fuel injected is determined only as a function of injection pressure and injection duration for engine calibrations and accurate fuel metering. Critical conditions are that the injection pressure has to be approximately twice the cylinder pressure to guarantee critical conditions (choked flow). Therefore, operating strategies with early DI require injection pressures in the range of approximately 5 and 20 bar, late injection strategies up to 100 bar, and multiple injection strategies with injection pulses during the actual combustion event of 100–300 bar (Verhelst and Wallner 2009).

9.10.4 Preignition Problems in Hydrogen Engines

Hydrogen burns cleanly and operates efficiently as it has the characteristics to operate in ultra-lean combustion conditions. This will drastically reduce the NO_x formation and have better performance at low load conditions.

However, at high loads there is a tendency for preignition and an increase in NO_x emissions also.

Preignition is defined as combustion prior to spark discharge that results from surface ignition at engine hot spots, such as spark electrodes, valves, or engine deposits. The limiting effect of preignition is that a preignition event will advance the start of combustion and produce an increased chemical heat release rate and rapid rate of pressure rise, high peak pressures that lead to higher in-cylinder surface temperatures.

In hydrogen engines, premature ignition is a problem because of its lower ignition energy, wider flammability range, and shorter quenching distance. Premature ignition occurs when the fuel mixture in the combustion chamber ignited before ignition by the spark plug and results in an inefficient rough running engine. If the premature ignition occurs near the fuel intake valve the resultant flame travels back into the induction system thus causing backfire. Crankcase oil enters into the combustion chamber through blow-by and in suspended form, or in the crevices just above the top piston ring, and acts as a hot spot that may contribute to preignition (White, Steeper, and Lutz 2006). Preignition in hydrogen vehicles, can be avoided by

- Injecting cold exhaust gases into the cylinder
- Injecting cold gaseous hydrogen
- Injecting water into the cylinder
- Increasing the compression ratio
- Using lean burn carburetor

9.11 Hydrogen in Compression Ignition Engines

Hydrogen can be used in CI engines in two ways: dual fuel mode and surface ignition.

9.11.1 Dual Fuel Mode Operation

In this method, hydrogen is inducted into the combustion chamber along with fresh air and compressed and then ignited by a spray of injected liquid fuel (diesel). Lkegami, Miwa, and Shioji (1982) investigated hydrogen combustion in a conventional swirl chamber type diesel engine. It has been reported by the investigators that hydrogen-fueled diesel combustion could be achieved to a limited extent because of the autoignition characteristics of the fuel. The problems associated with this method are: when an insufficient quantity of fuel is injected then incomplete combustion will occur and when hydrogen-gas mixture becomes rich combustion becomes uncontrollable.

An excessive introduction of the preliminary fuel may cause autoignition by itself thus giving rise to rough combustion. About 30% hydrogen can be replaced by this method. Introduction of a higher percentage of hydrogen leads to uncontrolled pressure rise.

Das and Polly (2005) conducted investigations on CI engines to achieve knock-free operation in various proportions by using diluents such as nitrogen, helium, and water. A comparative assessment among the three diluents shows that conventional diesel engines can be converted to operate on hydrogen-diesel dual mode with up to about 38% of full-load energy substitution without any sacrifice on the performance parameters such as power and efficiency. A long-term endurance study on this system showed that there was no problem related to material compatibility on such configurations and as such they can be safely adopted.

Kumar, Ramesh, and Nagalingam (2003) reported that with hydrogen induction, due to high combustion rates, the NO level was increased at full output. They reported that ignition delay, peak pressure, and the maximum rate of pressure rise were increased in the dual-fuel mode of operation. Combustion duration was reduced due to higher flame speed of hydrogen. Moreover, a higher premixed combustion rate was observed with hydrogen induction. Welch and Wallace (1990) conducted investigations on hydrogen combustion by its autoignition with glow plug assist in a reciprocating engine at a compression ratio of 17. Hydrogen-fueled diesel engines can produce higher power than an ordinary diesel engine with reduction in NO_x and nil smoke emissions.

9.11.2 Direct Injection

In this method, hydrogen is directly introduced into the cylinder at the end of compression. The gas impinges on the glow plug in the combustion chamber and hence surface ignition burning of fuel occurs. Moreover, it is possible to introduce lean hydrogen–air during intake of an engine and then inject the bulk of hydrogen toward the end of compression. The advantage of this method is that surface ignition is not dependant on the compression ratio for ignition. Literature shows that a very small difference in brake thermal efficiency was observed in the compression ratio 12–18.

With hydrogen directly injected into the combustion chamber in a CI engine, the power output would be approximately double that of the same engine operated in the premixed mode. The use of hydrogen DI in a diesel engine has given a higher power to weight ratio when compared to conventional diesel-fueled operation, with the peak power being approximately 14% higher. The power output of such an engine would also be higher than that of a gasoline engine, since the stoichiometric heat of combustion per standard kilogram of air is higher for hydrogen (approximately 3.37 MJ for hydrogen compared with 2.83 MJ for gasoline). Higher diffusivity and

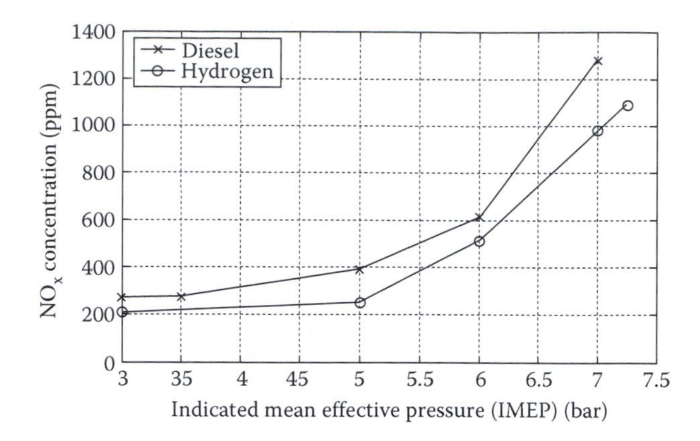

FIGURE 9.12
NO_x emissions of hydrogen engines. (From Antunes, J. M. G., Mikalsen, R., and Roskilly, A. P., *International Journal of Hydrogen Energy*, 34, 6516–22, 2009. Reprinted with permission from the International Association of Hydrogen Energy and Elsevier Publications.)

the lower inertia of the injected hydrogen gas compared with diesel fuel, enhances the fuel-air mixing process after injection. This reduces the local peak temperatures in the combustion chamber. The DI of hydrogen allows much better control of engine operation compared to when operating in the port injected, HCCI mode (Figure 9.12; Antunes, Mikalsen, and Roskilly 2009).

The DI of hydrogen allows much better control of engine operation compared to when operating in the port injected HCCI mode. Consideration must be given to the control of injection timing and duration, as these variables heavily influence factors such as the rate of pressure rise and maximum combustion pressure. Direct injection offers the possibility to control and limit excessive mechanical loads while this is virtually uncontrolled in the HCCI mode of operation (Antunes, Mikalsen, and Roskilly 2009).

9.12 Hydrogen–CNG Engines

A good opportunity in the short-term can be represented by the utilization of blends of hydrogen with other fuels, first of all with natural gas (HCNG). When used in an ICE, even the addition of a small amount of hydrogen to natural gas (5–30% by volume that means ~1.5–10% by energy) leads to many advantages, because of some particular physical and chemical properties of the two fuels.

Methane has a slow flame speed while hydrogen has a flame speed about eight times higher (Figure 9.13); therefore, when the equivalence ratio (lambda) is much higher than for the stoichiometric condition, the combustion of methane is not as stable as with HCNG.

As a consequence of the addition of hydrogen to natural gas an overall better combustion had been verified, even in a wide range of operating conditions (lambda, compression ratio, etc.), finding the following main benefits: a higher efficiency; and lower CO_2 production and emissions. Because of hydrogen's chemical and physical properties, HCNG, despite its higher LHV per kg, has a lower LHV per Nm^3 (Figure 9.14), depending on the hydrogen content. Therefore, a natural gas engine, when fueled with HCNG, shows a lower power output, while maintaining better efficiency.

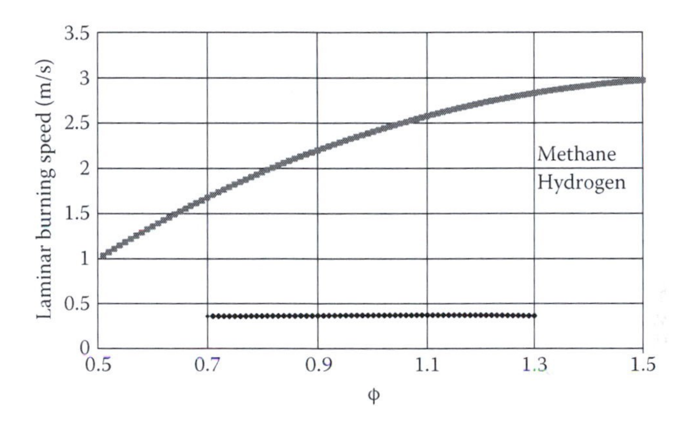

FIGURE 9.13
Laminar burning speed for methane and HCNG15. (From Ortenzi, F., Chiesa, M., and Conigli, F., *Experimental Tests of Blends of Hydrogen and Natural Gas in Light Duty Vehicles,* HYSYDays, -Turin, 2nd World Congress of Young Scientists on Hydrogen Energy Systems, 2007.)

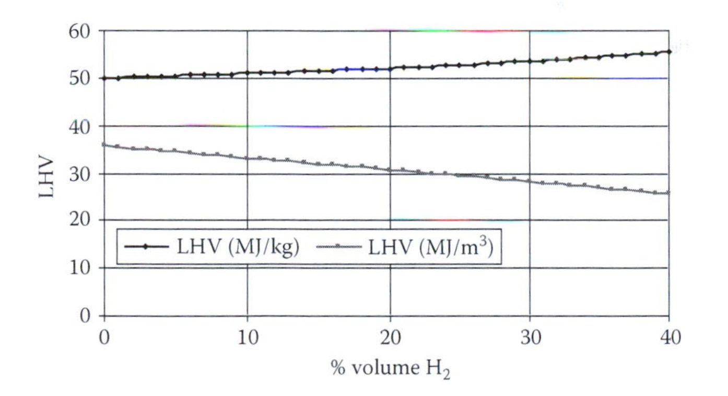

FIGURE 9.14
LHV for different percentages of hydrogen. (From Ortenzi, F., Chiesa, M., and Conigli, F., *Experimental Tests of Blends of Hydrogen and Natural Gas in Light Duty Vehicles,* HYSYDays, -Turin, 2nd World Congress of Young Scientists on Hydrogen Energy Systems, 2007.)

To restore a good value of power output, especially for lean burn mixtures (for lambda = 1.4, a naturally aspirated engine loses 50% of its power), a good solution could be represented by a turbocharged engine with a higher charging pressure.

Additionally, CO_2 emissions can be reduced as a result of the substitution of CNG by hydrogen. The special properties of hydrogen as a combustion stimulant can produce leverage factors much greater than 1 by improving fossil fuels and not just displacing them.

Hydrogen leverage is defined as the following ratio:

$$\text{Hydrogen leverage} = \frac{\%\text{ Emissions Reduction}}{\%\text{ Energy Supplied as Hydrogen}}.$$

The increased efficiency makes this value higher than 1. An obvious benefit of the leverage effect is that a CO_2 reduction is possible even if the hydrogen used is produced by natural gas without any sequestration of CO_2.

Experiments of application of blends of hydrogen and natural gas in ICEs started in 1991, in the framework of a research program financed by DoE and NREL, called the "Denver Hythane Project." The results are shown in Table 9.6.

In the same years, the University of Pisa and ENEA carried on some activities with the following interesting results. During the last 15 years many experiments have been conducted all over the world. All the experiments had mainly examined the reduction of the emissions of NO_x with respect to different air/fuel ratios (λ values) and percentages of hydrogen by volume (Figure 9.15). All the experiments had shown that the blends of hydrogen and natural gas reduce the exhaust emissions of both regulated pollutants and CO_2 and increases the efficiency of a spark ignition engine. During the last years, a number of fleet tests have been carried out.

The recent Hythane® (24.8% vol. Hydrogen, Frank Lynch, Hydrogen Components, Inc., HCI), bus demonstration project at Sunline transit in California used a 7% hydrogen by energy formula and the NO_x emissions were reduced by 50%. Based on success with Hythane® buses, and the cost-effectiveness of Hythane® compared to available fuel cell technology, a

TABLE 9.6

Denver Hythane Project

Fuel	NMHC (g/mile)	CO (g/mile)	NO_x (g/mile)
Gasoline	0.59	14.1	2.2
ULEV	0.04	1.7	0.2
Natural gas	0.01	2.96	0.9
Hythane	0.01	0.7	0.2

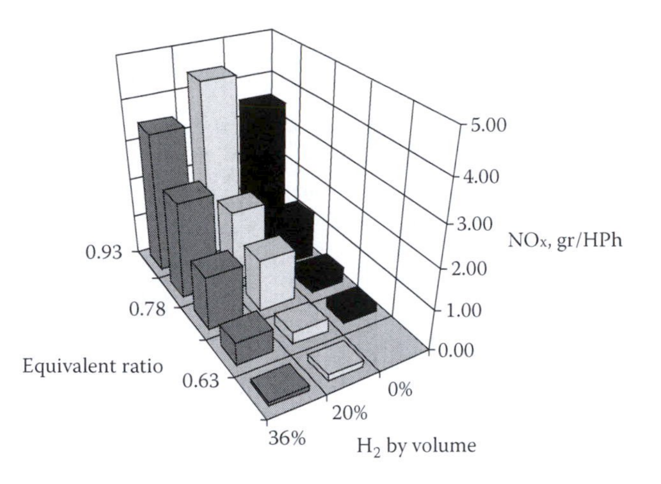

FIGURE 9.15
NO_x emissions in function of both the equivalent ratio and the H_2 percentage by volume.

number of projects are currently being carried out around the world, like the Beijing Hythane Bus Project, whose demonstration phase will be to adapt 30 natural gas engines for Hythane operation before 2010. At a European level, the most significant example of application of blends of hydrogen and natural gas in ICEs is given by the tests still ongoing in Malmo (Sweden) on urban buses.

The experimental results are available for blends with a hydrogen content of 8% by volume for their use on real driving cycles, while the data relative to a blend of 25% by volume are available just for the engine tests, in specific functioning points. The available data show an increase of the engine performance with the increase of the hydrogen quantity in the blend. Furthermore, depending on the λ values associated with the combustion of the blends, an overall environmental benefit can be noticed for λ values higher than 1.

From Figure 9.16 an efficiency increase from 31.2–35% for a λ value of 1.61 is illustrated. For inferior values of λ the efficiency is higher; nevertheless, since the first target related to the use of blends is the reduction of atmospheric emissions of urban pollutants and CO_2 the tests have to optimize the combustion and try to obtain the maximum reduction of total hydrocarbons (HC), nitric oxides (NO_x), and carbon monoxide (CO).

From Figure 9.17, it emerges that for $\lambda = 1.61$ there is a reduction of all the pollutant emissions with the exception of nitric oxides whose emissions do not decrease with respect to the use of pure NG. Nevertheless, for the blends of hydrogen and natural gas the combustion remains stable even for air/fuel ratio values that exceed those relative to natural gas, equal to 1.6, over which the combustion remains unstable.

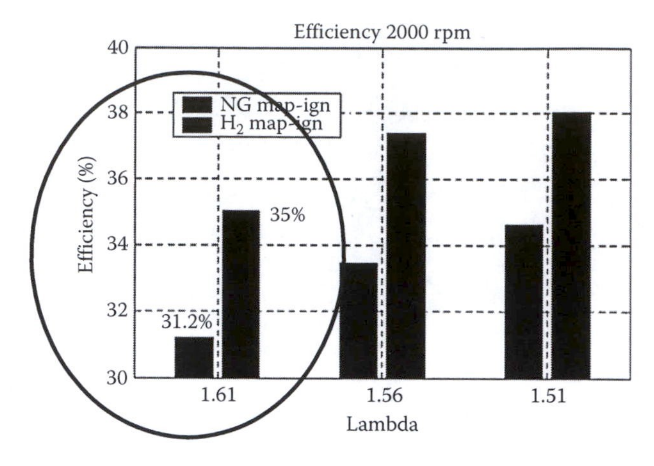

FIGURE 9.16

Efficiency and emissions of an ICE fueled with a HCNG8 blend.

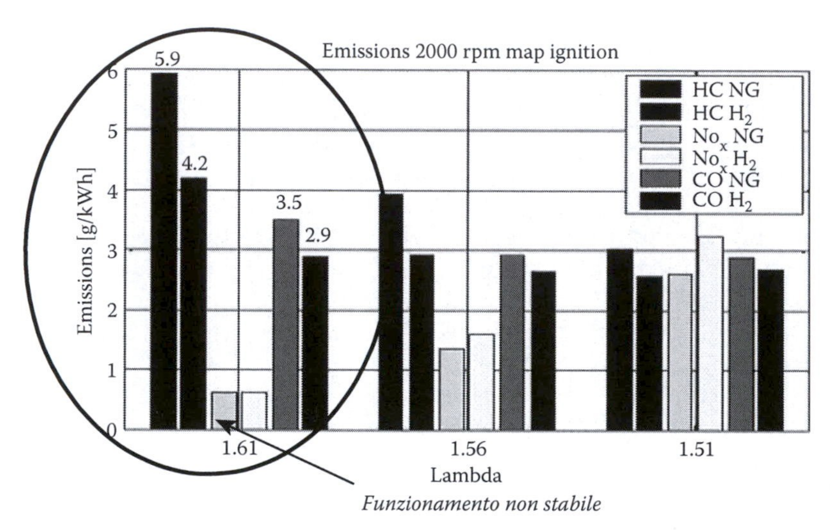

FIGURE 9.17
Emissions of an ICE fueled with a HCNG8 blend.

9.12.1 HCNG Engine Tests at ENEA

In an application of 15% of HCNG at an ENEA research center on a naturally aspirated light duty vehicle (Figure 9.18), two configurations have been tested: stoichiometric combustion and lean burn. In stoichiometric configuration, without any modification of the injection control map, the NO_x emissions increase as a consequence of the increasing combustion speed. A spark advance reduction (which means a little retard compared to the case

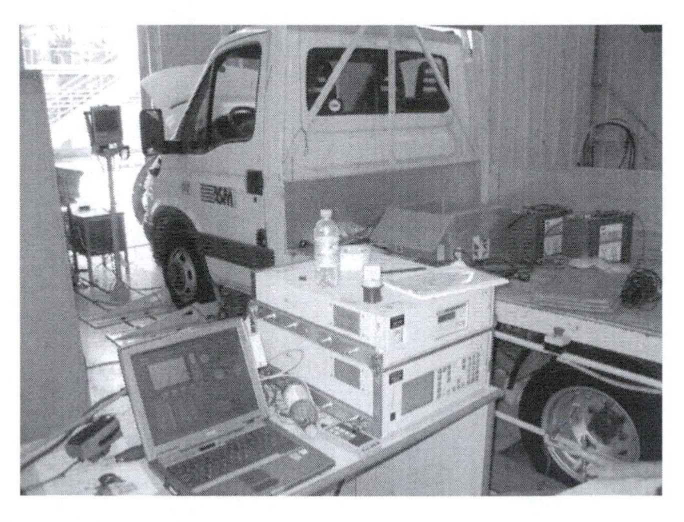

FIGURE 9.18
Iveco Daily CNG fueled with 10% and 15% of HCNG at ENEA research laboratories.

of pure methane) brings a large decrease of NO_x emissions, without torque reduction. For a lean burn blend, research of better λ (we wanted to reach a value of 100 ppm, the same of pure methane) had been limited from $\lambda = 1$ to $\lambda = 1.45$. Furthermore, the increase of λ values had caused very important power losses.

Therefore, $\lambda = 1.45$ had been the maximum value initially fixed (this value substantially reduces NO_x) and a series of tests had been produced changing the spark ignition advance to optimize the control strategy in order to increase the performance. Unfortunately, NO_x had grown in an exponential way, while power gain had not been significant. Therefore, it had been decided that it is more convenient to adopt a lower λ value without changing the spark advance instead of setting a high λ together with optimal advance timing.

Values (for λ) greater than 1.45 could be interesting to analyze for their better efficiency and consumption, but the engine will need substantial modifications such as a different compression ratio, the addition of a turbocharger, and so on. In the following pictures (Figure 9.19), the obtained values for emissions produced on the ECE15 driving cycle for the configurations analyzed are represented and compared with the original CNG values. For the stoichiometric configuration, it can be seen that the simple fuel substitution (without engine tuning) can give bad results especially for NO_x emissions, for the reasons explained above. Using the optimized maps, the emissions levels are even lower than the original CNG ones, especially for NO_x, while for CO and HC there are improvements caused by a better combustion quality (for HC) and a less carbon presence in the fuel (for CO). However, with the HCNG10 the vehicle presents the lowest emissions levels.

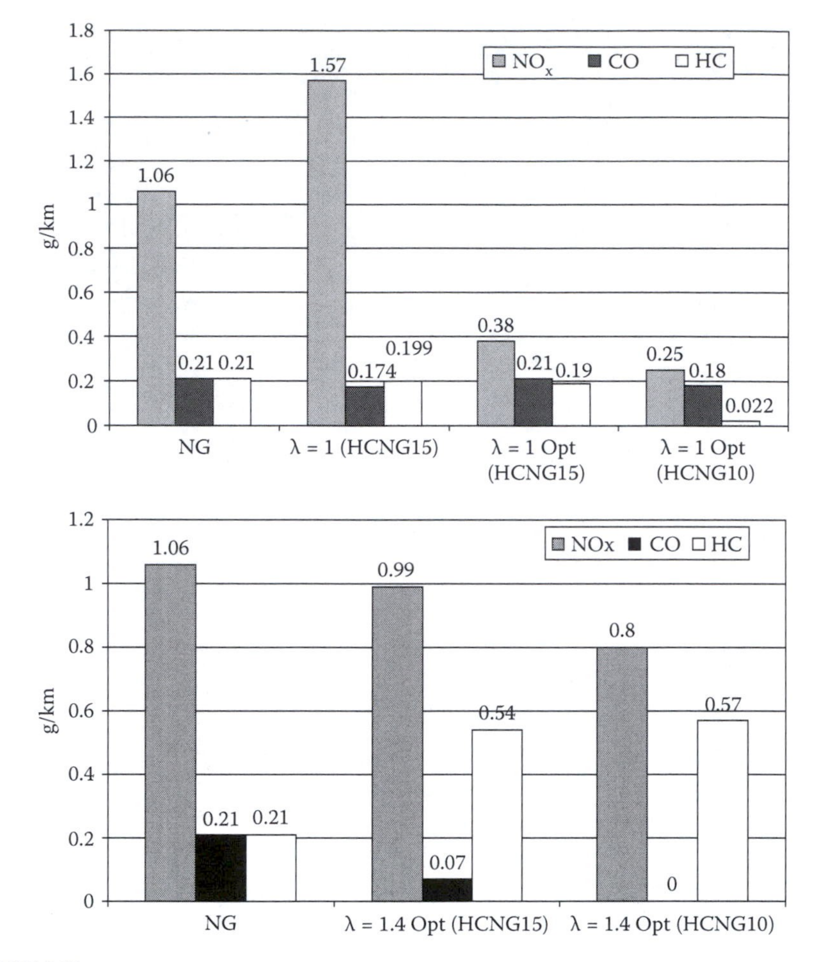

FIGURE 9.19
Emissions produced on the ECE15 driving cycle with stoichiometric mixtures (left) and lean (right) for HCNG10 and HCNG15. (From Ortenzi, F., Chiesa, M., and Conigli, F., *Experimental Tests of Blends of Hydrogen and Natural Gas in Light Duty Vehicles,* HYSYDays-Turin, 2nd World Congress of Young Scientists on Hydrogen Energy Systems, 2007.)

For lean burn mixtures the emissions present two different behaviors: the CO emissions are always lower for blends with 10% and 15% of H_2 by volume. Nevertheless, concerning the NO_x emissions, the values are lower than those using pure CNG, but not as good as stoichiometric values. Furthermore, HC emissions increase for the lower combustion quality due to lean mixture.

It is interesting to note that operating the vehicle with the O_2 sensor disabled, the fuel system had not been able to adjust the blend to an "objective" value: in all the unsteady conditions (especially in acceleration phases) any variation toward leaner mixtures had led to higher HC emissions and richer

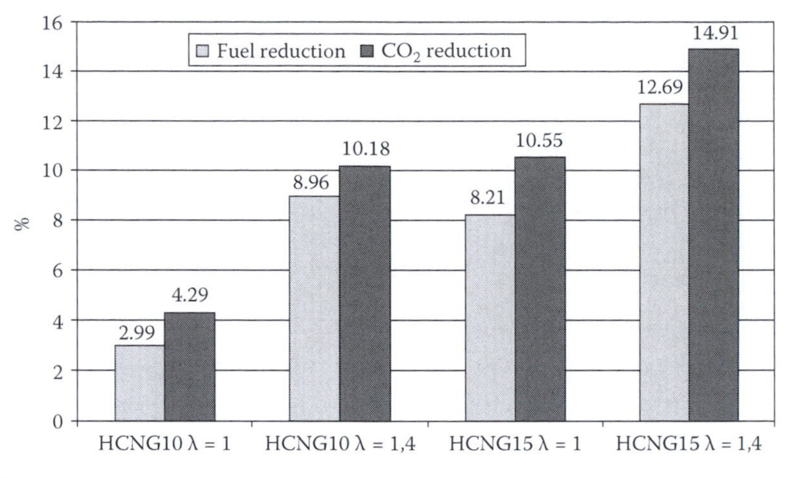

FIGURE 9.20
Fuel and CO_2 reduction for different blends and different lambda.

blends had increased NO_x emissions. In order to close "the loop" this fact has to be taken into consideration as a future development.

The advantages in terms of consumption are reported in Figure 9.20, leaner mixtures and moving toward higher percentages of hydrogen give better results. This is not the same for the emissions. The improvements of HCNG10 in stoichiometric conditions are about 3% for consumption and 4.3% for CO_2 emissions reduction, while for HCNG15 and $\lambda = 1.5$ are about 12.7 and 14.91%. The values for HCNG10 with $\lambda = 1.4$ and HCNG15 with $\lambda = 1$ are similar.

9.13 Hydrogen Storage

9.13.1 Liquid Hydrogen

Liquid hydrogen is a high energy content chemical and is widely used in space applications. Liquid hydrogen systems are much lighter and often more compact than hydride systems providing an equal range. For liquefaction processes, it requires an energy equivalent of approximately 30% of combustion energy of hydrogen that is liquefied. Moreover, engine application systems should have the facility to pump the cold liquid hydrogen to DI engines.

Liquid hydrogen storage is not significantly heavier than gasoline storage but it is bulkier. Hydrogen is stored in liquid form in the temperature of 20 K at 2 bar in double walled insulated cylinders. The liquid hydrogen may be delivered in liquid form or gaseous form based on our requirement.

The weight of the liquid hydrogen system is slightly more than that of conventional gasoline/diesel storage systems. The liquid hydrogen system has a volume about 10 times higher than that of gasoline systems for carrying an equal amount of energy. Hydrogen can be delivered to the engine in cold so that it increases the engine volumetric efficiency, which leads to increase in power output also. Cold hydrogen reduces acts like a thermodynamic sink and reduces the combustion chamber temperature. This leads to a reduction in NO_x emissions.

The maximum temperature excursions that a storage system would undergo from an ambient temperature of about 40°C to the storage temperature of about 13 K amounts to a difference of about 300°C. This places high thermal and hence mechanical loads on the pipe-work, flanges and vessels, and the cool-down has to be controlled to minimize the potential for leakage due to uneven contraction of the system. Liquefied hydrogen has the potential problem of causing hydrogen embrittlement of metallic components in the system. As the liquid is extremely cold, any spillage rapidly evaporates and cools the ground onto which it falls. This produces a rapid "puff" of expanding vapor that rises rapidly and disperses. The very high turbulence generated by the sudden expansion of vapor from the liquid mixes the hydrogen gas with the air, and the gas cloud typically expands by a factor of five times. Low clouds of dense, cold gas will spread along the ground, rather than immediately rise and dissipate (Astbury 2008).

9.13.2 Compressed Hydrogen

In general, hydrogen is produced in the form of gas and also its end use is in the form of a gaseous state. Hence, it is obvious that most of the industrial storage is in the form of compressed gas. Natural gas is stored at around 200 bar in strong composite pressure vessels, typically aluminum cylinders wrapped with fiberglass. However, hydrogen is eight times lighter than methane per mole, and thus requires a much higher storage pressure to attain a given volume. As the hydrogen obeys gas law ($PV = mRT$) at lower pressure, increase in pressure increases the storage density. However, there is a slight deviation in the law at higher pressures (Das 1996a).

Hydrogen can be stored as any gas in compressed form in high strength gas cylinders. One kilogram of hydrogen occupies 11 m^3 of volume at ambient pressure and temperature. Hence it requires a large volume of gas to store in its uncompressed form. Hydrogen is stored in compressed gas cylinders made of aluminum or other high strength materials. Hydrogen can be compressed up to 550 bar. Hydrogen cylinders (200 bar) weigh nearly three times more than that of comparable liquid hydrogen. However, the size of the system can be reduced by increasing the pressure of hydrogen in cylinders. Compressed hydrogen cylinders require about two times the volume of gasoline.

A standard pressure vessel of 0.05 m^3 at atmospheric pressure would hold approximately $(0.050/22.4) \times 2 = 0.00446$ kg of hydrogen, so at 1.4 MPa gauge (200 psi) a vessel would hold about 0.0625 kg of hydrogen; that is, there is an increase in storage capacity of about 50-fold compared with storing just the compressed gas (Guo, Shang, and Aguey-Zinsou 2008).

High strength steel/austenitic steel/aluminum cylinders wrapped with fiberglass are used as cylinder materials. For the high-pressure cylinders the cost of materials and safety features are critical issues. The hydrogen cylinders can be used for hydrogen vehicles as well as fuel cell vehicles. Hydrogen has a positive Joule–Thompson coefficient and hence a large reduction in pressure that occurs during a leak from a high-pressure system results in a rise in temperature, rather than a fall, due to the expansion.

Though compressed hydrogen was suitable for many applications, the use in automobile applications is limited because of high volume, weight, cost of compression, and safety aspects. Das (1996b) reported that for 164 atm. compressed hydrogen gas cylinder compression energy of 12.65 kJ/mol or 4100 kJ for a full cylinder is required. Hence, a lot of energy will be released under catastrophic rupture situations. This rapid release of hydrogen may cause backfire.

9.13.3 Solid Hydrogen

Storing hydrogen in metal hydrates has a number of advantages over compressed gas or liquid hydrogen. Solid state hydrogen fuel storage, either absorption in the interstices of metals and metallic alloys, or adsorption on high surface area materials such as activated carbons, has been receiving attention for on board hydrogen storage. It is a similar process to the adsorption of solvent vapors on activated charcoal, in that the adsorption is accompanied by a release of heat, and the desorption is driven by heating the metal to release the hydrogen as gas. The heat required for the working of metal hydrates (about 600 K) can be met by exhaust gases. As the heat of adsorption would need to be removed, the process of refueling requires the metal matrix to be connected to an external coolant supply during the filling process. Refueling hydride vehicles is relatively simple. It appears that hydride vehicles can be refueled in 10 minutes or less. Metal hydride storage is a safe, volume-efficient, storage method for onboard-vehicle applications. Metal hydrates storage is suitable for vehicle applications to avoid space constraints. Metal hydrate should possess the capacity of discharging the required level of hydrogen to cater to the fluctuating need of the vehicle. Hence, the proper choice of hydrate is important. The hydrate should be of high storage capacity with good absorption/desorption characteristics. The hydrate should be light weight and should have long life with the ability to withstand engine operation over a prolonged time.

Some of properties of metal hydrates are pressure inside metal hydrates, hydrogen capacity, change in volume, kinetics of reaction, rate of depreciation of hysterisis, and safety features. Guo, Shang, and Aguey-Zinsou (2008) advocates that for optimum H-storage, an ideal storage system should possess: a high H-storage capacity (>6 or 9 mass% hydrogen, or 5.5 kg H_2 in a tank that can be fitted under the car seat), a low thermodynamic stability (leading to a low desorption temperature <150°C, under a moderate pressure), high kinetics for hydrogen absorption/desorption (<1–2 mass%/minute, e.g., 5 minutes re-filling time), high stability against O_2 and moisture (hence a long cycle life >500 cycles), good thermal conductivity (for rapid conduction of sorption heat), and low cost <£2.5/kWh.

Hydrogen forms metal hydrides with some metals and alloys leading to solid-state storage under moderate temperature and pressure that gives them the important safety advantage over the gas and liquid storage methods. Metal hydrides have a higher hydrogen-storage density (6.5 H atoms/cm^3 for MgH_2) than hydrogen gas (0.99 H atoms/cm^3) or liquid hydrogen (4.2 H atoms/cm^3). Metal hydrates hydrogen storage systems include Mg-based metal hydrides, complex hydrides, alanates, and intermetallic compounds.

Magnesium and its alloys for onboard-hydrogen storage is more attractive due to their high hydrogen storage capacity by weight and low cost. Mg-based hydrides possess good-quality functional properties, such as heat-resistance, vibration absorbing, reversibility, and recyclability. Magnesium hydride has the highest energy density (9 MJ/kg Mg). The main disadvantages of magnesium hydrates is the high temperature of hydrogen discharge, slow desorption kinetics, and a high reactivity toward air and oxygen. The increased surface contact with catalyst during ball-milling leads to fast kinetics of hydrogen transformations. Complex hydrides are known as "one-pass" hydrogen-storage systems, which means that hydrogen evolves upon contact with water. Low weight complex hydrides include alanates [AlH4]–, amides [NH2]–, imides, and borohydrides [BH4]–. Hydrogen content of these complex hydrides can reach 18 mass% for LiBH4 and hydrogen release can occur at temperatures as low as 150°C for LiAlH4 (Sakintuna and Lamari-Darkrim 2007). Physic and energetic data concerning hydrogen is given in Table 9.7.

According to Energy Conversion Devices Inc. (Ovonic Hydrogen Solutions, 2004), the hydrogen is stored in a "... lightweight fiber-wrapped vessel with an internal volume of 0.05 m^3 that stores 3 kg of hydrogen as a metal hydride

TABLE 9.7

Physic and Energetic Data Concerning Hydrogen

	Gravimetric Capacity (mass)			Volumetric Capacity (vol.)
Fuel	**wt.% H_2**	**kWh/kg**	**MJ/kg**	**gH_2/L**
CGH2 (350b)	5.8	1.93	6.95	19
CGH2 (700b)	4.2	1.2	4.3	22
LH2	5	1.66	6	36
Chemical hydride	3.3	1.1	3.96	30

at an operating pressure of typically less than 200 psi … ." This is 50 times higher as compared to compressed air storage of the same capacity.

The metal hydrates are usually insulated with noncombustible materials so that in the event of tank crack, the major portion of hydrogen still remains stored in the hydrogen block, thus avoiding explosion. The hydrate need not be replaced frequently and the process of charging and discharging can be repeated an infinite number of times provided the metal hydrate does not get contaminated. In case of contamination of metal hydrates, the hydrogen can be reactivated by heating.

9.13.4 Onboard-Hydrogen Storage

An important alternative to hydrogen storage is onboard hydrogen production: using devices called *reformers* hydrogen can be extracted from a fossil hydrocarbon, typically natural gas. This obviously implies a larger onboard complexity and some gaseous polluting emissions by reformers, but has the advantage of not requiring an infrastructure for distributing hydrogen. Currently, two leading hydrogen supply systems onboard are envisaged: POX and methanol steam reformer (MSR; Ogden, Steinbugler, and Kreutz 1999). A detailed description of reforming systems could be found in Dicks (1996) and Ahmed and Krumpelt (2001).

However, it is worth remarking in which ways fuel processors affect fuel cell behavior and system performance. PEM FCs prefer to be fed with pure hydrogen, since fuel reformer by-products, namely CO, CO_2, and S, can poison the platinum catalyst obstructing reaction sites. Particularly, platinum attracts carbon monoxide that may be removed adding 2% oxygen to a fuel stream with 100 ppm of carbon monoxide. Unfortunately, this technique experiences different drawbacks since hydrogen stream from reformers contains 1000 ppm, while oxygen not reacting with CO will consume fuel. Obviously, reformers increase fuel cell system cost, size and weight, and introduces other issues related to their control and subsystems. Hydrogen streams leaving the fuel reformer have to be prefiltered before approaching fuel cells whereas hydrogen molar fraction effectively reacting at anode sites achieves only 0.3% against 75% assured fueling of pure hydrogen (Ogden, Steinbugler, and Kreutz 1999). Moreover, reformer efficiency lies in a range between 62 (POX) and 69% (MSR), thus further reducing PEMFC system performance. In real applications, an accurate control is required since fuel processors experience slow start-up and inadequate transient response.

To be used in a vehicle, hydrogen must be stored onboard or produced onboard. Main characteristics of the major Fuel cell typologies is given in Table 9.8. Hydrogen can be stored in four different ways:

- In compressed form, CGH2: in high-pressure tanks or gas cylinders made of ultra-light composite materials. They are used in prototype fuel cell automobile and buses.

TABLE 9.8

Main Characteristics of the Major Fuel Cell Typologies

Parameter	PEMFC	DMFC	AFC	PAFC	MCFC	SOFC
Electrolyte	Polymeric Membrane	Polymeric Membrane	Potassium Hydroxide	Liquid Phosphoric Acid	Liquid Molten Carbonate	Ceramic
Operating Temperature (°C)	60–100	30–100	65–220	150–220	600–700	600–1000
Efficiency (%)	35–50	20–40	45	40	45–50	45–60
Start-up	Fast (1 min)	Fast	Fast	Medium	Slow	Slow (20 min)
Charge carrier	H^+	H^+	OH^-	H^+	$CO_3^=$	$O^=$
Catalyst	Platinum	Platinum	Platinum	Platinum	Nickel	Perovskite

Source: Courtesy of Professor Cesare Pianese, University of Salerno, Italy, 2007.

- In liquid form, LH2: in super isolated tanks as cryogenic liquid (at –253°C, or 20K).
- In a bonded form, in a solid compound by: adsorbing (carbon structures), absorbing (simple crystalline hydrides), or chemically reacting (complex and chemical hydrides).
- in a fluid compound form by chemically reacting (e.g., methanol, ammonia, etc.).

Basic hydrogen storage criteria are:

- Weight Density (kg H_2/storage system weight)
- Volumetric Density (kg H_2/storage system volume)
- Reliable
- Compatible with available/planned infrastructure
- Safety: normal operation and emergency situations
- Service Life 15 years
- Cost (purchase and operative) High capital cost, low operating cost
- Refueling (fast filling time)

Each of these technologies has its own advantage, but only the second and third one seem to be viable for onboard storage, because of complexity and continuous energy requirements for the cryogenic solution.

Nickel metal hydride systems as hydrogen reservoirs assure high volumetric energy density (Figure 9.21) while excessively weighting vehicle power train (Burke et al., 2005). Volume reduction is appealing but metal hydride technology has to be further developed to reduce risk and costs. Similarly, carbon nanofibers-tubes reduce tank volume but the technical maturity is unsuited to practical applications.

Pressurized tank storage (Figure 9.22) has the advantage of rapid charge/discharge times and of relatively lower complexity; metal hydrides have the advantage of reduced safety constraints (their pressure is very low) and higher energy per unit of volume, but they are more complicated to manage because of the need of thermal management and have longer recharging times.

The design of the vehicle is simplified when pressurized vessels store hydrogen onboard. High-pressure tanks, similar to what was used in the Cute project, assure long operating life (20 years and more) and exhibit moderate costs and straightforward control, thus being a mature, reliable, and low risk technology. However, hydrogen production and distribution costs are high so that an accurate WTT analysis is required to evaluate economic and energetic impact of pressurized vessel technology.

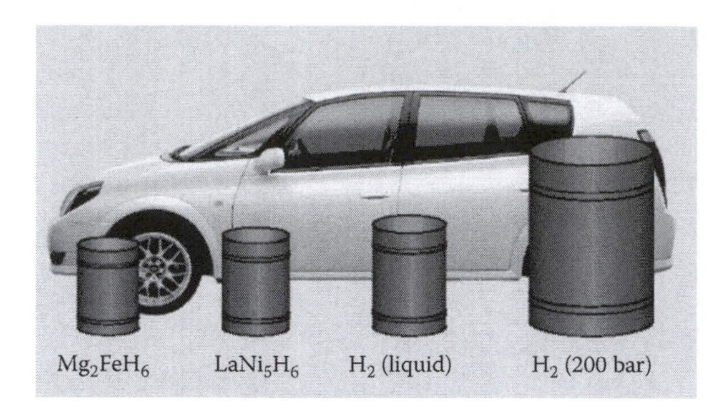

FIGURE 9.21
Hydrogen storage system comparison.

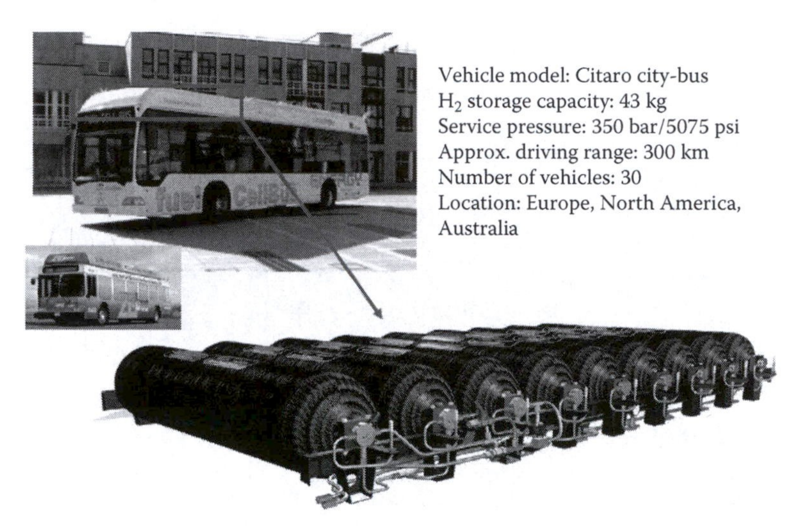

FIGURE 9.22
Hydrogen storage systems. (From Fraser, D., *Solutions for Hydrogen Storage and Distribution*, Dynetek Ind. Ltd, The PEI Wind-Hydrogen Symposium, June 22–24, 2003.)

9.14 Hydrogen Benefits

Karim (2003) narrated the positive aspects of hydrogen for its engine applications that include

1. Environmentally clean automobile fuel suitable for I.C. engines. Fuel leakage to atmosphere is not a pollutant.
2. It can be produced from an abundant raw material, water.

3. Hydrogen is an excellent additive in relatively small concentrations to fuel such as methane.
4. Hydrogen high burning rates make the hydrogen-fueled engine performance less sensitive to changes to the shape of the combustion chamber, level of turbulence, and the intake charge swirling effects.
5. Hydrogen ignition limits are much wider than gasoline so it burn easily and gives higher efficiency.
6. Requirement of less spark advance contributes to better efficiencies and improved power output.
7. Hydrogen burns nearly 10 times faster than gasoline mixtures.
8. Moderately high compression ratio operation is possible with lean mixtures of hydrogen in air that permits higher efficiencies and increased power output.
9. The exhaust heat can be used to extract hydrogen from the hydride.
10. The thermodynamic and heat transfer characteristics of hydrogen tend to produce high compression temperatures that contribute to improvements in engine efficiency and lean mixture operation.

9.15 Hydrogen Barriers and Challenges

Switching from diesel to hydrogen provides a number of environmental benefits, avoiding the local pollution (NO_x, CO_2, and particles) in any case. If the electricity is produced by renewable energy sources it can be justified to name the electrical trains as well as the hydrogen trains "ZEV, zero emission vehicles." With hydrogen and fuel cells, the energy efficiency is lower due to the conversion losses in electrolysis (0.8), hydrogen storage (0.9), fuel cell (0.45), and electric motors (0.9). Hydrogen fuel produced from renewable energy will still have comparable overall energy efficiency to diesel traction—roughly 30% before taking into account the considerable energy efficiency gains that will be achieved when adding regenerative breaking.

The overall energy gap is likely to be filled out by a number of different energy sources, primarily renewable energy. Due to the unpredictable nature of wind power, electricity prices are likely to fluctuate to a higher extent than what we experience today. This means that the electrolyser in more periods than today can purchase relatively cheap electricity.

Apart from expected price reductions on electrolyser hardware, another factor that speaks in favor of increased competitiveness for hydrogen produced by electrolyses in the future is that expected rising electricity prices are not reflected to a full extent in the price of hydrogen since depreciation of the electrolyser is included in prices.

If a country is to increase the dependency of wind power in the future further than 15–20%, some sort of electricity storage option is required in order to balance the electricity grid. A part of the solution to this problem is to establish electrolysis plants and H_2 storage facilities and use the hydrogen in the transport sector. Even if the general development of hydrogen and fuel cell technology will bring lower prices, longer cell lifetimes, better and cheaper solutions for hydrogen storage, production and other benefits, some areas of development are specific for railway use, and the technology needs to be tested, adapted to and demonstrated in railway vehicles to ensure feasible solutions.

Hydrogen has the potential for use as a fuel source for engines as well as fuel cells. However, like other fuels it has its own problems.

- Poor engine volumetric efficiency
- Higher NO_x emissions because of its higher flame temperature
- Higher fuel cost
- Infrastructure for distribution network
- Storage of hydrogen in vehicles
- Low energy density
- Flame trap and flash back arrestors are necessary for hydrogen systems
- Hydrogen requires 1/50th of energy gasoline-air mixtures to ignite

Use of hydrogen as a fuel in the transport sector would require significant changes in infrastructure. Distribution of hydrogen and local fueling of cars could not be done in the same way gasoline is handled today. Therefore the infrastructural problems must be given careful consideration, both concerning economy and safety, in relation to possible utilization of hydrogen as an energy carrier.

Hydrogen will emerge as the dominant fuel when climate change or greenhouse effects are strong and conventional fuel prices become very expensive. Many of the centralized systems focus on hydrogen use in road transport, and envisage local hydrogen pipeline grids linking early demonstration projects and fleet vehicle refueling depots, creating hydrogen corridors in areas of high demand. But the decentralized hydrogen production overcomes the infrastructural barriers associated with the hydrogen economy. This allows the distributed generation, home refueling, and authorizing the public to have control over the energy. For decentralized systems, the major technological challenge is the expense of hydrogen from small-scale natural gas reformers and electrolysers, while centralized systems rely on the viability of a large-scale hydrogen distribution infrastructure, and prospects for centralized systems are greatly enhanced by cost-effective coal gasification or nuclear–thermal water splitting (McDowall 2006).

However, the commercialization of hydrogen associated with fuel cost, infrastructure, and viability of technology depends on: high costs with the onboard hydrogen storage or hydrogen production and fuel cells vehicles as there is limited infrastructure for hydrogen distribution, it will be difficult to establish markets for hydrogen vehicles or fuel cell vehicles and vice-versa; technological advancement on development of fuel cell vehicles and more mileage; and public awareness about safety and acceptance. These problems will need to be overcome by government policies and the necessary actions to protect the environment as well.

References

Ahmed, S., and M. Krumpelt. 2001. Hydrogen from hydrocarbon fuels for fuel cells. *International Journal of Hydrogen Energy* 26:291–301.

Antunes, J. M. G., R. Mikalsen, and A. P. Roskilly. 2009. An experimental study of a direct injection compression ignition hydrogen engine. *International Journal of Hydrogen Energy* 34 (15): 6516–22.

Astbury, G. R. 2008. A review of the properties and hazards of some alternative fuels. *Process Safety and Environment Protection* 86:397–414.

BP Statistical Review of World Energy June 2004. Available at http://www.bp.com/liveassets/bp_internet/globalbp/STAGING/global_assets/downloads/S/statistical_review_of_world_energy_full_report_2004.pdf

Burke, A. and M. Gardiner. 2005. *Hydrogen Storage Options: Technologies and Comparisons for Light-duty Vehicle Applications,* UCD-ITS-RR-05-01, Hydrogen Pathways Program Institute of Transportation Studies, January 2005, University of California-Davis.

CONCAWE. *Well-to-wheels analysis of future automotive fuels and power trains in the European context.* A joint study by EUCAR/JRC/CONCAWE: Summary of Results. Heinz Hass, FORD, www.ies.jrc.ec.europa.eu/uploads/media/ WTW_ Report _220104.pdf

Carcassi, M. N., and N. Grasso. 2004. Safety, standards and regulations invited lecture. *Proceedings of hydrogen-power theoretical and engineering solutions international symposium,* September 7–10, 2003 Porto Conte, Italy, eds. M. Marini and G. Spazzafumo. Servizi Grafici Editoriali Padova, ISBN 88-86281-90-0; 569–79.

Das, L. M. 1996a. Hydrogen-oxygen reaction mechanism and its implication to hydrogen engine combustion. *International Journal of Hydrogen Energy* 21 (8): 703–15.

Das, L. M. 1996b. On-board hydrogen storage systems for automotive application. *International Journal of Hydrogen Energy* 21 (9): 789–800.

Das, L. M. 2002. Hydrogen engine: Research and development (R&D) programmes in Indian Institute of Technology (IIT), Delhi. *International Association for Hydrogen Energy* 27:953–65.

Das, L. M., and M. Polly. 2005. Experimental evaluation of hydrogen added natural gas (HANG) operated SI engine. *Symposium on international automotive technology. SAE* 2005-26-29.

Dicks, A. L. 1996. Hydrogen generation from natural gas for the fuel cell systems of tomorrow. *Journal of Power Sources* 61:113–24.

Eichlseder, H., T. Wallner, R. Freymann, and J. Ringler. 2003. The potential of hydrogen internal combustion engines in a future mobility scenario. *SAE* 2003-01-2267.

Fraser, D. 2003. Solutions for hydrogen storage and distribution, Dynetek Ind. Ltd. *The PEI Wind-Hydrogen Symposium,* June 22–24, Charlottetown, Prince Edward Island.

Guo, Z. X., C. Shang, and K. F. Aguey-Zinsou. 2008. Materials challenges for hydrogen storage. *Journal of the European Ceramic Society* 28:1467–73.

Holladay, J. D., J. Hu, D. L. King, and Y. Wang. 2009. An overview of hydrogen production technologies. *Catalysis Today* 139:244–60.

Kahraman, E., S. C. Ozcanli, and B. Ozerdem. 2007. An experimental study on performance and emission characteristics of a hydrogen fuelled spark ignition engine. *International Journal of Hydrogen Energy* 32:2066–72.

Karim, G. A. 2003. Hydrogen as a spark ignition engine fuel. *International Journal of Hydrogen Energy* 28:569–77.

Kumar, M. S., A. Ramesh, and B. Nagalingam. 2003. Use of hydrogen to enhance the performance of a vegetable oil fuelled compression ignition engine. *International Journal of Hydrogen Energy* 28:1143–54.

Lkegami, M., K. Miwa, and M. Shioji. 1982. A study of hydrogen fuelled compression ignition engines. *International Journal of Hydrogen Energy* 7:341–53.

Lynch, F. E. 1983. Parallel induction: A simple fuel control method for hydrogen engines. *International Journal of Hydrogen Energy* 8:721–30.

McDowall, W., and M. Eames. 2006. Forecasts, scenarios, visions, backcasts and roadmaps to the hydrogen economy: A review of the hydrogen futures literature. *Energy Policy* 34:1236–50.

Norbeck, J. M., J. Burden, B. Tabbara, J. W. Heffel, and T. D. Durbin. 1996. *Hydrogen fuel for surface transportation.* Warrendale, PA: SAE International.

Ogden, J. M., M. M. Steinbugler, and T. G. Kreutz. 1999. A comaprsion of hydrogen, methanol and gasoline as fuels for fuel cell vehicles: Implications for vehicle design and infrastructure development. *Journal of Power Sources* 79:143–68.

Ortenzi, F., M. Chiesa, and F. Conigli. 2007. *Experimental tests of blends of hydrogen and natural gas in light duty vehicles.* HYSYDays-Turin 2007, 2nd World Congress of Young Scientist on Hydrogen Energy Systems.

Ovonic Hydrogen Solutions. 2004. Hydrogen hybrid vehicle powered with ovonic (solid state hydrogen storage). Available from: http://ovonic-hydrogen.com/news/20040309.htm. Accessed on November 26, 2007.

Pant, K. K., and R. B. Gupta. 2009. Fundamental and use of hydrogen as a fuel. *Hydrogen fuel—Production, transportation and storage,* ed. R. B. Gupta. Oxford: Taylor & Francis 3–32.

RAPSDRA-4PCRD Program. 1995. *Reliability of advanced power device for railway applications* Available at http://www.aramis.admin.ch.

Rigas, F., and S. Sklavounos. 2009. Hydrogen safety. *Hydrogen Fuel: Production, Storage, and Transport.* Eds. R. B. Gupta. Oxford: Taylor & Francis.

Sakintuna, B., and F. Lamari-Darkrim, and M. Hirscher. 2007. Metal hydride materials for solid hydrogen storage: A review. *International Journal of Hydrogen Energy* 32:1121–40.

Saxena, R. C., D. Seal, S. Kumar, and H. B. Goyal. 2008. Thermo-chemical routes for hydrogen rich gas from biomass: A review. *Renewable and Sustainable Energy Reviews* 12:1909–27.

Verhelst, S., and T. Wallner. 2009. Hydrogen-fueled internal combustion engines. *Progress in Energy and Combustion Science* 35:490–527.

Welch, A. B., and J. S. Wallace. 1990. Performance characteristics of a hydrogen-fueled diesel engine with ignition assist. *SAE* 902070.

White, C. M., R. R. Steeper, and A. E. Lutz. 2006. The hydrogen-fueled internal combustion engine: A technical review. *International Journal of Hydrogen Energy* 31:1292–1305.

Zittel, W. 2005. *Alternative world energy outlook 2005: A possible path towards a sustainable future.* Ottobrunn, Germany: L-B-Systemtechnik GmbH.

10

Electric Vehicles

Nallusamy Nallusamy, Paramasivam Sakthivel, Abhijeet Chausalkar, and Arumugam Sakunthalai Ramadhas

CONTENTS

10.1 Introduction

In recent years, there has been increasing concern about global environmental issues and the problem in the balance of supply and demand for fossil fuels. Electricity is one of the clean energies at user point. Before electricity generation began over 125 years ago, houses were lit with kerosene lamps, food was cooled in iceboxes, and rooms were warmed by wood-burning or coal-burning stoves. Currently electricity is one of the prime energy sources in our day-to-day activities; that is, cooking, heating, light, computers, and other industrial and residential purposes. Electricity is an important tool for the industrial growth of the world. Various forms of other energy sources; that is, coal, petroleum, nuclear, and biomass can be converted into electric energy in its clean energy form. Electricity is produced in thermal power plants or nuclear power plants in large amounts. Stored electricity can be used for many applications including space applications, residential purposes, and automobiles. The storage of electricity is one of the prime movers for development of electric vehicles (EVs). The EVs charge the battery by grid power supply hence there is no pollution on the road.

Currently EVs have been gaining acceptability in the transportation sector at a global level. Also, the need to introduce EVs into the transportation sector as replacements for fossil fuel driven vehicles has been the subject of world-wide debate during the last few years. Originally, interest in EVs arose mainly from the concern over atmospheric pollution attributable to exhaust emissions from petroleum-powered cars. Compared with oil-run vehicles, battery-powered work vehicles offer low maintenance costs as well as low-running costs due to the lower cost of electricity and the use of off-peak power for recharging. The heart of an EV is its battery and the key to the success of the EV industry is a better battery with less weight, more compactness, ability to store more energy, longer durability, recharged more rapidly, and costs less than existing ones. The performance of the vehicle depends mainly on the performance, efficiency, and reliable operation of the battery.

However, the economic feasibility of EVs has yet to be proved. The performance of present EVs is limited mainly by the low energy and limited power densities of the lead–acid battery. The weight penalty thus imposed, limits the range of a four-seat passenger car to about 150 km and the cruising speed to about 55 km/h.

10.2 Principle of Electric Vehicles

The EVs consist of battery, motor, and controllers in addition to the normal components of the automobile. The electricity power stored in the battery drive the motor that runs the vehicle. An EV is operated by an electric motor, which draws electricity from a battery bank. Different types of batteries for EVs and other applications are being developed for better performance. Electric vehicles should match with comparable petrol or diesel driven vehicles in terms of performance, reliability, durability, and cost.

In order to properly assess the use of the electrochemical energy conversion and storage systems (storage batteries, super capacitors, and fuel cells) to power EVs, it is mandatory to quantitatively estimate the power and energy required for propelling a modern car (Shukla et al. 2001).

$$P_{\text{traction}} = P_{\text{grade}} + P_{\text{accel}} + P_{\text{tires}} + P_{\text{aero}} + P_{\text{inertial}},$$

where, P_{grade} is the power required for the gradient;
P_{accel} is the power required for acceleration;
P_{tires} is the rolling resistance power consumed by the tires;
P_{aero} is the power consumed by the aerodynamic drag; and
P_{inertial} include inertial losses of rotating components.

Loss in electrical devices occurs when energy is consumed by the battery management system and the cooling device that controls the battery's temperature. The battery management system monitors the state of the batteries, delivers the proper charging current, and maintains safety.

During battery pack charging, part of the energy is converted to heat and lost due to the battery's internal resistance when a lead–acid battery is charged over the rated capacity by a certain percentage to ensure a full charge. While driving, part of the energy is converted to heat and lost due to the battery's internal resistance (Ogura 1997).

10.3 Construction of Electric Vehicles

The EV consists of all the parts similar to conventional automobiles, however the internal combustion engine power train is replaced by motor and battery. Modern EVs are built with the original body and frame to meet the structural requirements and to provide the wide flexibility of electric propulsion. Figure 10.1 shows the major parts of a basic EV power train. A typical configuration of modern EVs include energy source, electric motor propulsion, and control system (Ehsani et al. 2005).

10.3.1 Motor

The electrical energy available in the battery is in the form of direct current. Direct current (DC) motors are used to convert electrical energy into mechanical energy at the driving wheels. It is based on the principle that

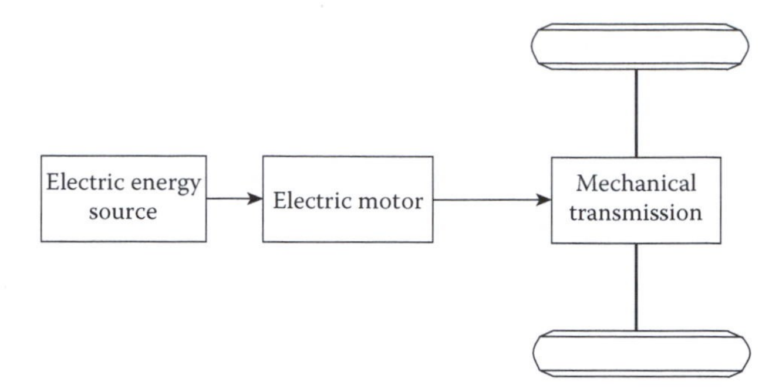

FIGURE 10.1
Block diagram of a basic electric vehicle drivetrain.

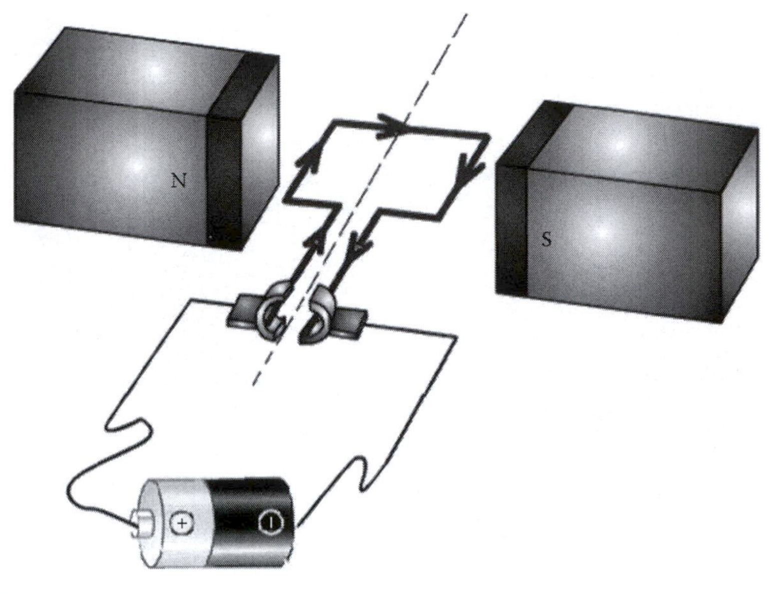

FIGURE 10.2
Schematic of working principle of DC motor. (From http://www.ncert.nic.in/html/learning_basket/electricity/animations/dc-motor.html.)

when a coil carrying current is held in a magnetic field, it experiences a torque that rotates the coil. Figure 10.2 shows the schematic of DC motor working principle.

The motor works on the principle of Flemming's left-hand rule. One side of the coil experiences a force directed inward and perpendicular to the plane of the coil and the other side experiences it in the outward direction and perpendicular to the plane of the coil. These two forces being equal, unlike and parallel form a couple, which rotates the armature coil in a counter-clockwise direction. After the coil rotates through 180° the direction of the current is reversed in both sides of coil and so the forces are experienced by each. Thus the armature coil continuously rotates in the same direction; that is, counterclockwise.

Motor efficiency is the ratio of back e.m.f (electromagnetic force) to applied e.m.f. The back e.m.f is due to changes in magnetic flux linked with the coil and it opposes the battery current in the circuit.

$$\text{Motor efficiency} = \frac{\text{back e.m.f.}}{\text{applied e.m.f.}}.$$

The constructional detail of a DC motor is illustrated in Figure 10.3

- The rotor is surrounded by stator magnets. The rotor has a number of coils wound on it perpendicular to the axis of the rotor. The rotor coils are free to rotate in between magnets. The rotor consists of windings (generally on a core), the windings being electrically connected to the commutator.
- The stator is the stationary part of the motor that includes the motor casing, as well as two or more permanent magnet pole pieces. The movement or rotation of the rotor is caused by the electromagnetic interaction between the rotor and the magnets.
- Rotor coils are connected to the fixed commutator on the rotor shaft. These are two halves of the same ring. The ends of the armature coil are connected to these halves, which also rotate with the armature.

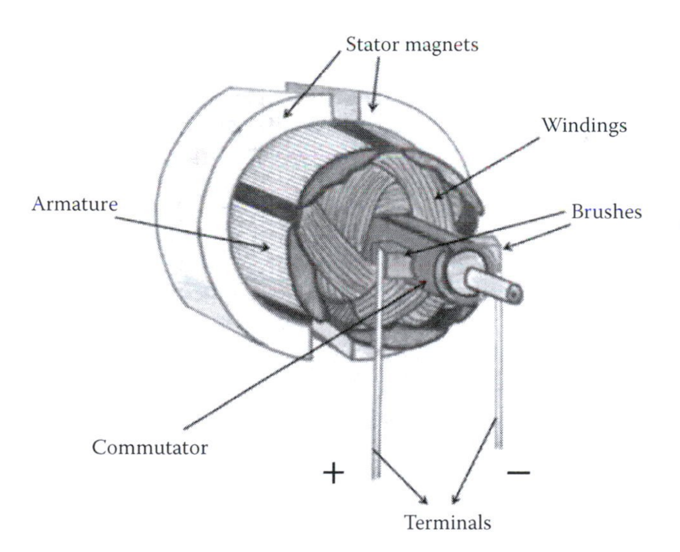

FIGURE 10.3
Constructional details of DC motor. (From http://www.cvel.clemson.edu/auto/actuators/motors-dc.html.)

- Brushes are two flexible metal plates or carbon rods, which are fixed that they constantly touch the revolving rings. The fixed brushes are connected alternatively to opposite ends of the rotor for every 180° of shaft rotation. The direction of current flow in the rotor is reversed twice for every turn of the rotor to maintain the rotation of the rotor.
- The battery is connected to the brushes, which convey current to the rings where it is carried to the armature.

The advantage of using an electric motor as a propeller are: quieter and a more efficient propulsor, easier to install the regenerative braking system, more efficient transmission through the possibility of direct connection to the wheels, fewer moving parts, and lower operating/maintenance costs.

In general, DC motors are classified into four types based on the windings configuration:

1. DC series motor
2. DC shunt motor
3. DC compound motor
4. Separately excited DC motor

10.3.1.1 DC Series Motor

In this type of motor, the field coil is in series with the rotor/armature coil (Figure 10.4). The amount of current flow is controlled by the resister. The DC motor has a high torque at zero or nearer to zero speeds and the torque falls as the speed increases. The speed of the motor is controlled by adding

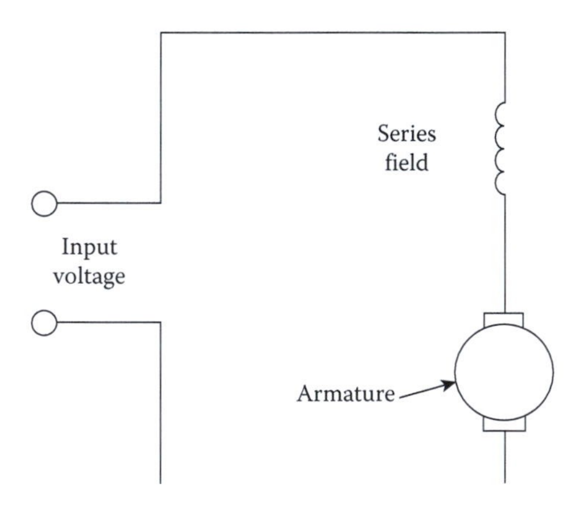

FIGURE 10.4
DC series motor.

additional resistance in series or parallel with the field. The series motor is suitable for EVs and it gives excellent acceleration.

10.3.1.2 DC Shunt Motor

In the shunt motor, the field coil is in parallel with the rotor coils (Figure 10.5). The current flows through the motor and is controlled by field resistance and armature resistance. At higher speeds, the DC shunt motor takes the full load. With increasing speed, the load decreases at small percentages.

10.3.1.3 DC Compound Motors

By combining series and shunt coils, it is possible to get the wide range of speed–load conditions to its best operating condition (Figure 10.6). In this condition, the majority of its field is provided by the shunt coil and the series coil provides any supplementary requirements. The arrangement can be tailor-made according to the requirement of the vehicle applications.

Figure 10.7 illustrates the speed and torque characteristics of various types of DC motors. The series wound motor has one distinct advantage of producing high torque even at very low speeds, which is a primary requirement of a motor to be used in EVs. Starting torque is of 300% to as high as 800% of full load torque. Moreover, the torque would be infinite at zero speeds if there was no limitation on the current available and the magnetic circuit had zero reluctance. In actual case, the current is limited by the series resistor and armature coil resistance to the maximum that the field windings, rotor windings, and brushes can withstand without overheating. The series motor characteristic is particularly suitable for an EV as it gives an excellent acceleration from the rest combined with a controlled slowing down on hills and a constant high speed on the flat.

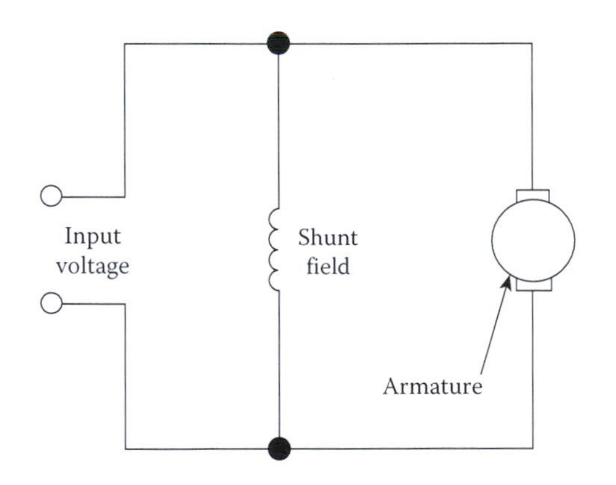

FIGURE 10.5
DC shunt motor.

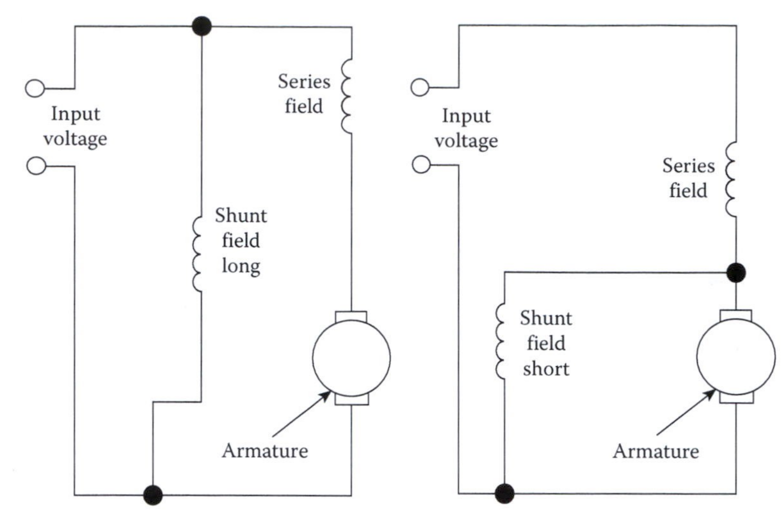

FIGURE 10.6
DC compound motor.

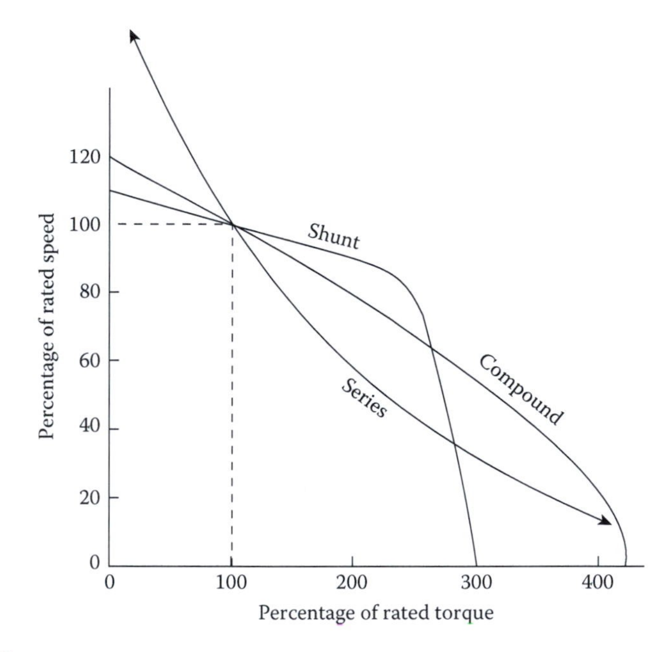

FIGURE 10.7
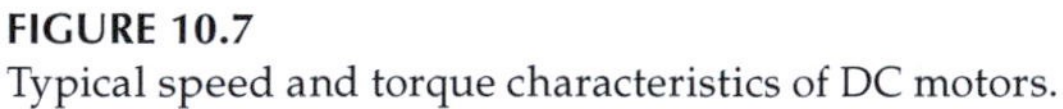
Typical speed and torque characteristics of DC motors.

In case of a shunt wound DC motor, the torque almost remains constant as the speed increases until the maximum point is reached. In general, starting torque is of 125–200% full load torque. Efficient thyristor or transistor controls capable of controlling both field and armature current by turning it on or off at high speed makes the efficient use of shunt motors for EVs. The distinct advantage of shunt motors is the ability to reverse the motor by only reversing the relative low-current field connections, instead of reversing the full armature current like the series wound motor. As increased use of electronic controls improves the efficiency and controllability of shunt wound motors, their use has spread although series motors still have a place in some low-cost vehicles.

Combining the series and shunt field coils in a motor having a compound wound field provides a possibility to obtain a wide range of characteristics between the extremes of the series and shunt wound motors. Two arrangements can be made; keeping the shunt long or short, with the series field supplementing the shunt field. This cumulative compounding can make a way to use compound wound motors to act as a series like motor or a shunt like motor, getting both advantages. The characteristics can therefore to some extent be tailored to the requirement of a particular vehicle design (Wakefield 1998).

10.3.1.4 Separately Excited DC Motor

Use of power electronics with the capability of controlling higher current and relatively higher voltages in recent years makes the DC motor's armature current and field current to be controlled independently by separate excitation. This makes the possibility of achieving any required variable combination of series, shunt, and compound characteristics. But the major drawback lies in sending the feedback of rotor speed to the electronic power control system, which necessitates the use of commutator and brush to carry the armature current. Hence the motor speed is limited and causes wear and consequent unreliability.

DC motor drives have been widely used in applications requiring adjustable speed, good speed regulation, frequent starting, braking, and reversing. Various DC motor drives have been widely applied to different electric traction applications due to the simplicity, low cost of control systems, and the maturity of the technology. This will remain in service at least in low cost EV markets.

10.3.1.5 AC Motor Drives

Commutatorless motor drives offer a number of advantages over conventional DC commutator motor drives for the electric propulsion of hybrid EVs. AC induction motor drives have additional advantages such as being lightweight in nature, small volume, low cost, and high efficiency. Generally the AC motors fall under three categories: induction motors, synchronous motors, and switched or variable reluctance motors.

10.3.2 Battery Storage Systems

Batteries are energy storage systems that are used to store electrical energy. Electrochemical batteries, more commonly known as batteries that convert electrical energy into chemical energy during charging, and converts chemical energy into electrical energy during discharging. The battery is as important as the combustion engine of a conventional vehicle and considered as the essential power plant of EVs.

Batteries can be classified into two types: primary batteries and secondary batteries. Primary batteries irreversibly transform chemical energy into electrical energy. When the initial supply of reactants is exhausted, energy cannot be readily restored to the battery by electrical means. The primary batteries have higher energy densities than secondary batteries. The primary batteries supply energy immediately; however these batteries have to be discarded once the energy is completely used. Typical primary batteries (disposable batteries) are alkaline batteries and zinc–carbon batteries.

Secondary batteries are known as rechargeable batteries. The chemical reactions can be reversed by supplying electrical energy to the cell and it can be restored to its initial position. Battery chargers are used to recharge the battery by supplying electricity. Rechargeable batteries are used in EVs.

10.3.2.1 Requirement of Electric Vehicle Batteries

The important requirement of EV traction batteries is to have cycle-life as high as possible. During usage, it undergoes charging and discharging many times and the performance of the battery comes down after a particular period of life. Generally, the degenerative stages of batteries during service are interactive and accumulative. So, when the performance starts to decline, it soon accelerates and the battery becomes unusable. Despite these problems, modern-day batteries with advanced technologies are able to fulfill requirements for more than 1000 cycles.

The batteries should have

1. A stable voltage output over a good depth of discharge
2. High energy capacity for the given battery weight and size
3. High peak power output per unit mass and volume
4. High energy efficiency
5. Able to function with wide ranges of operating temperatures
6. Good charge retention on open-circuit stand

7. Ability to accept fast recharge
8. Ability to withstand overcharge and over discharge
9. Reliable in operation
10. Maintenance free
11. Rugged and resistant to abuse
12. Safe both in use and accident conditions
13. Made of readily available and inexpensive materials with environmental friendliness
14. Efficient reclamation of materials at the end of service life

10.3.2.2 Electric Vehicle Batteries

A battery is composed of several cells stacked together in a single container. A cell is an independent and complete unit that possesses all the electrochemical properties. Depending on the requirement of battery voltage, the number of cells will be decided and connected in series to get the desired voltage.

Every cell consists of three important elements: positive electrode, negative electrode, and electrolyte as shown in Figure 10.8. The number of positive plates and negative plates to be connected with respective electrodes will be decided based on the requirement of battery capacity. The positive plate and negative plates will be separated by insulators. One half-cell includes

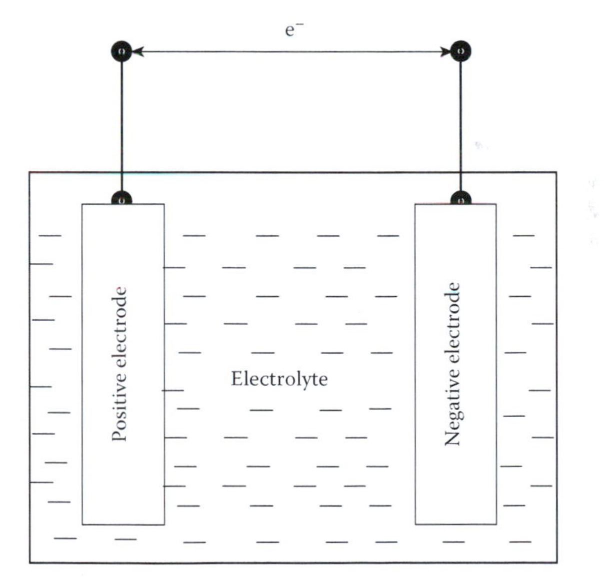

FIGURE 10.8
Typical electrochemical battery cell.

the electrolyte and the electrode to which negatively charged ions migrate; that is, negative electrode or anode. The other half-cell includes the electrolyte and the electrode to which positively charged ions migrate; that is, the positive electrode or cathode. In the battery, addition of electrons occurs at the cathode, while removal of electrons occurs at the anode. The electrodes are not physically touching each other but are electrically connected by the electrolyte, which can be either solid or liquid.

Usually, the batteries are specified with their capacities in terms of Ampere-Hours (Ah), which is defined as the number of Ah gained when discharging the battery from a fully charged state until the terminal voltage drops to its cut-off voltage as shown in Figure 10.9.

The various types of batteries for automotive applications are classified as follows:

1. Lead–acid batteries
2. Nickel-based batteries
    a. Nickel–iron battery
    b. Nickel–zinc battery
    c. Nickel–cadmium battery
    d. Nickel-metal hydride battery
3. Lithium-based batteries
    a. Lithium-solid polymer battery
    b. Lithium-ion battery
4. Other types
    a. Sodium–sulfur battery
    b. Sodium–nickel chloride battery

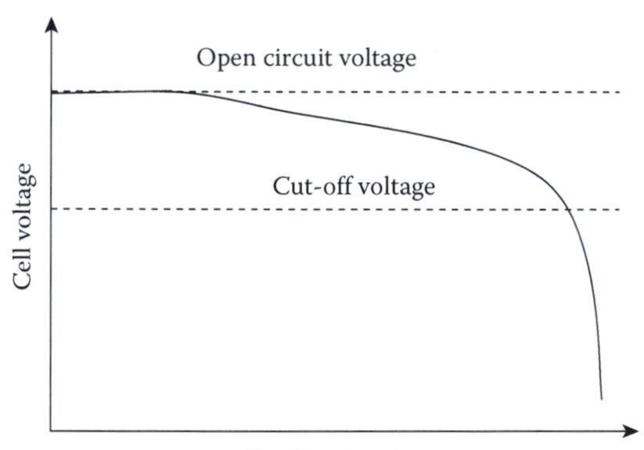

FIGURE 10.9
Characteristics of a typical battery.

c. Aluminum–air battery
d. Zinc–air battery

10.3.2.3 Lead–Acid Batteries

The lead–acid battery has been a successful commercial product for over a century and is still widely used as electrical energy storage devices in the automotive field and other applications. During discharge, the cathode is positive and the anode negative, and the reverse is the case during charging.

Figure 10.10 shows the major components of a lead–acid battery. It consists of positive and negative terminals connected with respective plates, electrolyte, cell connectors, cell dividers, vent caps, and a container.

In the charged state, each cell contains electrodes of elemental lead (Pb) and lead dioxide (PbO_2) in an electrolyte of approximately 33.5% v/v sulfuric acid (H_2SO_4). In the discharged state, both electrodes turn into lead sulfate ($PbSO_4$) and the electrolyte loses its dissolved sulfuric acid and becomes primarily water. The reaction is shown in Figure 10.11. Due to the freezing-point depression of water, as the battery discharges and the concentration of sulfuric acid decreases, the electrolyte is more likely to freeze during winter weather.

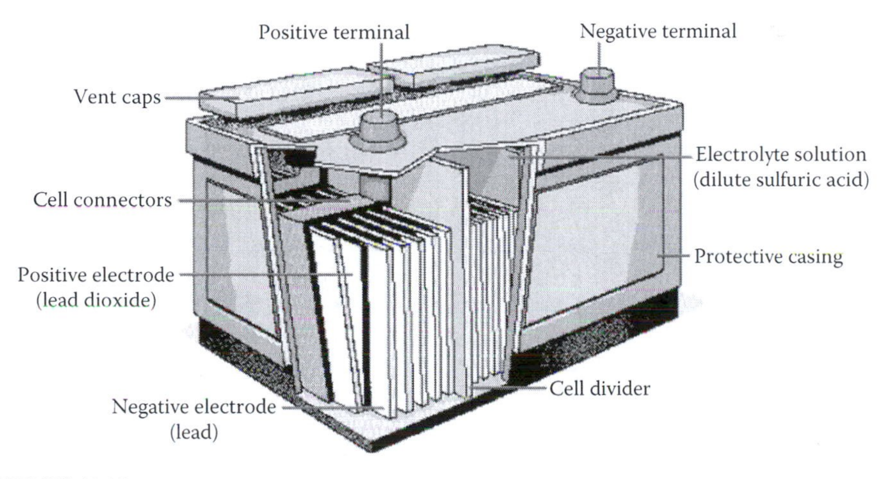

FIGURE 10.10
Schematic diagram of a lead–acid battery. (From www.reuk.co.uk/Lead-Acid-Batteries.htm.)

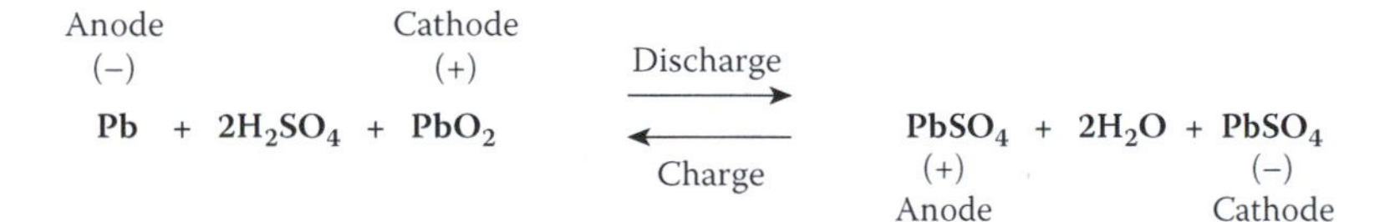

FIGURE 10.11
Chemical reactions in a lead–acid battery.

The electric passenger vehicles being tested today operate mostly on the lead–acid system and will remain dominant in the near-term. The major advantages of a lead–acid battery include low cost, mature technology, and relative high power capability. The performance of present EVs is limited mainly by the low-energy density of the lead–acid battery. The other drawbacks are slow recharging time, the need for careful maintenance, and poor performance under cold climates where the temperature gets less than 10°C. Modern technologies enable the use of lead–acid batteries without these major drawbacks, but of course, with an increase in cost.

Lead–acid batteries constructed with tubular positive electrodes, give a significant improvement in life. A lifetime can be obtained as much as 5 years under well controlled charge and discharge conditions. Gassing during charging produces hydrogen that requires venting to the atmosphere and in most vehicle installations positive extraction is used to avoid the buildup of a potentially explosive gas mixture.

Advanced lead–acid battery technologies offer improved retention of positive plate material during high discharge operations. This includes an increase in maximum energy density from 35 to 45 Wh/kg and a substantial improvement in maximum power capacity from 150 to 250 W/kg.

In sealed bipolar technology, the electrical resistance of the lead grids and the connectors between them is minimized by the use of a conducting plastic in the cell wall that permits low loss transmission of current while maintaining the seal between the cells.

The major breakthrough came with valve regulated lead acid (VRLA) batteries that allow fast charging typically 50% in 5 minutes and 80% in 15 minutes. This is done using computers to control the charging rate while monitoring the battery temperature and internal resistance. These batteries have pressure relief valves, which will activate when the battery is recharged with high voltage. The charging characteristics of a typical VRLA battery charged to an interactive pulsed current/constant voltage algorithm are one second bursts of high current with a ten millisecond pause between each pulse to measure internal resistance.

10.3.2.4 Nickel-Based Batteries

Nickel is a lighter metal than lead and has very good electrochemical properties desirable for battery applications. The different types of nickel-based battery technologies include nickel–iron, nickel–zinc, nickel–cadmium, and nickel-metal hydride.

10.3.2.4.1 Nickel–Iron

This battery uses nickel as the positive electrode and iron as the negative. Usually potassium hydroxide is used as an electrolyte. The self discharge of these batteries are high and eventually this problem is reduced by adding sulfur to the electrode, or lithium sulfide ions to the electrolyte. With recent advancements, the battery can perform satisfactorily down to –20°C. The power density of these batteries is around 100 W/kg, which makes it adequate for vehicle acceleration. The battery has a long lifetime with up to 2000 deep discharge cycles.

10.3.2.4.2 Nickel–Zinc

These batteries have an energy density of 70 Wh/kg and a power density of 150 W/kg. The fundamental problem lies with the dendrite growth, which is common to all zinc-based batteries as it limits the maximum number of deep discharge cycles to 300. Though a number of attempts have been made in the past to increase the lifetime, the research and development on zinc-based batteries slowed down significantly in recent years.

10.3.2.4.3 Nickel–Cadmium

A nickel–cadmium battery uses the same positive electrodes and electrolyte as the nickel–iron battery, in combination with metallic cadmium negative electrodes. This technology has seen enormous technical improvement, due to high specific power over 220 W/kg, long cycle life in the order of 2000 cycles, and low-discharge rate. The disadvantages are the high initial cost, relative low-cell voltage, and carcinogenicity of cadmium. There are two types of nickel–cadmium batteries used currently, one is the vented type and other is the sealed type. The vented sintered plate is a more recent development, which has a high specific energy but is more expensive. Sealed type incorporates a specific cell design feature to prevent a build-up of pressure in the cell caused by gassing during overcharge. As a result, the battery requires no maintenance.

10.3.2.4.4 Nickel-Metal hydride

These batteries have been in use since 1992. The characteristics are similar to those of the nickel–cadmium batteries. The principle difference between them is the use of hydrogen, absorbed in a metal hydride, for the active negative electrode material in place of cadmium. The overall reaction in Ni-MH battery is given below

$$MH + NiOOH \longleftrightarrow M + Ni(OH)_2$$

When the battery is discharged, the metal hydride in the negative electrode is oxidized to form a metal alloy; and nickel oxyhydroxide in the positive electrode is reduced to nickel hydroxide. During charging, the reverse reaction occurs. The nickel-metal hydride batteries have distinct advantages over nickel–cadmium batteries such as superior specific energy, environmental friendliness, and quick recharging.

10.3.2.5 Lithium-Based Batteries

Lithium is the lightest of all metals and presents very interesting characteristics from an electrochemical point of view. It allows very high thermodynamic voltage that results in high specific energy and specific power. The two major types of lithium-based batteries are lithium-polymer and lithium-ion.

10.3.2.5.1 Lithium-Polymer Battery

These batteries use lithium metal as the negative electrode and transition metal intercalation oxide as the positive electrode. A thin solid polymer is used as an electrolyte that offers improved safety and design flexibility. On discharge, lithium ions formed as the negative electrode migrates through the solid polymer electrolyte and are inserted into the crystal structure at the positive electrode. On charging, the process is reversed. The major advantage of these batteries is the very low self-discharge rate and the drawback is the relatively weak low-temperature performance due to the temperature dependence of ionic conductivity.

10.3.2.5.2 Lithium-Ion Battery

Lithium-ion batteries use a lithiated intercalation material for the negative electrode instead of metallic lithium, a lithiated transition metal intercalation oxide for the positive electrode, and a liquid organic solution or a solid polymer as the electrolyte. On discharge lithium ions are released from the negative electrode, migrate via the electrolyte and are taken up by the positive electrode. On charging the process is reversed. Lithium-ion batteries are considered to be the most promising rechargeable batteries of the future. Though this technology is at the developmental stage, it has gained wide applications in EVs.

10.3.3 Motor Controllers

For the efficient operation of the EV it is necessary to control the vehicle system components effectively so that energy is available all the time for all the elements of the engine at the required level. In older vehicles, electric motor speed and torque were controlled by the variation of field and rotor resistance of the DC motor. However, in recent days, to meet the tag of zero emission vehicles, it is necessary to utilize the full advantages of EVs by employing suitable control systems. To achieve that, every part of the vehicle needs to be controlled by sophisticated electronics/computer control systems.

Motor controller is a device or group of devices that serves in some predetermined manner the performance of an electric motor in an EV. It includes manual or automatic means for starting and stopping the motor, selecting forward or reverse rotation, selecting and regulating the speed, regulating or limiting the torque, and protecting against overloads and faults. Motor controller can be DC or AC current operated therefore it is called DC or AC controller based on the current it operates. Generally electric motors used in EVs are DC operated as it is simple to use and control.

A method of controlling motor response in an electrically powered vehicle having a motor and a manually operable accelerator, comprises these steps: monitoring the temperature of the motor; determining the maximum available power output based upon the current motor temperature; and adjusting the motor response to manipulation of the accelerator based upon determining the maximum power by increasing the gain of the motor control signal as the determined heat increases, wherein the motor control signal corresponds to the manipulation of the accelerator.

In an EV a simple DC controller is connected in between the batteries and the DC motor (Figure 10.12). The EV controller is the electronics package that operates between the batteries and the motor to control the EV's speed and acceleration much like a carburetor does in a gasoline-powered vehicle. Unlike the carburetor, the controller will also reverse the motor rotation and convert the motor to a generator so that the kinetic energy of motion can be used to recharge the battery when the brake is applied. If the driver floors the accelerator pedal, the controller delivers the full voltage from the batteries to

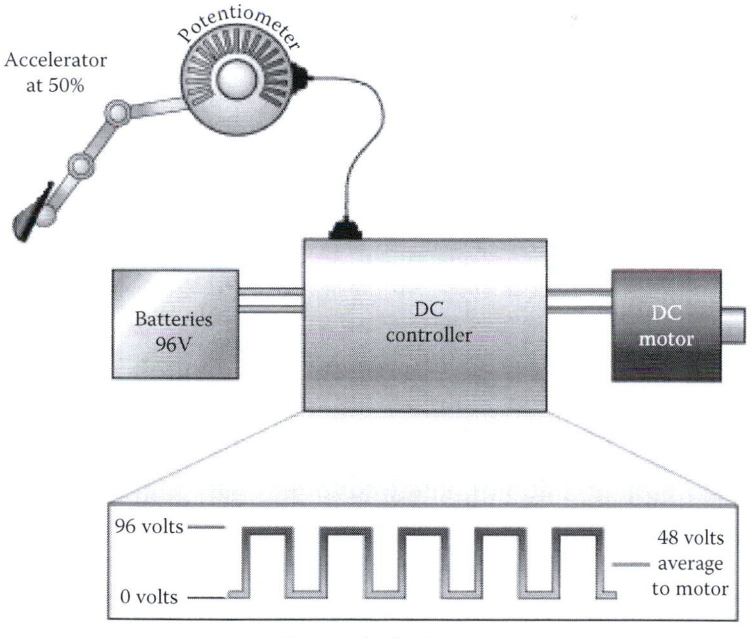

FIGURE 10.12
Working of motor controller. (From www.evsociety.ca.)

the motor. If the driver takes his/her foot off the accelerator, the controller delivers zero volts to the motor. For any setting in between, the controller "chops" the full volts thousands of times per second to create an average voltage somewhere between 0 and full volts.

On pushing the accelerator, a cable from the accelerator connects to the potentiometers. The potentiometer sends signals to the controller determining the amount of power to be delivered to the electric car's motor. There are two potentiometers in the vehicle and the motor controller reads both potentiometers to confirm that the received signals are equal. The controller will not operate if the received signals are unequal. A DC controller is a big on/off switch wired to the accelerator. The operating mechanism of the DC controller is simple and easy to understand. When the accelerator is pushed, it would turn the switch on, and when the foot is taken off the accelerator, it would turn it off. This mechanism of On/Off approach works well but the task for the driver becomes difficult if he is supposed to manually control the speed therefore the motor controller generates the pulse required to carry on/off operation. This helps in regulating the power generated and also the speed of the vehicle. The controller reads the setting of the accelerator from the potentiometers and regulates the power accordingly. For example, if the accelerator is pushed halfway down the controller reads that setting from the potentiometer and rapidly switches the power to the motor on and off so that it is on half the time and off half the time. If the accelerator pedal is 35% of the way down, the controller pulses the power so it is on 35% of the time and off 65% of the time (www.howstuffworks.com).

Modern controllers adjust speed and acceleration by an electronic process called pulse width modulation. Switching devices such as silicone-controlled rectifiers, rapidly interrupt (turn on and turn off) the electricity flow to the motor. High power (high speed and/or acceleration) is achieved when the intervals (when the current is turned off) are short. Low power (low speed and/or acceleration) occurs when the intervals are longer. The process is shown in Figure 10.13.

The controllers on most vehicles also have a system for regenerative braking. Regenerative braking is a process by which the motor is used as a generator to recharge the batteries when the vehicle is slowing down. During regenerative braking, some of the kinetic energy normally absorbed by the brakes and turned into heat is converted to electricity by the motor/controller and is used to recharge the batteries. Regenerative braking not only increases the range of an EV by 5–10%, it also decreases brake wear and reduces maintenance costs.

In an AC controller, the controller creates three pseudo-sine waves. It does this by taking the DC voltage from the batteries and pulsing it on and off. In an AC controller, there is the additional need to reverse the polarity of the voltage 60 times a second. Therefore, six sets of transistors are required in an AC controller, while only one set in a DC controller is required. In the

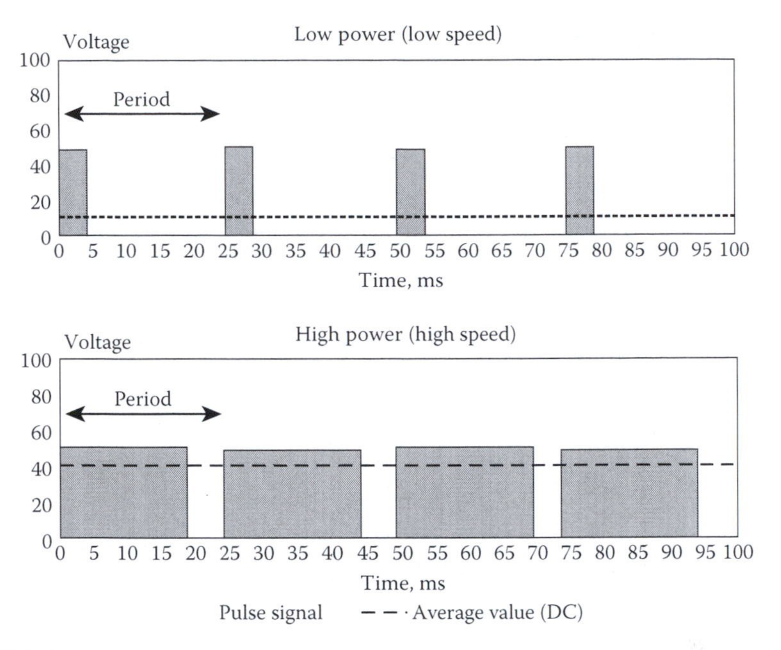

FIGURE 10.13
DC pulse width modulation. (From www1.eere.energy.gov/vehiclesandfuels/avta/light_duty/fsev/fsev_ev_power.html.)

AC controller, for each phase one set of transistors is required to pulse the voltage and another set to reverse the polarity.

10.4 Charging of Electric Vehicle Batteries

The charging system is a critical part of the energy cycle in an EV. The system should be able to charge or recharge the traction batteries as quick as possible so as to enable the batteries to supply the necessary power for traction. Electric vehicle batteries have to be charged in a most convenient way either at home using a domestic power supply or at a charging station depending on the journey (Figure 10.14).

At the earliest stage of use of EVs battery charging systems were simple. If an AC supply was available a transformer was used to supply the voltage according to the battery in use. Then being full wave rectified and the resulting DC smoothened by a capacitor, fed to the battery through a rheostat to adjust to a suitable current and a ballast resistor to limit the charging current. Batteries were also charged by DC supply from a motor–generator

FIGURE 10.14
Electric cars being charged at a charging station. (From www.mlive.com/news/baycity/index.ssf/2009/08/bay_city_downtown_development.html.)

setup. The lower battery voltage enables a high voltage DC supply to be used for DC-to-DC step-up conversion.

10.4.1 Charging Techniques

AC domestic and industrial supply systems are found universally for vehicle battery charging. The current into the battery is controlled using inductive reactance electronically by the continuous feedback of battery critical parameters. In the initial phase of charging a lead–acid battery, the voltage raises steadily from 2.1 to 2.35 V per cell, at which point the battery is about 80% charged and hydrogen and oxygen evolution (gassing) at the electrodes commences. After this voltage rises rapidly to a maximum value of about 2.45 V and as a consequence the charging current drops, its reduction is determined by the internal impedance of the charger.

More sophisticated chargers are being used to charge modern lead acid batteries like valve regulated batteries in which the hydrogen and oxygen evolution within the battery cannot be vented to the atmosphere under normal conditions. In this case, both constant current phase and constant voltage phase are used depending on the temperature of the battery sensed by the probe. In the initial phase, the maximum constant current that the battery will accept or that the charger can provide is maintained until gassing starts, without raising the battery temperature above an acceptable level. In the next phase, the battery will be charged at constant voltage, while the charge current decreases.

Aqueous batteries are developed and used in EV applications. For competitive traction applications, high performance batteries are being developed. An EV with an advanced Li-ion battery could in principle achieve a 400–480 km (250–300 miles) range, but these batteries would take up 450–600 liters of space (equivalent to a 120–160 gallon gasoline tank).

10.5 Vehicle Tests

Testing of batteries for EVs has two distinct focuses. For batteries, testing concentrates on optimizing battery performance and cycle life. It normally uses constant current or constant power and simple cycles. For EV systems, testing focuses on the total vehicle including the battery. This testing is multifaceted and usually involves a complex driving life cycle. These tests are demanding and require much higher physical and electrical performance than a constant-current cycling regime. Tests commonly used today are FUDS, SFUDS, and GSFUDS life cycle profiles. These tests have shown to be a much better predictor of battery life cycle performance than constant-current or constant-power cycling. Driving life cycle tests are one facet of the six characteristics considered when evaluating EV batteries and EV systems.

In testing batteries for potential application in EV systems, one commonly starts with cells. These are put through the characterization steps of Table 10.1. These cell tests are then repeated by varying the environment as detailed in Table 10.2. The objectives of this testing are primarily to characterize and optimize the cell.

The testing of EV systems represents the testing of a complete vehicle platform that includes (Brandt, 1992)

- Complete platform
- Propulsion system

TABLE 10.1

Steps for Cell Characterization Tests

S. No.	Cell Characterization Tests	Unit
1	Ah capacity	3 b rate
2	Specific power	W/kg
3	Specific energy	Wh/kg
4	Cycle life	80% DOD
5	Utilization of active material	%

Source: From Brandt, D. D., *Journal of Power Sources*, 40, 73–79, 1992. Reprinted with permission from Elsevier Publications.

TABLE 10.2

Environments for Cell Characterization Tests

S. No.	Cell Characterization Tests
1	Cell environment variables
2	Discharge (vary the rate and type)
3	Recharge (constant voltage or pulsed)
4	Cycling at various states-of-discharge
5	High and low-temperature performance
6	Variation in cell geometry and electrolyte
7	Variation of raw materials and separators
8	Operation with mechanical vibration

Source: From Brandt, D. D., *Journal of Power Sources*, 40, 73–79, 1992. Reprinted with permission from Elsevier Publications.

- Regenerative braking
- Rolling resistance
- Battery system
- Range (km)
- Acceleration

The major consideration in testing an EV system is its range and acceleration. The specific energy (Wh/kg) of the battery primarily determines an EV's range. This is determined by the battery's specific power (W/kg; peak power). Electrical tests have been developed to simulate real-life driving conditions in the laboratory. These tests have become the standard of the EV industry as a cost-effective means of comparing performance of cells and battery systems. They provide hard data without going to the dynamometer or the test track. The electrical tests that have been used to simulate real-life driving conditions have evolved from simple constant current or constant-wattage cycles to stepped constant-wattage cycles, to the sophisticated average power integer.

10.6 Solar Electric Vehicles

A solar vehicle is an EV powered by solar energy obtained from solar panels on the surface of the vehicle. Photovoltaic (PV) cells convert the sun's energy directly into electrical energy. PVCs are the components in solar paneling that convert the sun's energy to electricity. A solar array is the combination of various PV cells.

10.6.1 Design of Solar Vehicles

Designing solar vehicles is a multistage process where many parts need to be carefully designed and assembled. The various stages and steps involved are:

1. Designing chassis and basic framework
2. Designing and selection of suspension, braking system, and steering system
3. Designing and selecting motor and electric drive train
4. Selecting motor controller
5. Designing solar array with PV cells
6. Selecting proper batteries
7. Selecting electrical systems and instruments to display speed, load, temperature, and so on.

Solar vehicle design is done keeping in mind many factors. Reliability and operational efficiency forms the most important criteria for a successful design of the solar vehicle. Various factors to be considered during the design are:

1. Efficient PV cells
2. Good aerodynamic structure
3. Use efficient long running durable batteries
4. High performance motor
5. Light weight (200–350 kg)
6. Reliable chassis

10.6.2 Photo Voltaic Cells

Solar car's power storage capacity depends on the efficiency of PV cells to convert sunlight into electricity. While the sun emits 1070 ± 3.4% watts per square meter of energy, 51% of it actually enters the earth's atmosphere and therefore approximately 700 watts per square meter of clean energy can be obtained. Silicon is the most common material used for PV cells and has an efficiency of about 15–20%. The sunlight's energy then frees the electrons in the semiconductors, creating a flow of electrons. That flow generates the electricity that powers the battery or the specialized car motor in solar cars. Schematics of a single solar cell is shown in Figure 10.15. Some solar cars use gallium arsenide solar cells, with efficiencies around 30%. Other solar cars use silicon solar cells, with efficiencies around 20%.

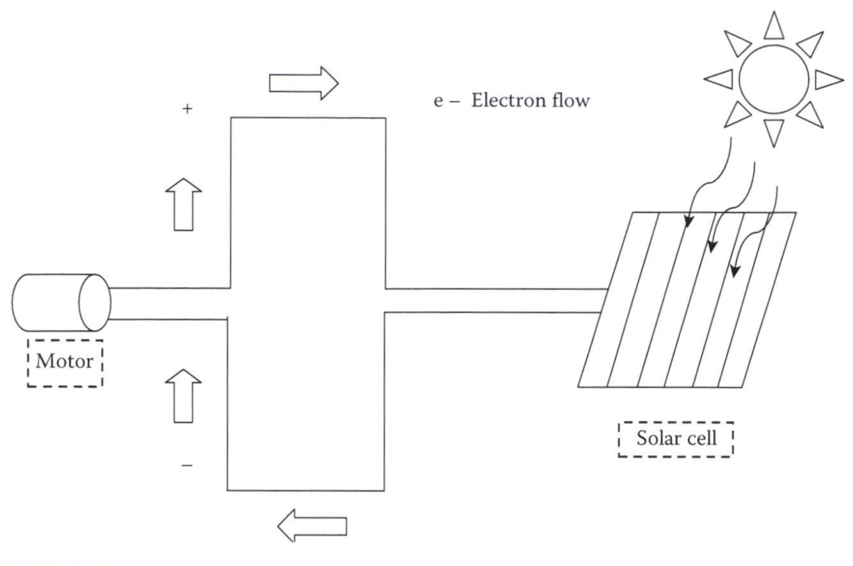

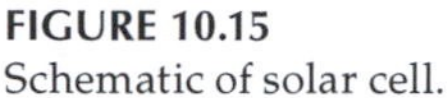

FIGURE 10.15
Schematic of solar cell.

10.6.3 Solar Arrays

To increase their utility, many numbers of individual PV cells are interconnected together in a sealed weatherproof package called a module. When two modules are wired together in series, their voltage is doubled while the current stays constant. When two modules are wired in parallel, their current is doubled while the voltage stays constant. To achieve the desired voltage and current, modules are wired in series and parallel into what is called a PV array. Solar arrays comprising these PV cells are mounted on solar cars. Solar arrays on solar cars are usually mounted using industrial grade double-sided adhesive tape right onto the car's body. It should kept in mind during the assembly process that the solar cells are fragile and can easily get damaged therefore extreme caution should be taken in handling these cells.

The larger arrays in use can produce over 2 kilowatts (2.6 hp). The power produced by the solar array varies depending on the weather, the sun's position in the sky, and the solar array itself. On a bright, sunny day at noon, a good solar car's solar array will produce well over 1000 watts (1.3 hp) of power. The power from the array is used either to power the electric motor or stored in the battery pack for later use.

The cost of solar cells can range from $10 up to $400 each. PV cells do have a defined lifespan. The lifetime of a solar module is approximately 30 years. However, in automotive purposes they need to be sealed well if meant to go for decades in all weather conditions. There is a problem of driving the vehicle in the night and rainy season when there is no sunlight. Some technical breakthroughs are required to overcome this problem. With a big solar panel on the roof the space available for the user is less therefore designing the car for commercial applications is a big design challenge. The solar panel adds weight to the car weight thereby affecting the car performance in terms of speed and load carrying capacity.

10.7 Applications of Electric Vehicles

Depending on the special requirements of a particular operation in specific industries, the following are some of the varieties of materials handling equipment (Sivaramaiah and Subramanian 1992):

- Electric truck for use in steel plant soaking pit
- Electric truck for frame and channel handling and mounting stud planter and stud puller in aluminum industries
- Fork-lift truck for bale handling
- Electric truck for batch trolleys in textile industries
- Electric truck for die handling and for use as an order picker
- Electric truck for inter-bay movements
- Electric locomotive for mines
- Electric tow truck for handling passenger luggage at airports
- Electric vehicle for passenger movement at airports

10.8 Benefits of Electric Vehicles

1. Electric vehicles do not emit greenhouse emissions and toxic exhaust gases. The EVs are zero emission vehicles if the energy required to operate the vehicle (i.e., charging of battery) is produced from renewable energy sources or onboard hydrogen fuel cells.
2. Electric motor is much more efficient than conventional internal combustion engines.
3. Electric vehicles are noiseless.
4. Electric vehicles reduce the dependency of fossil fuels if they run on alternative fuels.

10.9 Challenges of Electric Vehicles

1. The cost of EVs is high as compared to contemporary gasoline and diesel cars. Recharging time of the batteries is high and research is going on to reduce the recharge time. It takes hours to recharge the batteries, which is more than the refueling time of gasoline and diesel vehicles.
2. Driving range of the EVs depending on the power of the batteries. To date the driving range is limited and is around 100 miles in one recharging.
3. Limited seating space availability as increase in the space will increase load of the passengers affecting the driving range as increased load puts stress on the battery performance and life.
4. Less customer acceptance and satisfaction. This is a marketing challenge as it will be a difficult task to change customer perception with so many limitations of EVs and when other options that are far better than EVs are available in the market.

References

Brandt, D. D. 1992. Battery cycle testing of electric vehicle batteries and systems. *Journal of Power Sources* 40:73–79.

Ehsani, M., Y. Gao, S. E. Gay, and A. Emadi. 2005. *Modern electric, hybrid electric and fuel cell vehicles.*, Boca Raton, FL: CRC Press.

Ogura, M. 1997. Development of electric vehicles. *JSAE Review* 18:51–56.

Shukla, A. K., A. S. Arico, and V. Antonucci. 2001. *Renewable and Sustainable Energy Reviews* 5:107–55.

Sivaramaiah, G., and V. R. Subramanian. 1992. An overview of the development of lead/acid traction batteries for electric vehicles in India. *Journal of Power Sources* 38:165–70.

Wakefield, E. H. 1998. *History of the electric automobile.* Warrendale, PA: Society of Automobile Engineers.

11

Fuel Cells

Parthasarathy Sridhar, Sethuraman Pitchumani, and Ashok K. Shukla

CONTENTS

11.1 Introduction

Air pollution and global warming arising due to the ever expanding use of energy have become problems of global concern. Global atmospheric warming is now a generally accepted fact, although there is still debate on the true nature of its origin. Concerns about climate change and the impact of exhaust emissions and energy security have given the impetus to look for alternative fuels/power sources for automotive applications. According to the U.S. Environmental Protection Agency (EPA) Report, vehicles in the United States account for about 75% of carbon monoxide (CO) emissions, about 45% of nitrous oxide (NO_x) emissions, and nearly 40% of volatile organic compound emissions (Emadi et al. 2004). European countries contribute about 20% of the greenhouse gases from vehicular transport (Adcock, Kells, and Jackson 2008).

The growing concerns on environmental issues have been constantly demanding cleaner and energy-efficient vehicles without compromising the convenience of the conventional internal combustion engine vehicles (ICEVs). To meet these demands, major automotive manufacturers have been attempting to introduce electric vehicles (EVs), hybrid electric vehicles (HEVs), and plug-in hybrid electric vehicles (PHEVs) as alternatives to ICEVs. However, such vehicles have several techno-economic issues that need to be addressed. Among zero emission vehicles (ZEVs), battery-powered electric vehicles (BEVs) are the most viable alternatives. But BEVs have several short-comings, namely limited electrical energy storage, long charging time, small operating temperature range, low-cycling capability, and high cost.

Fuel cells that can overcome the major technical limitations of the storage batteries, namely their energy density, have shown promise for automotive applications. A fuel-cell vehicle (FCV) uses a fuel cell stack as the source of electric power to drive an electric-traction motor. Unlike internal combustion engines (ICEs), fuel cells are not limited by Carnot cycle and hence can operate highly efficiently.

Various types of fuel cells are: alkaline fuel cells, phosphoric acid fuel cells, polymer electrolyte fuel cells (PEFCs), molten carbonate fuel cells, and solid oxide fuel cells. Among these, PEFCs are seen to be most attractive for automotive applications due to their quick start-up and low-temperature operation. A PEFC uses hydrogen as fuel and oxygen from air as the oxidant. Ironically, however, hydrogen is not available freely in nature and hence needs to be produced from other hydrogen containing fuels. It is projected that the use of FCVs will help reduce the emission of carbon dioxide gas by about 33% and there will be virtually no other polluting gases. In this article, we estimate the power and energy consumption of a modern car, and examine the feasibility of the PEFCs for realizing a viable FCV.

11.2 Power and Energy Requirements of the FCV

To assess the energy and power requirements of an electric automobile, it is appropriate to quantitatively estimate the power and energy required for driving a modern ICEV (Shukla 2005; Shukla, Aricò, and Antonucci 2001) as follows. Neglecting relatively minor losses due to road camber and curvature, the power required at the drive wheel ($P_{traction}$) may be expressed as

$$P_{traction} = P_{grade} + P_{accel} + P_{tires} + P_{aero}, \quad (11.1)$$

where P_{grade} is the power required for the gradient,
P_{accel} is the power required for acceleration,
P_{tires} is the rolling-resistance power consumed by the tire, and
P_{aero} is the power consumed by aerodynamic drag.

The first two terms in Equation 11.1 describe the rates of change of potential (PE) and kinetic (KE) energies associated during climbing and acceleration, respectively. The power required for these actions may be estimated from the Newtonian mechanics as follows:

$$P_{grade} = d(\mathrm{PE})/\mathrm{dt} = Mg\nu \sin\theta, \quad (11.2)$$

and

$$P_{accel} = d(\mathrm{KE})/\mathrm{dt} = d(1/2M\nu^2)/\mathrm{dt} = Mav. \quad (11.3)$$

In Equations 11.2 and 11.3, M is the mass (kg) of the car, v its velocity (m/s), a its acceleration (m/s^2), and $\tan\theta$ is the gradient. The potential and kinetic energies acquired by the car as a result of climbing and acceleration represent reversibly stored energies and, in principle, may be recovered by appropriate regenerative methods wherein the mechanical energy is converted and stored as electrical energy.

The last two terms in Equation 11.1 describe the power required to overcome tire friction and aerodynamic drag that are irreversibly lost, mainly as heat and noise and cannot be recovered. The power required here may be estimated from the following empirical relations:

$$P_{tires} = \mathrm{C_t}Mg\nu, \quad (11.4)$$

and

$$P_{aero} = 0.5\mathrm{dC_a}A(v + w)^2 v. \quad (11.5)$$

In Equations 11.4 and 11.5, C_t and C_a are dimensionless tire friction and aerodynamic drag coefficients, respectively, d the air density (kg/m^3), w the headwind velocity (m/s), g (= 9.8 m/s^2) the gravitational acceleration, and A the frontal cross-sectional area (m^2) of the car.

From the parameters associated with a typical modern medium-size car; that is, $M = 1400$ kg, $A = 2.2$ m^2, $C_t = 0.01$, $C_a = 0.3$, $d = 1.17$ kg/m^3, its power requirements may be estimated from Equations 11.2 through 11.5. For the irreversible losses, Equations 11.4 and 11.5 show that while P_{tires} is linearly dependent on velocity, P_{aero} varies as the third power of velocity and although negligible at low velocities, the latter becomes the dominant irreversible loss at high speed. As an example, for these parameters, for a car traveling at about 50 km/h, tire friction is twice the aerodynamic drag and together amount to about 3 kW. At 100 km/h highway cruising, aerodynamic drag increases considerably to over twice the tire friction, increasing the total power requirement to about 12 kW. It is noteworthy that for both these estimates, the wind speed (w) has been taken to be zero for the sake of simplicity. But, in practice, the effect of wind speed on the performance of the car could be quite substantial. For example, P_{aero} at a favorable tailwind speed of 30 km/h will be as low as 0.8 kW but would amount to 4 kW at a similar opposing tailwind velocity. Accordingly, the energy performance of the car will drop from 40 km/kWh to 15 km/kWh (Wicks and Marchionne 1992). Taking the example of a hill with a substantial 10% gradient, climbing at 100 km/h requires about 50 kW, including tire friction and aerodynamic drag. Acceleration is more demanding, particularly at high velocities. For example, acceleration at 5 km/h/s requires 30 kW at 50 km/h but increases to 66 kW at 100 km/h.

The above estimates are for the power supplied to the wheel of the car and do not include the losses incurred in delivering that power to the wheels. At this time in the development of electric traction systems, a precise estimate of this is difficult to obtain but anecdotal information suggests that the efficiency of the power conditioning electronics together with the electrical and mechanical drivetrain is likely to be about 0.85. Additional power may also be required to power the accessories like radio, lights, steering, air-conditioning, and so on, which is likely to add about 5 kW to the total power demand of the car.

An analysis of this kind indicates that the power plant of a modern car must be capable of delivering about 65 kW of sustained power for accessories and hill climbing, with burst-power requirement for a few tens of seconds to about 105 kW during acceleration. For a car with these performance characteristics, this sets the upper power limit, but in common usage rarely exceeds 20 kW while cruising.

The heating value of gasoline-fuel is 32.5 MJ/l but a heating value of only 6.5 MJ/l will be available with an ICEV of near 20% well-to-wheel efficiency. This is about 1.82 kWh/l of gasoline-fuel and considering the average drive range of the car with the parameters listed above as ~10 km/l, it would amount to 182 Wh/km. The heating value of diesel fuel is 35.95 MJ/l and accordingly the estimated energy will be 201 Wh/km for diesel-driven cars, which have well-to-wheel efficiency of about 30% and a drive range of about

15 km/l. It is mandatory that EVs meet these power and energy requirements. Various aspects of FCVs are discussed in the following section.

11.3 Fuel-Cell Vehicles

Fuel-cell technology is not new and has a long history. The first fuel-cell was invented in 1839 by Sir William Grove and an experimental vehicle powered by an alkaline fuel-cell appeared in 1970. As indicated earlier, among the fuel cells based on hydrogen as fuel, PEFC has been broadly accepted for automotive applications. PEFC comprise a solid polymer electrolyte membrane sandwiched between an anode and a cathode, and end plates. A PEFC stack has several cells stacked in series with electrically conducting bipolar plates having flow fields to distribute hydrogen fuel and oxidant air at the anodes and cathodes, respectively. Hydrogen fuel can be supplied either as compressed gas from a cylinder or from a cryogenic tank, but both of these have certain disadvantages. To overcome the difficulties of hydrogen storage and lack of hydrogen-distribution network, a methanol or gasoline-fuel processor is incorporated to produce a hydrogen-rich gas stream onboard the FCV.

Ballard in May 2005 emphasized four areas critical to commercial adaptation of automotive PEFC-stack technology; namely, durability, cost, freeze start, and volumetric power density. The technical specifications for the PEFC system were set as: power density ~2500 W/l, endurance of ~5000 h in test driving cycle, freeze start capacity of 30 s to 50% power at –30°C, and fuel cell stack cost of $\$30/kW_e$ net at a volume of 500,000 units (Wee 2007).

The advantages of using fuel cells for automotive applications are

1. High-energy density subject to the limitation of the size of fuel-storage tank
2. Constant power, unlike batteries where power density varies with the state-of-charge
3. Fast refueling

The limitations presently faced are

1. Low-power density (kW/kg and kW/l)
2. High cost
3. Problems with fuel production and storage

A recent advancement in fuel-cell technologies has been the development of direct methanol fuel cells (DMFCs). Although DMFCs have made substantial advances in recent years, they still require significantly higher amounts of precious metal catalyst than the direct-hydrogen PEFCs. Another issue is to mitigate the problem of methanol crossover through the membrane.

Hydrogen can also be produced electrolytically from water using renewable sources of energy that can reduce pollution and emission of greenhouse gases (Hocevar and Summers 2008). Hydrogen can also be used in IC engines. But the efficiency of IC engines is quite low. Onboard storage of hydrogen, albeit being cost and volume intensive in relation to gasoline or diesel tanks is much closer to automotive costs and performance figures. Fuel cells have not had the advantage over combustion engines or batteries for a long time in spite of the demonstration of various FCVs such as Allis–Chalmers Fuel-Cell Tractor in 1959 and GM Electrovan in 1966 due to their limited power-density. The advent of new electrolyte materials has offered the possibility for compact and lightweight fuel cells. Though the power density of fuel cells has increased over the years, fuel cells remain heavier and bulkier than conventional drivetrains. Accordingly, further technological developments focusing on additional size reduction and increased performance are desired (Helmolt and Eberle 2007).

A fuel-cell running on hydrogen looks attractive as a long-term option for passenger cars as it eliminates emissions on the tank-to-wheel path. Hydrogen can be produced from many sources, and has attractive efficiencies. It is noteworthy that a fuel-cell reaches its highest efficiency at part loads and there is little advantage against the ICE at full loads. Interestingly, passenger vehicles are mostly operated at part loads, significantly below their rated power, such that the efficiency gain offered by fuel cells could be maximum. It is noteworthy that at low-power output even the fuel-cell system (FCS) efficiency drops as many of the balance of plant units needed to be operated even at idle power. Accordingly, the system has to be optimized for low-power consumption to capitalize on the part load efficiency advantage of the fuel-cell. Performance at high-power density and efficiency in terms of high cell voltage has been the mainstay of fuel-cell development. In recent years, these are complemented by other factors, namely lower cost at high volume production with increased reliability and durability. The targets are derived from competing conventional automotive propulsion systems, which are designed for 5500 h of operational lifetime at a cost of U.S. $50/kW, including fuel storage.

Durability is a major issue for the fuel cell stacks, especially with thinner membranes and lower catalyst loadings for cost effectiveness. Among the factors limiting the lifetime of PEFCs, chemical degradation has been identified as a major problem. However, there has been remarkable progress in system reliability owing to the continuous improvements in engineering and operation of complete FCSs and vehicles.

PEFCs are the most widely used fuel cells in transport applications. It is estimated that more than 90% of the FCVs on the road are equipped with PEFCs (Cropper, Geiger, and Jolli 2004). PEFCs exhibit high-power density at operating temperatures of about 70°C. There are still some drawbacks with the current state-of-the-art PEFCs: one issue being the low CO tolerance. Most PEFC stacks require CO levels well below 10 ppm to mitigate losses in performance. A second factor is the operating temperature of the PEFCs that may require an unacceptably large radiator area for cooling the fuel cell stack (Mallant 2003). These two factors have led to research efforts in proton conducting membranes at elevated temperatures (Li et al. 2003; Savogado 2004). It is projected that these developments will take a long time before they will find a use in systems for practical applications (Bruijn 2005). This type of high-temperature PEFCs can at best be regarded as second-generation systems. The conventional low-temperature PEFC stacks will be the first products in practical applications requiring relatively pure hydrogen. PEFC stacks in transport applications will be used in a hybrid configuration with batteries/super capacitors.

For auxiliary power units (APUs) in FCVs, SOFCs are also being considered as a possible candidate. SOFCs are characterized by their high-operating-temperature; the state-of-the-art ytria stabilized zirconia based SOFC has an operating temperature of about 1000°C. There are two basic configurations of SOFCs, tubular and planar. Tubular SOFCs are regarded as suitable for large scale stationary applications while planar SOFCs are preferred for automotive applications due to their higher power density (Singhal and Kendall 2003). SOFCs offer two major advantages over PEFCs, namely no need for a platinum catalyst and greater fuel flexibility without any external reforming. However, their high-operating temperature and consequent lower durability and long start-up times, pose significant hurdles for use in automotive applications (Ormerod 2003). Advances in materials have brought in intermediate temperature (IT) SOFCs operating between 500 and 750°C (Brett et al. 2008). The lower operating temperature for IT-SOFCs, and consequent faster start-up time, greater durability, and lower cost materials, make them a possible candidate for automotive applications. The IT-SOFCs offer high-fuel flexibility, efficiency, and more tolerance to impurities than PEFCs and, unlike PEFCs, do not require external fuel reforming. However, they exhibit reduced activity toward oxygen reduction reactions at the cathode in relation to high-temperature SOFCs (Shao and Haile 2004). But, IT-SOFCs might be promising only as APUs in heavy-duty FCVs (Steele and Heinzel 2001). The use of SOFCs in automotives would best be limited to APUs and is not seen viable for automotive purposes in general due to the specific requirements associated with traction, and in particular start-up behavior and dynamic load changes (Oosterkamp et al. 2006). In light of the foregoing, only PEFCs are considered suitable for vehicular applications by automotive manufacturers.

11.4 Polymer Electrolyte Fuel Cell

11.4.1 PEFC Operating Principle

A PEFC comprises an anode and a cathode separated by a proton conducting polymer electrolyte membrane. When hydrogen is supplied through the porous anode, the catalyst dissociates hydrogen into protons (H^+) and electrons (e^-). Since only protons can flow through the electrolyte, the electrons pass through the external circuit to the cathode forming water on combining with protons and oxygen from air. It is noteworthy that fuel cells have the advantage over both the combustion engine and the battery. Like a combustion engine, a fuel-cell runs as long as it is provided fuel, and like a battery, fuel-cell converts chemical energy directly into electrical energy.

The main constituents of a PEFC stack are an electrolyte membrane, bipolar plates and a catalyst. Figure 11.1 shows the operating principle of a PEFC. Accordingly, to realize a viable system for vehicular application, innovations in several disciplines such as polymer chemistry and catalysis in conjunction with electrochemistry, mathematical modeling, and the integration of PEFC with the vehicle drive system are desired (Arita 2002).

The bipolar plates act as the current conductors between cells, provide conduits for flow of reactant gases, and constitute the backbone of the power stack. They are commonly made of graphite composite with high-corrosion resistance and good surface contact resistance; however, their manufacturability, permeability, and durability for shock and vibration are not as good

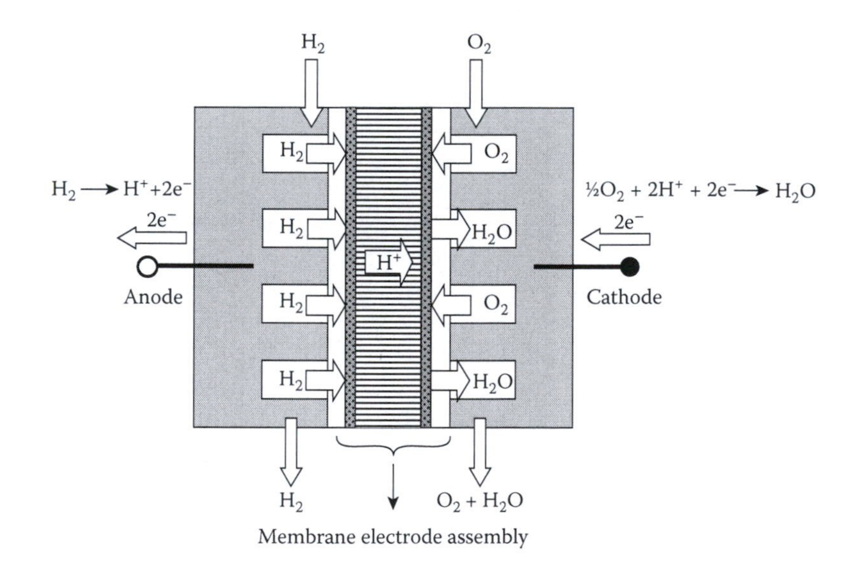

FIGURE 11.1
Operating principle of a PEFC.

as metals. Various methods and techniques must be developed to combat metallic corrosion and eliminate the passive layer that causes unacceptable reduction in contact resistance and possible fouling of the catalyst and the ionomer (Tawfik, Hung, and Mahajan 2007). The schematic of a simplified direct hydrogen PEFC system (Yang 2000) is shown in Figure 11.2.

A PEFC system comprises the following four subsystems, namely the PEFC stack, stack fuel delivery, stack air delivery, and stack water/thermal management subsystems.

11.4.1.1 PEFC Stacks

The fuel cell stack that drives the balance of plant and the bulk of the trade-offs is the heart of the system. Key system metrics include stack size, active area, cell voltage, current density, operating temperature, gas pressures, anode and cathode stoichiometries, and gas prehumidification. Typical PEFC operating temperatures range between 60 and 80°C but there is also interest in operating them at higher temperatures with operating pressures ranging from near-ambient to over 3 bar while keeping stoichiometries of the fuel and air as ~1 and 1.4, respectively. Prehumidification, which impacts the water and thermal management subsystems, has been demonstrated between 0% and 100% relative humidity both for the air and fuel streams.

The power density of fuel cell stacks developed for automotive applications with pure hydrogen are significantly higher than those developed for stationary applications as shown in Table 11.1. A power generating fuel cell stack requires four major subsystems: hydrogen supply, air supply, water

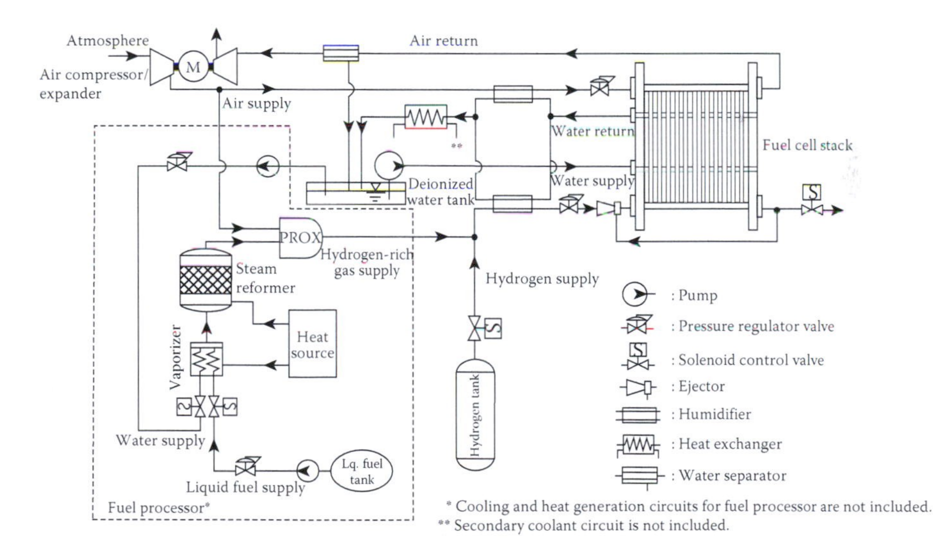

FIGURE 11.2
Schematic of a fuel-cell system.

TABLE 11.1

Commercial PEFC Stacks for Automotive Applications

Vehicle	Power (kW)	Power Density kW l^{-1}	kW kg^{-1}	Conditions
Ballard Mark 902	85	1.13	0.88	H_2–air/1–2 bar g/80°C
GM HydroGen 3	94	1.60	0.94	H_2–air/1.5 bar g/80°C

management, and heat management. A typical fuel-cell power plant for automotive applications is shown in Figure 11.2. For a direct hydrogen FCS, liquid or compressed hydrogen is stored in a tank installed in the vehicle. Hydrogen is supplied to the anode manifold through a deionized water humidifier as the electrolyte membrane needs to be water saturated to conduct protons. Hydrogen recirculation at a certain level is necessary to keep the system flooded with gas and to improve the power demand transient responses. It is noteworthy that while using dilute hydrogen-gas from a fuel processor, the exhaust-gas cannot be recirculated owing to its low-hydrogen content.

Unlike a naturally aspirated ICE, fuel cell stack requires forced air supply at a specific system pressure. It is important that the flow through each cell in the fuel cell stack be evenly distributed, especially when dilute reactant gas such as air or reformate gas is used. To ensure the flow through each cell at an optimum operating-pressure, the air supply system requires a coordinated pressure and flow control system. An air compressor/super-charger with speed control can be used to supply variable amounts of air, and, to recover the energy from the compressed exhaust air, an expander can be incorporated with the air compressor. One of the most important vehicle performance criteria is the transient response on power demand. The fuel-cell itself has a fast dynamic response. However, the response times for the fuel and air supply system depend on the system design and may affect the overall system transient-performance.

Three major functions of the water management system are humidification of the reactant gases to properly hydrate the membrane, removal of the product water in the stack, and cooling of the stack to control the operating temperature. Ionic conductivity of the membrane is highly dependent on the degree of membrane hydration and is critical to the stack performance, especially at high-power density operations. A conventional design with a water saturated gas stream is shown in Figure 11.2. Proper removal of the product water in each cell is important to prevent the airflow passage from being blocked by the accumulated water leading to cell degradation. Most of the power losses in the fuel-cell are converted into thermal energy producing significant amounts of heat requiring substantial heat removal from the fuel cell stack. In most of the PEFC-stack designs, a cooling circuit using deionized water as a cooling medium is incorporated as shown in Figure 11.2. For automotive applications, a second stage heat exchange circuit is incorporated using a high thermal capacity coolant and a radiator akin to ICEs. The removed heat can also be used to heat the vehicle compartment.

As an alternative to the direct hydrogen supply to the fuel cell stack, a fuel processor can be incorporated onboard the vehicle to produce a hydrogen-rich gas stream from a liquid fuel such as methanol or gasoline. Although major technical advances for onboard vehicle applications are still desired, the concept is well proven and has been tested on a small size passenger vehicle using methanol as the fuel. One of the most significant drawbacks of the onboard fuel processors in vehicular applications, besides the packaging issues, is the system time constant; slow dynamic response is mainly due to the slow heat transfer processes, mass transfer, and mixing delays.

Three major subsystems currently requiring improvements are the air supply, thermal management, and water management. The most efficient operating pressure for current PEFC systems is about 207 kPa. This requires an air supercharging mechanism at the cost of parasitic loads. Current superchargers used in FCSs have excessive parasitic loads, approaching 10–15 kW at a peak power of 50 kW net, and are heavy, big, and costly. As pressure drops in most of the stacks are only of the order of several psi, significant energy is still contained in the cathode exhaust-gas. To recover this energy and to reduce the total parasitic load on the supercharger, expanders are being developed as part of the air supply subsystem. The present air supply subsystem requires further improvement for successful applications of fuel cells in automotives.

Thermal management subsystems also require innovation and improvement. Two major difficulties facing thermal management are: (a) the large radiator and fan required for removing the low-temperature waste heat from the stack and (b) development of nonconducting coolants capable of operating at subfreezing temperatures. Due to the low-stack temperature of about 90°C at 3 atm operation, the radiator size for a 60 kW net power system will be 1.5 times larger than that for an ICEV.

The automotive market will require higher temperatures and drier operating conditions (Beuscher, Cleghorn, and Johnson 2005). This requirement is driven by the available radiator technology used to transport excess heat away from the fuel cell stack (Oosterkamp et al. 2006). To maintain the current size of automotive radiators in PEFC vehicles, temperatures above 100°C are desired for the fuel cell stack to increase the heat transfer efficiency of the radiator. An additional advantage of higher temperatures is lower sensitivity to fuel impurities, particularly CO. System size and weight constraints associated with water management require lower humidity conditions. Operation under automotive drive cycle, with frequent stops and starts including freeze tolerance, makes materials selection for FCVs even more difficult. Proposed future operating conditions are challenging for current PEFC membrane materials. High temperature and low humidity have detrimental effects on both performance and durability of currently used membranes. New membrane materials must be developed to reach the targets set by the fuel-cell industry. To this end, the ionomer needs to be optimized to provide sufficient conductivity at low humidity. A mechanical reinforcement will be useful to sustain the integrity of the membrane under high-mechanical stress during fuel-cell operation. It will be necessary to understand the location and transport of the water inside the fuel-cell to assess the influence of material properties on performance and durability. Careful fuel-cell experiments with mathematical simulations will be valuable for the development of new and successful membrane materials.

Current fuel cell stack designs require use of deionized water for the stack coolant. If the ionic conductivity of the coolant is too high, shunting currents can arise that lower the stack efficiency. Since deionized water is corrosive to aluminium radiators, a separate liquid–liquid heat exchanger is required. Deionized water flows through the stack at one side of the heat exchanger, while glycol coolant flows between the radiator and the other side of the heat exchanger. Use of an intermediate heat exchanger further exacerbates removal of waste heat from the stack. Development is underway on alternative radiator designs, nonconductive coolants, and stack designs capable of using water/glycol coolants. Operation at subfreezing temperatures is an issue as pure water freezes at 0°C. Interestingly, the membrane may not be a significant problem at subfreezing temperatures as the membrane retains significant proton conductivity even at temperatures of < –20°C. However, the presence of deionized water will require significant system level developments to match the performance of current ICEs at low temperatures (Yang 2000).

11.4.1.2 Stack Fuel Delivery Subsystem

For a direct hydrogen subsystem, the fuel delivery subsystem is considerably more simple than the fuel processor based system (Masten and Bosco 2003). Hydrogen sources include compressed gas, cryogenic liquid, and metal or chemical hydrides. In all cases, hydrogen is delivered from the source to the anode, often with external humidification and temperature preconditioning. In direct hydrogen applications, where the hydrogen is nearly pure, hydrogen utilization can be near unity, markedly enhancing system efficiency. Hydrogen is often recirculated with a pump or ejector to maintain flow and distribution within the anode. Alternatively, the anode may be operated "dead-ended" where there is no continuous exhaust flow and only an occasional purge is required to alleviate contaminant build-up. In the standard case, humidification is achieved through deionized, water-fed membrane humidifiers wherein a small pump recirculates the anode flow with a solenoid valve and residual hydrogen is consumed in a small catalytic combustor.

11.4.1.3 Stack Air Delivery Subsystems

A compressor or blower delivers air in excess to the cathode, typically with external humidification and temperature conditioning. Delivery pressure is either controlled with a back-pressure valve and/or expander or is left unconstrained in ambient systems. Systems have been designed where the stack pressure drop is used as a method for controlling the operating pressure. Typically the exhaust air is fed into a tail-gas combustor to serve as the oxidant. In pressurized systems, either the cathode or combustor exhaust may be fed to an expander for additional power generation. Common compressor types include twin screw, scroll, centrifugal turbo machines, and roots blowers. In the system shown in Figure 11.2, discharge air from the compressor is conditioned to stack temperature with a heat exchanger. As with the anode, cathode humidification is performed by a deionized, water-fed membrane humidifier.

11.4.1.4 Stack Water and Thermal Management Subsystems

Although a FCS is the most efficient means of converting fuel to energy onboard a vehicle, it produces considerable amounts of thermal energy. Thermal loads are of the same magnitude as in conventional automotive systems but at lower temperatures that require enhanced thermal rejection. Heat exchangers are employed to carry out several key functions within the FCS, namely, to reject fuel-cell heat, condition compressor discharge air to stack temperature, condense water for stack humidification, cool power electronics, and to enhance energy recovery in expanders.

Water management is often reduced to the practice of providing the gas-fed humidification required to maintain the performance of the fuel-cell membrane during its life. Gas-feed humidification has been accomplished by a number of means, including liquid water-fed membrane humidifiers, direct cathode water injection, compressor water injection, humidification wheels, porous bipolar plates, and water vapor transport membranes. Water for humidification is typically collected from the stack at the cathode exhaust. Depending on pressure and temperature, a cathode exhaust condenser may be required to recover product water to maintain system water neutrality.

Thermal management must also address the conductive and corrosive nature of automotive coolants. Stack coolants must be nonconductive, corrosion cum freeze resistant, and should have a high-heat capacity and thermal conductivity without being excessively viscous. Deionized water is a traditional coolant but glycol/water-based coolants have also been employed with promise.

The aforesaid requirements present a significant challenge to system design and hence FCV commercialization. In the system shown in Figure 11.2, the stack coolant supplies the water and vaporization energy for the anode and cathode inlet humidifiers. A cathode exhaust condenser is used to supply make-up water to the humidifier.

11.4.1.5 PEFC System Efficiency

System efficiency is defined as

$$\eta_{system} = \frac{P_{net}}{P_{fuel}},$$

where P_{net} = FCPS net power (W); and P_{fuel} = fuel power, which is equal to fuel flow rate × fuel heating value (W). The automotive standard is to use the lower heating value, which is 242 kJmol^{-1} for hydrogen. Stack efficiency is defined as

$$\eta_{stack} = \frac{P_{gross}}{P_{H_2consumed}} \frac{P_{net}}{P_{gross}},$$

where P_{gross} = fuel cell stack gross power (W); and $P_{H_2consumed}$ = hydrogen fuel power consumed electrochemically.

$$\eta_{stack} = \frac{V_{cell}}{1.23V},$$

where V_{cell} is in volts. The system efficiency can be written as:

$$\eta_{system} = \eta_{H_2util}\ \eta_{stack}\ \eta_{parasitic},$$

where $\eta_{H_2util} = P_{net}/P_{gross}$.

For a typical direct-hydrogen system, η_{H_2util} is near unity and $\eta_{parasitic}$ is normally dominated by the compressor auxiliary power and may vary between 80 and 95%. For operating cell-voltages between 1 V at open circuit and 0.6–0.7 V at peak power, η_{stack} varies from nearly 80% at open circuit to 50–60% at peak power.

11.4.1.6 Stack Sizing/Packaging

Stack sizing is perhaps the most important system design component. As the stack happens to be an expensive part in FCV, it needs to be small in size. This results in reduced peak efficiency, which adversely impacts the balance of plant. The actual design choice will in part depend on the system efficiency target, but is likely to be driven by factors such as cost, packaging, and thermal management.

After stack size, the most fundamental system trade-offs are stack operating pressure, temperature, humidification, stoichiometry, and their interactions. Higher temperature operation is desirable for effective heat rejection.

Moreover, higher temperature enhances stack efficiency through improved cathode electrocatalyst activity. Higher temperature operation increases water requirements, influencing water and thermal management. Higher temperature operation also adversely impacts the durability/performance of proton exchange membrane materials. High pressure, low-cathode stoichiometry operations facilitate water management by reducing water content required for membrane and feed gas saturation. High pressure also enhances cell voltage through reduction in cathode overpotential with increased oxygen partial pressure and improved mass transport. Conversely, high pressure increases parasitic compressor power consumption, often as much as 15–20% of gross stack power. Reductions in cathode stoichiometry reduce humidification requirements and compressor parasitic power but at the cost of cell voltage and stack gas distribution related stability.

To understand the push to higher temperature PEFCs for FCVs, it is desirable to compare the thermal requirements of FCVs with ICEVs. As a rule of thumb, one-third of the system energy is removed each by the coolant and the exhaust in ICEVs. Conversely, in a FCS, typically less than 10% of the heat is rejected with the exhaust gases. In FCVs, waste heat is approximately equal to the power generated. Thus, albeit being more efficient than ICEVs, FCVs have similar or larger, coolant loads. In a PEFC stack, coolant temperature is significantly lower, typically 60–80°C, as compared to 120°C in ICEVs. Since radiator performance is proportional to the initial temperature-difference, the driving force between the coolant and ambient temperature in an ICEV has an approximately 2–4 times higher heat rejection capability during operation at elevated temperatures. Even a modest sized, FCS requires a cooling system that taxes the bounds of normal automotive design. While heat exchange does not present an insurmountable technical obstacle, considerable work remains to be done in optimizing thermal management hardware for FCSs.

Stack temperature and operating pressure are linked to the PEFC's water requirement. At elevated temperatures, the water feed required to achieve saturation increases. At higher temperatures and lower pressures, the humidification energy duty and associated condensate water requirement may affect the water and thermal management capabilities of an automotive thermal system. Increased system pressure can offset the water requirement but increased complexity and cost potentially reduce the efficiency. Alternative humidification schemes, such as water vapor transport membranes or adsorption/desorption in a desiccant wheel, are promising. It is hoped that PEFC development will lead to higher temperature, lower humidification materials circumventing the need to transport large quantities of water through the system.

Presently, packaging constraints imposed by the system thermal requirements are a limiting step. Accordingly, either novel heat rejection concepts need to be employed or stack temperatures need to be increased. Humidification loads will also increase assuming the proton exchange mechanism to be constrained by current material characteristics are within the fuel cell stack. These will require the stacks with high temperature MEAs at reduced humidification, and the system and application side where emphasis needs to be placed on novel heat exchanger designs and development of advanced fuel-cell air machinery. Fuel-cell specific membranes, compressors, expanders, and humidification components are all in development. In the past decade alone, the power density of the fuel cell stacks has increased by a factor of 10. With the rapid pace of technological innovation and advancement in fuel-cell materials, more elegant system solutions are expected to emerge.

11.5 Fuel-Cell Vehicles Versus Hybrid Vehicles

It is documented that PEFC-Ni/MH and PEFC-Li-ion battery powered hybrid cars are more efficient than the FCVs (Britsche and Gutmann 2004). Demirdoven and Deutch (2004) have stressed on the priority to deploy hybrid cars against FCVs. Unlike ICEs, FCS exhibit higher efficiency at part loads. This is particularly advantageous for automotive applications since the vehicles are mostly operated under part load conditions; the average demand on standard U.S. drive cycles is <20% of the rated power of the engine. A recent study suggests that the fuel economy of hydrogen FCVs could be ~three times the fuel economy of the ICEVs (Ahluwalia et al. 2003).

Since the fuel cells are more efficient at part load than at full rated power, hybridizing a FCV is different from hybridizing the ICEVs. One motivation for hybridizing the FCVs is to improve their fuel economy by recovering a portion of the braking energy. Hybridization can also help if the energy storage device has higher specific power ($kWkg^{-1}$) at lower cost ($\$kW^{-1}$) in relation to the FCS to make the hybrid system lighter and less expensive. Owing to the higher part load efficiency, even in a hybrid configuration, it appears advantageous to preferentially operate the FCS in a load-following mode and to use the power from the battery on demand.

The automotive industry is pinning its hopes on two major automotive technologies; namely, HEVs and FCVs. Fuel-cell system design is heavily influenced by automotive requirements that are generally application specific. Engine peak-power is usually determined by the degree of acceleration, grade, or top speed performance targets. Maximum rated power may be either a continuous requirement or for a finite duration lasting tens of seconds. For the latter, hybridization, combining a down sized FCS designed for continuous maximum-power with an energy buffer, typically a battery bank, to augment the peak power may be an option. To date, early prototype light-duty passenger vehicles are designed between 50 and 80 kW, both with and without hybridization. Full size bus applications are also focused

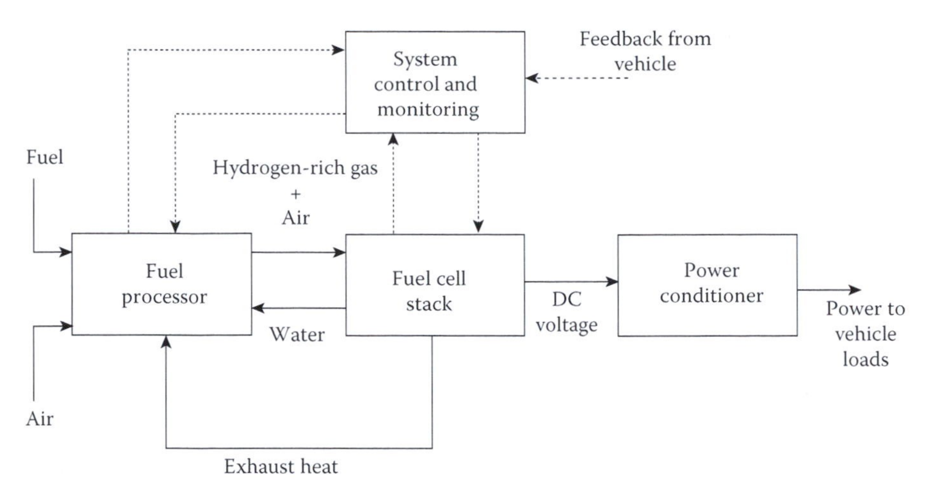

FIGURE 11.3
Schematic diagram of a fuel-cell system with fuel processor.

on 100–200 kW both with and without hybridization. GM has developed both hybrid ("HydroGen 1" and "Gasoline Fuel-Cell S-10") as well as nonhybrid ("HydroGen2" and "HydroGen3") FCVs. Market forces, vehicle target requirements, and FCS trade-offs will determine both power and hybridization levels for FCVs.

A low-voltage DC is produced as the fuel-cell output, which is applied to an electric machine through a suitable DC/DC or DC/AC converter. The electrical machine drives the wheels of the vehicle. The schematic diagram of a basic FCV power system with an onboard fuel processor is shown in Figure 11.3 (Emadi and Williamson 2004).

There are three major steps involved in the generation of power from a fuel-cell. The first step is to achieve purity of the available hydrogen from a fuel processor wherein suitable hydrocarbon fuel is fed that, in turn, produces a hydrogen-rich gas stream at its output. This hydrogen-rich gas stream is then fed to the anode of the fuel-cell. The generation of the DC voltage through the fuel-cell makes up the second stage of the power-processing unit. Lastly, the power output needs to be properly harnessed, which is achieved through an appropriate power conditioner. Ideally, the power conditioner must have minimal losses for higher efficiency. Power conditioning efficiencies can typically be higher than 80%.

In addition, there exists a system controller, which is a vital link between the FCS and the load. This controller receives feedback from the vehicle and sends control signals to the fuel cell stack. The power-conditioning unit (PCU), in turn, provides the appropriate power to the load. The control methodology of the PCU generally involves an advanced power electronic controller.

11.6 Fueling Options for Fuel-Cell Vehicles

Besides the fuel cell stack, another subsystem influencing the vehicle cost and performance is the fuel storage. The key issue in the use of fuel cells for automotive applications is the availability of hydrogen fuel. Without a widespread network of hydrogen stations, no large scale introduction of fuel-cell cars is feasible and, without a substantially large network of fuel-cell cars, large investments in the hydrogen infrastructure is unlikely. Onboard storage of hydrogen is critical to the success of FCVs to address the issues concerning driving ranges comparable to gasoline cars, low-storage volume, low weight, and low cost.

The state-of-the-art hydrogen storage options include compressed hydrogen, liquid hydrogen, and metal hydrides. U.S. DoE has published targets for onboard hydrogen storage devices (Ross 2006). The energy storage density for hydrogen (in MJ/l) amounts to 2.8 for hydrogen as gas at 345 bar to a value between 7 and 12 for metal hydrides, 8.4 for liquid hydrogen, and 31.1 for gasoline.

DoE has established a series of targets as indicated below.

- Gravimetric energy density: 2 kWh/kg
- Volumetric energy density: 1.5 kWh/l
- H_2 storage capacity (mass fraction) of 6 wt% (on a system basis)
- Operating temperature: –30°C to + 50°C
- Refueling time <5min
- Refueling rate: 1.5 kg H_2/min
- Recoverable amount of hydrogen: 90%
- Cycle life: 500 times (requirements for the physical properties of storage materials)
- Cost targets: U.S. $5/kWh (storage materials without peripheral components)

There are four options for onboard hydrogen storage; namely (a) liquid hydrogen at 20–30 K, 0.5–1 MPa, (b) compressed gaseous hydrogen at 35–70 MPa and at room temperature, (c) solid-state storage using hydrides or high-surface materials, and (d) porous solid adsorption of molecular hydrogen. The first two methods have reached the engineering prototype stage while with the last two methods much remains in finding the optimum system for further development.

Liquid hydrogen can be stored onboard the vehicle as demonstrated by BMW. In these vehicles, most of the hydrogen is supplied to an ICE with a part going to a fuel-cell that provides the electrical power for air-conditioning, and so on. This approach yields a gravimetric storage density of 100% H_2 without the weight of the container. However, if the mass of the tank is included, it is reduced to about 10% of gravimetric storage density of H_2. This has two major disadvantages, namely an inevitable boil-off rate that is currently ~1%/day and there is high-energy loss due to the refrigeration process that amounts to 30% of the energy available by burning hydrogen. Until very recently, liquid hydrogen has also been considered to be a technologically viable option for automotive application. But its drawbacks concerning efficient thermal insulation needs to be addressed satisfactorily. Also, due to the low-operating temperatures between 20 and 30 K, heat flow looks inevitable. This heat input comprises three fractions, namely thermal conduction, convection, and thermal radiation. Among these, the thermal conduction through pipes and cables to the inner storage vessel and the heat radiation from the environment to the cryogenic liquid predominate. To achieve the low overall values of aforesaid heat transfer, it is important to work with cylindrical tank structures as this geometry is closer to the optimal surface-to-volume ratio. Additionally, it is required to implement an efficient multilayer vacuum super insulation with about 40 layers of metal foil. Wrapping the metal foil around the storage vessel in general and around the dome area in particular is time consuming and highly demanding. The remaining heat input leads to enhanced evaporation of the liquid hydrogen stored inside that eventually causes a pressure rise. Typically, when a system pressure of about 1 MPa is reached, a valve needs to be opened for venting hydrogen. The complexity of the liquid hydrogen storage system together with the challenge to reduce evaporation adds to liquid hydrogen storage system costs that are not favorable over compressed gaseous hydrogen storage systems. Also, the energy required to liquefy hydrogen is about 30% of the chemical energy stored compared to just 15% for 70 MPa compressed hydrogen with reference to the net calorific value of 120 MJ kg^{-1} hydrogen.

The second approach is to store hydrogen at high pressures. Since the volumetric storage density of the compressed hydrogen tank is rather low, the packaging of such a fuel system into the existing mass production vehicle architecture remains a challenge. Because of the comparatively high-operating pressure of these vessels, a cylindrical design is mandatory. Despite limitations of the compressed gaseous-hydrogen approach, this option yields the best overall technical performance and has the highest maturity for automotive applications.

A conventional steel cylinder stores about 1% gravimetric hydrogen. Recent developments of fiber-reinforced resin have reached pressures of 700 bar, corresponding to about half the density of liquid hydrogen. Hydrogen-fueled buses under Clean Urban Transport for Europe (CUTE) use hydrogen stored in conventional steel cylinders at 150 bar.

Solid-state absorbers of hydrogen offer an impressive volumetric hydrogen density. A short refueling-time requirement causes a significant engineering burden on the system. Considering a 6 kg H_2 tank system comprising a storage material M with the heat of formation ΔH ($M + H_2 \rightarrow MH_2 + \Delta H$) of about 25 MJ kg^{-1} H_2, a thermal load of 150 MJ is desired during refueling.

This leads to an average heat exchange of 800 kW. Such a high-performance device is unimaginable onboard a vehicle due to cost, volume, and weight considerations. A further constraint is that many solid-state absorbers have to be limited to 70°C to be practical, which could be reached by using the waste heat of the FCS. Hydrogen absorbers often consist of powder materials. The absorber accounts for 50% of the total weight, whereas 50% is due to valves, pipes, pressure vessels, heat exchanger, and so on. A 70 MPa compressed gaseous hydrogen storage system is currently the best-in-class option available for hydrogen storage onboard a vehicle.

The most important factor for the success of FCVs lies in the success of the hydrogen economy and related technologies. McNicol, Rand, and Williams (2001) reported that a FCV system equipped with a direct conversion fuel processor could compete successfully with conventional ICEVs. Buses and recreation vehicles have more space to house the fuel processor than passenger cars. Onboard fuel reforming is another option to utilize the existing liquid hydrocarbon fuel infrastructure and to produce hydrogen onboard. The desired reformer-FC system is obviously more complex than an onboard hydrogen storage FC system.

Reforming for automotive applications requires an air-assisted primary reforming called Catalytic Partial-Oxidation (CPO) or Auto-Thermal Reforming (ATR). During CPO, air is applied together with a hydrocarbon fuel while, during ATR, air is supplied together with steam and a hydrocarbon fuel. The gas treatment consists of shift conversion and CO removal.

The architecture of the reformate system is influenced by the type of the fuel used. There are many candidate fuels such as methanol, ethanol, naphtha, low sulfur gasoline, and dimethyl ether. It is advisable to use the existing infrastructure for fueling FCVs. Performance is also affected by the architecture of the reformate system. A gasoline FCV equipped with an ATR is estimated to emit almost the same level of CO_2 as a diesel HEV. Accordingly, performance should be improved to be competitive with HEVs in terms of fuel economy. If the target for volumetric power density is achieved, a reformate system with a volume of 60 l will be able to provide a power output of 100 kW by 2010. A system of that size will be easy to install under the cabin floor. A small volume reformate system will reduce heat losses from each component, resulting in improved system efficiency. In order to reduce the volume of each component, it will be necessary to enhance the performance of the catalyst used in ATR, shift reactors, and preferential oxidation (PrO_x) reactors, especially for activity under gas high space velocity (GHSV) conditions. It is noteworthy that using ceramic foam as a catalyst substrate helps reduce the volume of the ATR reactor.

Onboard reforming presents a number of technical and economical challenges that include volume targets, weight targets, and start-up requirements.

These requirements set a number of characteristics for the reformer system and in particular for the catalysts used in different reactors as indicated below (Oosterkamp et al. 2006):

- Higher catalytic activity than conventional palletized catalysts
- Good stability for dynamic operation and fast load changes
- Nonpyrophoric properties, ability to operate under air atmosphere
- Good sulfur resistance
- Small size and low weight GHSVs as shown in Table 11.2
- Low costs
- Fast and safe start-up/shut-down
- High turn-down ratio
- Low pressure (0–few bar)
- Minimum number of heat exchangers
- Integration of heat exchangers with reactors

Reducing start-up time is a difficult issue. A compact reformate system would be effective in achieving quicker start-up. However, an efficient start-up system needs to be developed. A start-up combustor coupled with a vaporizer might be useful to this end. ATR reformate systems appear promising for automotive applications in the near future. Although many companies have abandoned the idea of onboard reforming, Nuvera, together with Renault is developing a concept based on an onboard autothermal reforming for passenger cars.

To better understand the critical issues related to the commercialization of the FCVs, a brief discussion on the PEFC operation, related subsystems, and power-train configurations are introduced in the following subsections.

TABLE 11.2

Fuel Processor Targets and Industrial Practice

	Fuel Processor for Automotive Applications	Industrial Hydrogen Plant
Process	**GHSV (h^{-1})**	**GHSV (h^{-1})**
Reformer	200,000	1000
Water gas shift reactor (HTS)	60,000	1000
Water gas shift reactor (LTS)	60,000	3000
CO removal	150,000	3000
Pressure (bar)	1–10	>30

Source: From Oosterkamp, P., Kraaj, G., Laag, P., Stobbe, E., and Wouters, D., *WHEC* 16, June 13–16, 2006.

11.7 Fuel-Cell Vehicle Integration

The development of FCVs requires onboard integration of a FCS and electrical energy storage device with an appropriate energy management-system. In order to evaluate the FCV with onboard PEFC, it is important to drive test them according to a standard-duty drive cycle that includes reiterations such as start/stop, acceleration, cruising, and braking. The integration of the FCS into vehicles can be achieved in a manner akin to ICEVs. It has been demonstrated that sufficiently powerful and compact drivetrains are possible to realize. Cylindrical vessels are required for hydrogen storage. The existing vehicles lack the space for hydrogen storage for providing sufficient drive range. The system design and integration involve a choice of fuel cell stack sizing, operating conditions such as system pressures, temperatures, humidification levels, subsystem components, control strategies, and a myriad of other parameters to meet the requirements; namely, low mass, restrictive packaging, fast start, fast dynamic operation, large turndown ratio; that is, the ratio of maximum-to-minimum power, extreme environmental conditions, freeze tolerance, long lifetime, efficiency, and cost effectiveness. Detailed steady-state dynamic modeling and simulation tools as well as component, subsystem, and integrated vehicle testing are desired to optimize these parameters.

11.8 Technical Issues in Fuel-Cell Vehicles

Arita (2002) highlights the technical issues of FCSs for automotive applications. Arita added that achieving the targets for automotive applications will require technical breakthroughs in efficiency, compactness, water and thermal management, durability, and cost of PEFCs (Table 11.3).

In order to improve the total efficiency of the PEFC stacks, it is essential to improve the efficiency of the cells constituting it. Reducing the activation losses at the cathode is seminal for improving efficiency. One of the most promising components in this regard is the application of a Pt-alloy catalyst. In general, the gas diffusion layer (GDL) must be optimized to reduce the mass transfer loss, and reduction of membrane resistance is necessary to lower the ohmic loss and contact resistance between each component. Compactness is an important requirement for installing the FC stack in a vehicle. It is necessary to enhance the power density of the FC stack to achieve greater compactness. To attain high power density, the flow-field design of the bipolar plates should be optimized to improve the performance of the membrane electrode assembly (MEA). Another promising measure for achieving compactness is the utilization of thinner bipolar plates. The

TABLE 11.3

Current Status and Targets for PEFCs

Parameter	Part	Current Status	Target
Efficiency	Rated (A/cm^2)	About 50% (gross)	>60% (gross)
Power density	Stack	0.9–1.5 kW/l	>3 kW/l
Operating temperature	Coolant	80°C	>80°C
Operating pressure		Ambient – 2 bar	Ambient – 2 bar
Humidification		Necessary	No external humidification
Low-temperature start-up	Cold stability	>0°C	–30°C
	Storage temperature	>0°C	–40°C
Reliability		>1,000 h	>5,000 h
Cost		About $4,500/kW	About $10–15/kW

Source: From Arita, M., *Fuel Cells,* 2, 10–14, 2002. With permission from Wiley-VCH Verlag GmbH & Co.

flow channels are provided on the surfaces of the bipolar plates to a depth of around 0.5 mm. Therefore, it is difficult to reduce the plate thickness below 1 mm with channels on either side.

In a PEFC, the ionic conductivity of the membrane is proportional to its water content, and protons drag water molecules from the anode to the cathode during cell operation. Although water is essential for maintaining stable operation, it freezes below 0°C, but FCVs must be capable of starting and running at temperatures well below the freezing point of water. Water in the electrolyte membrane does not freeze at –20°C, so the stack can generate electricity at these temperatures. On the other hand, since water in the GDL and the catalyst layer freezes, it is necessary to operate the stack such that the generated heat energy can be used to prevent water from freezing in these components. The power output of the stack is limited at these temperatures. Therefore, the freezing problem is one of the critical issues that need to be solved before FCVs are put in the market. One solution to the problem is to eliminate the external humidifier in the FC stack. Several measures have been proposed to this end. Water back diffusion is shown to be a promising measure for facilitating nonhumidification operation. Intelligent Energy in the UK is targeting automotive applications using evaporative cooled stacks. The stack requires no external humidification, has low-pressure operation, and requires minimal balance of plant.

Another critical issue is to maintain the thermal balance of the FC system under all driving conditions. At present, operating temperature for PEFC stacks with Nafion membranes is around 80°C, which is lower in relation to the ICEVs. But lower temperatures are not desirable for automotive applications. Heat from the coolant is released to the air by using the temperature difference between the coolant and the air. In the case of a conventional ICE, the highest temperature of the coolant is about 110°C, so there is a temperature difference of 80°C when the ambient air is 30°C. In the case of a PEFC stack, however, the difference is only ~50°C. Therefore, the radiator of the PEFC-stack system must be 60% larger in size than that for an ICE. However, it is difficult to find space for installing such a large radiator in a vehicle. A high temperature membrane capable of withstanding temperatures between 110 and 120°C needs to be developed to resolve this issue.

The fuel supply infrastructure for FCVs is critical to their popularization. However, hydrogen supply infrastructure exists scarcely at present. FCVs need to have the same range as ICEVs. The tank mileage of FCVs should be at least 500 km. Current FCVs can travel almost 100 km/kg of hydrogen. Therefore, the hydrogen storage material and system must be capable of storing at least 5 kg of hydrogen. There are many candidate materials and systems for storing hydrogen. A high-pressure hydrogen gas tank might be promising at present. Capability of hydrogen gas containers has been rising from 25–30 MPa, and is expected to increase to 50 or 70 MPa in the near future. Carbon nanotubes could be a promising material, but only store 0.5 wt.% hydrogen. Metal hydrides can store around 2 wt.% hydrogen but their hydrogen storage capacity needs to be improved to > 5 wt.%. Another critical issue for the use of metal hydrides is the time required to store hydrogen. Several tens of minutes are required for refueling. Alkaline hydride compounds have good hydrogen storage capacity, but are difficult to recycle. It might be difficult to use existing hydrogen storage materials in automobiles. Recently, ammonia borane has been projected as a promising hydrogen carrier. The efficiency of peripheral systems such as the air supply system and the heat release capability of the cooling system also need to be improved to achieve a compact and high-performance FC system.

11.9 FCV Power Train Configurations

The challenge in designing automotive fuel-cell power systems is in converting the electrical output from the fuel-cell into usable power for varying system sizes (Williamson and Emadi 2002). Furthermore, the goal is to realize the full potential of the fuel-cell technology by using efficient methods to convert fuel-cell output to a useful electrical/mechanical energy. In addition, the conversion process must be cost effective.

The primary power required is a three-phase variable AC output for the traction motor of the vehicle. It is noteworthy that a high peak-to-average power ratio is generally required for the power electronics. If there is an auxiliary energy storage system, the fuel-cell also needs to maintain the critical power ratio to the maximum. Secondary converters would be used for the 12 V or 42 V auxiliaries, and about 140 V or 300 V for other loads such as pumps, air-conditioning, and power-steering applications. It is important, at the same time, to maintain high levels of efficiency, low-electromagnetic-interference, low-acoustic disturbances, and most importantly, low costs. It is critical for fuel-cell based automotive power electronic systems to last approximately 10–15 years. In addition, an alternative automotive application involves providing power for auxiliary loads, where the peak-to-average power ratio may be more suitable for fuel-cell operation.

A simple DC/AC inverter system is suitable only if the fuel cell stack can produce a voltage suitable for the inverter to operate without any additional converter requirements. Mostly, this is not the case and, hence, an additional DC/DC booster is required between the fuel cell stack and the DC/AC inverter. Another possible arrangement involves the usage of a higher current output inverter and a 60 Hz output transformer, when the DC link voltage is below the required inverter operation voltage, typically about 400 V.

The schematic for the power system of a fuel-cell based drivetrain for a FCV is shown in Figure 11.4. The arrangement shown in Figure 11.4 also includes a battery pack and has regenerative braking (Williamson and Emadi 2002). The secondary battery provides an input DC voltage to the DC/AC inverter during warm-up time of the fuel-cell. Once the fuel-cell warms-up, the battery is removed from service and the system is run on the fuel-cell and the dual power conversion circuitry. The battery system also provides voltage during transient conditions. The traction controller basically sends control signals to the fuel cell, DC/DC converter, and DC/AC inverter depending on the feedback speed and torque signals as well as drive commands. Thus, the speed and torque of the traction motor is controlled at all times, which, in turn is coupled to the vehicle transmission system.

A typical hybridized power system with a lead–acid or nickel/cadmium battery pack connected in parallel with the fuel-cell in order to meet the peak power demand to also take advantage of regenerative braking is shown in Figure 11.5. The DC/DC converter raises the level of the voltage from the fuel cell stack to the level of the main DC bus voltage. The initial peak power during transients, such as start-up and acceleration, is supplied by the battery pack. Varying DC and AC loads are fed from the main DC-bus through appropriate DC/DC converters and DC/AC inverters. Furthermore, the electric motor is controlled by a motor controller system, which, in turn, drives the wheels of the bus. On changing the speeding and braking commands, a suitable signal is fed back to the controller, which accordingly controls the speed and torque delivered by the motor.

Another configuration for fuel-cell buses includes a nonhybridized topology. In this case, direct hydrogen fuel cells are used to provide the entire power to the propulsion system. XCELLSiS Fuel Cell Engines, jointly owned by Daimler-Chrysler, Ford Motor Company, and Ballard Power Systems, are involved in the development of such fuel cell engines for automotive applications.

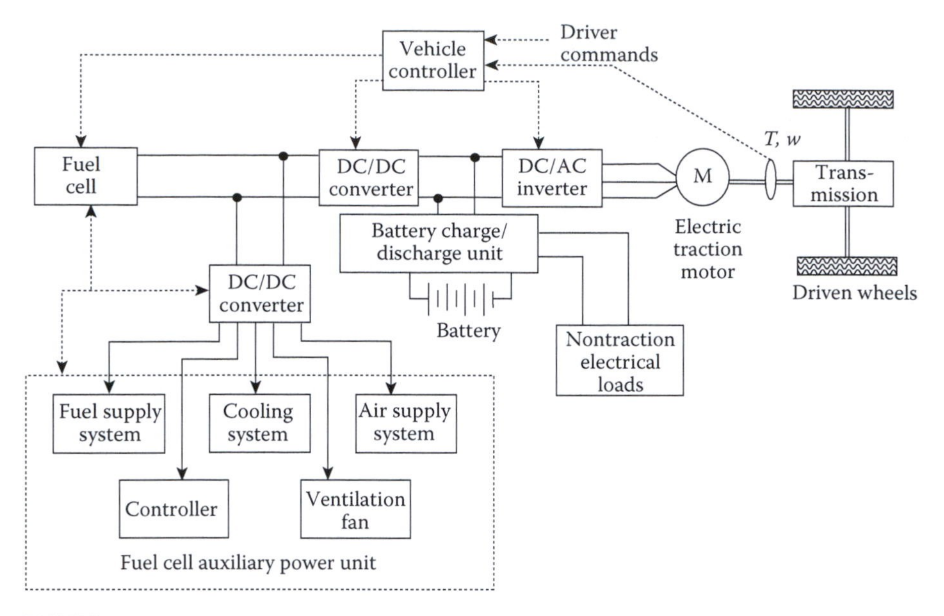

FIGURE 11.4
Schematic of a fuel-cell based power system for a passenger car.

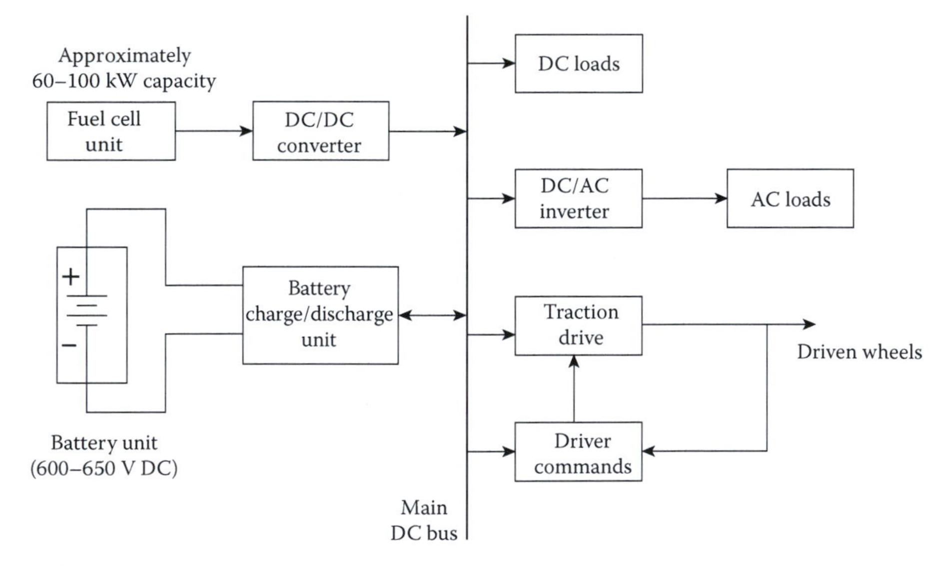

FIGURE 11.5
Typical block diagram of a heavy-duty, fuel-cell powered transit bus.

The drivetrain of a vehicle basically comprises the flow of power from the source to the ultimate sink, where it is used. A typical hybrid fuel-cell based drivetrain is shown in Figure 11.6.

In the case of direct hydrogen FCV, the output voltage from each cell is collected and sent to the power electronic converter stage, where it is converted to usable power by using a DC/DC converter. This power is then supplied to the motor based on the optimized control strategy. The power control unit processes acceleration commands of the driver and produces the necessary power for specific operation of the FCS.

In HEVs, there are two popular topologies. The first is the series hybrid configuration and the second option is the parallel hybrid topology. The difference between the two is presented schematically in Figures 11.7 and 11.8. A typical drivetrain configuration for a series HEV is shown in Figure 11.7 wherein the electric motor is connected to the wheels; the electric generator supplies electricity to the battery pack that, in turn, feeds the traction motor. Generally, the engine/generator set maintains the battery charge at around 65–75%. The engine starts-up when the battery reaches its lower limit, when shut off the battery reaches its upper limit (Moghbelli, Ganapavarapu, and Langari 2003).

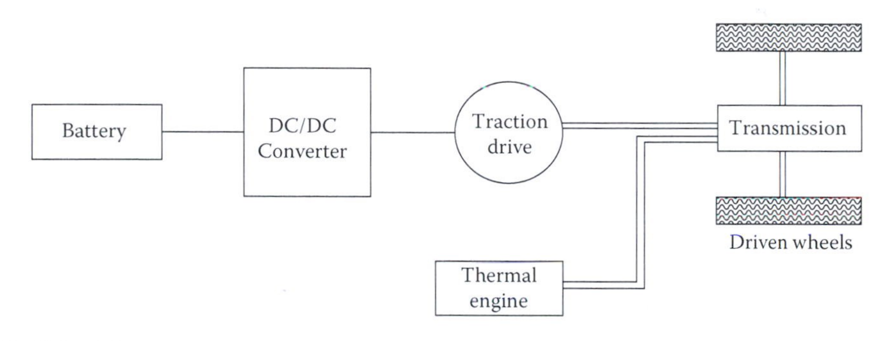

FIGURE 11.6
Typical hybrid fuel-cell based drivetrain.

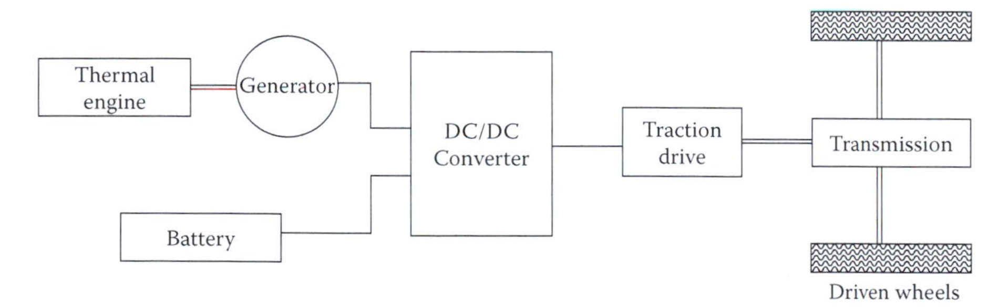

FIGURE 11.7
Schematic block diagram of a series HEV drivetrain.

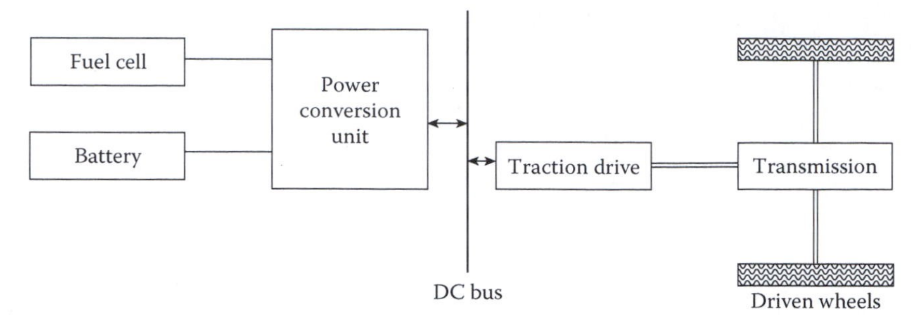

FIGURE 11.8
Schematic block diagram of a parallel HEV drivetrain.

A typical layout of a parallel HEV is shown in Figure 11.8 wherein the wheels are connected to both the electric motor and the heat engine. Parallel HEV topology does not need a generator since the traction motor itself serves this purpose.

Auto manufacturers have turned their attention to hybrid electric drivetrains as propulsion systems in order to solve the range problems of the electric drivetrains. The two popular topologies are the ICE/battery hybrid and fuel-cell/battery hybrid. Among these, ICE/battery hybrids are already in production.

Emissions from FCVs depend on the fuel used. Tailpipe emissions from hydrogen-fueled vehicles are zero. Daimler Chrysler Necar 5 is the fuel-cell equivalent of the Mercedes-A passenger car and runs on methanol. All tailpipe emissions of Necar 5 are lower than their ICE equivalent (Bruijn 2005).

A considerable number of FCVs are presently being tested and demonstrated on the road. These tests show the advancement of fuel-cell technology in terms of their robustness, compactness, and driving performance. Table 11.4 gives a summary of some FCV demonstrations (Bruijn 2005).

The majority of these vehicles are run on hydrogen. Daimler Chrysler, Toyota, and General Motors (GM) have demonstrated vehicles, which produced hydrogen onboard using fuel processors, mostly running on methanol. At present, most manufacturers are focusing on further development of vehicles with onboard hydrogen storage.

Fuel-cell technology has already demonstrated its feasibility and advantages as a power source for automotive applications. In order to become competitive, there needs to be improvements in performance, lifetime, and cost. To reach the goal, accurate modeling is desired. The fuel cell generator is a key component of the power train. The fuel-cell stack is subjected to rapid change in load currents. A good knowledge of dynamic behavior is essential. In this context, a model based on the electrochemical impedance is proposed by Garnier et al. (2003).

TABLE 11.4
Fuel-Cell Vehicle Demonstrations

Company	Vehicle Type	Fuel	Year	Accomplishment
Daimler Chrysler	Small passenger car: Necar 5	Methanol	2002	U.S. coast-to-coast trip, 4500 km
Volkswagen	Midsize passenger car: Bora HY power	Hydrogen	2002	Midwinter mountain trip across Simplon Pass at −9°C
General Motors	Midsize passenger car: HydroGen 3	Hydrogen	2004	10,000 km Journey in 38 days
Daimler Chrysler	Small passenger car: Mercedes A F-cell	Hydrogen	2004	60 cars in operation in Germany, Japan, United States, and Singapore
Honda	Passenger car: FCX Clarity	Hydrogen	2007	Named 2009 World Green Car
Nissan	SUV: Nissan X –Trail	Hydrogen	2009	Lease of a X-TRAIL Fuel-Cell Vehicle (FCV) to Sacramento Coca-Cola Bottling Co., Inc.
Daimler Chrysler	Bus: Citaro	Hydrogen	2003	30 Buses in 10 European cities in daily operation

11.10 Fuel-Cell Powered Buses

Fuel-cell powered buses have been in development since the 1990s (Figure 11.9). The advantage of fuel cells in buses are multiple. From a technical point of view, the availability of space has made it easy to integrate the system and hydrogen storage in a bus without sacrificing the available space for passengers. In 2003, Folkesson et al. (2003) evaluated hybrid urban FC buses for the CUTE project. The aim of the project was to design, build, and demonstrate hybrid-FC buses. The FCS had a maximum power output of 50 kW. The fuel was compressed hydrogen and the oxygen used in the fuel cell was derived from compressed ambient air. An integrated DC/DC converter adjusted the fuel-cell voltage with the voltage of a common power bus of 600 V. The propulsion system was located at the rear of the bus. The entire system was comprised of the FCS, battery, wheel motors, power electronics, and auxiliaries that could be easily removed from the rest of the bus in order to facilitate its servicing and other duties on the system. The heart of the FCS was a stack module containing two PEFC stacks, each comprising 105 cells. The stack assembly measured 58 cm in height, 42 cm in width and 57 cm in length, and had a total volume of 139 l. This system had a power density of about 0.2 kW/l. The mean-power consumption was about 17–24 kW during

FIGURE 11.9
Fuel-cell buses.

the test duty cycle. It was claimed that a FCS with a nominal power output of approximately 35–50 kW is adequate for a full size hybrid electric bus, even with a 20–25 kW air-conditioning system. The net efficiency of the FCS was approximately 40% and its fuel consumption was between 42% and 48%. In addition, bus subsystems such as the pneumatic system for door opening, suspension, brakes, hydraulic power-steering, water pump, and cooling fans consumed about 17% of the net power output from the FCS.

11.11 Fuel-Cell Vehicle Scenario

11.11.1 FCV Activities at GM

General Motors (GM; Rodrigues, Fronk, and McCormick 2003) demonstrated the first FCV in 1966. The vehicle comprised cryogenic hydrogen and oxygen tanks, a potassium hydroxide tank, and a 32-cell stack module. The system produced 32 kW continuously with a peak power of 60 kW. The driving range of the vehicle was 150 miles with a top speed of 70 miles per hour (Figure 11.10).

The development strategy of GM is to build stack technology to understand and address fundamental stack and system interfaces and architecture trade-offs. This includes reformate compatible stack modules as well as hydrogen compatible stack modules. The development strategy for fuel processing is to include gasoline reforming as well as reforming of other fuels such as natural gas. GM's fuel cell stacks have significantly improved over the years. Starting from the early large stack development in 1997 with Gen3 where power densities were 0.26 kW l^{-1} (0.16 kW kg^{-1}), volumetric density has increased almost seven times while the gravimetric density has increased about eight times. Although the fuel-cell program at GM initially

FIGURE 11.10
GM Chevrolet Equinox FCV.

concentrated on methanol-steam reforming, GM changed its focus to gasoline reforming because of the noncommercial viability of methanol as a fuel. Gasoline reforming activities at GM have made several accomplishments over the years in performance and power densities. The volumetric power densities of gasoline reformers (based on kW of H_2) have increased 2.5 times over a two-year period and the gravimetric power densities have increased three times over the same period. Also, start time of the fuel processor has decreased by about nine times, but it still needs further improvement. The Gen3 fuel processor has been successfully integrated and demonstrated in an S10 pick-up truck. The S-10 fuel-cell generates 25 kW of net electric-power. The Gen3 gasoline reformer in the S-10 reforms "clean" gasoline onboard and is capable of starting in less than three minutes. The reformer has a peak efficiency of 80%. The fuel processor and stack combine with a battery charger in the vehicle electric drivetrain. Onboard gasoline reforming is significant because all other fuel cells run on pure hydrogen or hydrogen from a methanol reformer. Developing a gasoline fed FCS would make the technology more flexible.

One of the earliest PEFCV demonstrations was Zafira, an open compact van with a 50 kW fuel-cell unit fed by onboard steam reformed hydrogen from methanol. The HydroGen1 is a five-seat concept vehicle based on Opel's Zafira compact van. It is powered by a 60 kW fuel-cell module using hydrogen and ambient air with the system efficiency of about 40% at 60 kW.

11.11.2 FCV Activities at Honda

Honda has extensive experience in the development of BEVs (Matsuo 2003). Based on similar vehicle structure and electric drivetrain technology, the

FCV was developed by replacing the battery box with a FCS and related components. This accelerated the early FCS development and helped Honda in realizing a high level of vehicle and system performance over a short period. FCX-V1, Honda's first FCV, had a simple hydrogen storage system with a metal hydride tank (Figure 11.11). The second, FCX-V2 experimental vehicle was fueled with methanol and incorporated a Honda-designed ATR system. It was powered by a first-generation, Honda-designed PEFC stack rated at 60 kW and propelled by a newly developed electric drive motor with increased output and efficiency. FCX-V1 and FCX-V2 were early proof-of-concept vehicles. Honda has also advanced the technologies for efficient vehicle energy management and control, electrical energy storage devices, and gaseous fuel storage systems. Technological development of FCVs at Honda is summarized in Table 11.5.

11.11.3 FCV Activities at Daimler Chrysler

New Electric Car (NECAR 1) truly demonstrated the feasibility of using fuel cells for automotive applications (Schmid and Ebner 2003). NECAR 2, a six-seater minivan, fueled with compressed hydrogen, has a downsized FCS with improvised power, maximum speed, and range. NECAR 3 was the first car with a methanol reformer, demonstrating the feasibility of hydrogen generation onboard a compact car. NECAR 4, introduced in 1999, uses liquid hydrogen as fuel with the tank-to-wheel efficiency at 37%. Daimler Chrysler developed an improved version of NECAR 4 with a new generation fuel-cell by storing hydrogen in three aluminium tanks wrapped with carbon foils and pressurized up to 350 bar. NECAR 5, an improved version of NECAR 3, combines new fuel-cell technology, the MARK 900 stack series, with more effective thermal management of the reformer with further reduction in size and weight of the system. NEBUS (New Electric Bus), the first Daimler-Benz

FIGURE 11.11
Honda FCX Clarity zero-emission hydrogen powered fuel-cell sedan.

TABLE 11.5

Technological Development of FCVs at Honda

Parameter	FCX-V1	FCX-V2	FCX-V3	FCX-V4	FCX	New FCX
Hydrogen supply	Hydrogen-absorbing metal alloy tank	Methanol reformer	High-pressure hydrogen tank (250 atm)	High-pressure hydrogen tank (350 atm)	High-pressure hydrogen tank (350 atm)	High-pressure hydrogen tank (350 atm)
Hydrogen storage capacity	—	—	100 L	137 L	156.6 L	156.6 L
Fuel-cell stack	Ballard	Honda	Ballard/Honda	Ballard	Ballard	Honda
Power assist	Battery	Battery	Ultra-capacitor	Ultra-capacitor	Ultra-capacitor	Ultra-capacitor
Motor max. output	49 kW	49 kW	60 kW	60 kW	60 kW	80 kW
Max. speed	—	—	130 km/h	140 km/h	140 km/h	150 km/h
Vehicle range	—	—	180 km	315 km	355 km	430 km
Number of occupants	2	2	4	4	4	4

fuel-cell bus introduced in 1997, was designed for city and regional transportation of up to 58 passengers. The power output from the 250 kW FCS is similar to conventional diesel engines.

11.11.4 FCV Activities at Nissan

Nissan developed a methanol reformer integrated FCV, R'nessa FCV, in 1999. In the year 2000, Nissan developed X-TRAIL FCV with a 63 kW UTC PEFC stack, compact Li-ion battery, and a 350 bar hydrogen-storage system (Figure 11.12). An improved version X-TRAIL FCV 2003 model comprised a fuel cell stack developed by Nissan delivering an enhanced power output of 90 kW, making the maximum speed and acceleration comparable to a gasoline engine. It was the first car commercialized for limited leasing. Nissan also owns the credit of releasing a fuel-cell powered car for hired service. Nissan is presently testing an advanced version with a new generation fuel-cell that is 25% smaller and 40% more powerful than the earlier 90 kW version. Its electrodes contain around 50% less Pt and it also has a new 700 bar hydrogen storage cylinder that increases its vehicular hydrogen storage capacity by approximately 30% in relation to earlier 350 bar cylinders without any change in the cylinder's external dimensions. The new high-pressure storage cylinder is made of an inner aluminum liner and an outer shell of several layers of a high strength, high elasticity, carbon fiber.

11.11.5 FCV Activities at Ballard

Ballard is recognized as the world leader in developing, manufacturing, and marketing zero-emission hydrogen fuel cells. Ballard is developing a sixth-generation fuel-cell module for the bus market, incorporating the most advanced fuel-cell technology available to date (Figure 11.13). The next generation, FC velocity-HD6, heavy-duty, fuel-cell module will power fuel-cell

FIGURE 11.12
Nissan X-Trail FCV.

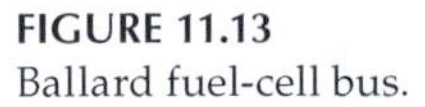
FIGURE 11.13
Ballard fuel-cell bus.

buses operated by British Columbia Transit (BCT) during the 2010 Olympics and Paralympic Winter Games. The BCT fleet of 20 buses will be the largest fleet of fuel-cell buses in the world.

11.12 Fuel-Cell Vehicle Market

No fuel-cell is yet ready for commercialization. In 2007, the total number of light-duty FCVs deployed worldwide was approximately 800, with an additional 3000 niche vehicles for powering forklifts, motorcycles, and marine craft (Adamson, Butler, and Hugh 2008). Schafer et al. predict that a market competitive, light-duty FCV would be available in approximately 15 years but a major fleet penetration could take more than 50 years (Schafer, Heywood, and Weiss 2006). During 2008, Daimler, Ford, GM, Honda, Hyundai, and Toyota had announced plans to commercialize FCVs anywhere between 2012 and 2025 (Martin 2002). To get to this stage, however, breakthroughs in materials and fabrication methods will be necessary.

The FCVs will get to the mass market only if they can be produced at an affordable cost of about U.S. \$50/kW. Projections suggest this to be a reasonable target, if FCVs are produced in high volumes. The best cost projection can be made for components where the construction materials are well established with proven manufacturing technologies. Components like air compressors and hydrogen recycling pumps, based on traditional manufacturing processes, need to be produced with existing equipment for the cost to become comparable with the automotive target. For new technology elements where there is no established production process, both production volume and manufacturing processes remain uncertain. For materials, for example, polymer electrolyte membranes that are used in small quantities, large production would be required. The cost of FCSs for mobile applications at present is estimated to be \$325/kW at a production level of 500,000 units per year (Adamson 2008). U.S. DOE has set technical as well as cost targets for mobile FCSs that have to be met in order to become competitive with ICEVs. The direct hydrogen fuel-cell power system has to have 60% electric efficiency at a cost of \$45/kW by 2010 and \$30/kW by 2015. Alternatively, a reformer-based-fuel-cell-power system operating on clean hydrocarbon or alcohol that meets emission standards has to have a 45% electric efficiency at a cost of \$45/kW by 2010 and \$30/kW by 2015. The start-up time for a reformer-based system should be less than 30 s. The total cost consists of \$5/kW for the membrane, \$5/kW for the electrodes, \$10/kW for the MEA, and \$10/kW for the bipolar plates. For PEFCs, the challenge is catalyst cost reduction.

The search for nonplatinum catalysts has been on for more than 40 years. At low temperatures and using hydrogen, platinum-based catalysts remain most active (Arico et al. 2005). One could potentially use a greater amount of a cheaper, less active catalyst, but the acidic environment of PEFCs rules out non-noble metals at present (Gasteiger et al. 2005). Another option is to reduce platinum loadings but this presents the problem of reducing catalyst durability (Zhang et al. 2006). The need for alternative catalysts is also driven by the availability of platinum. Platinum catalysts in automobiles already amount to approximately half of the platinum sold globally each year (Loferski 2008). Unless breakthroughs are made, low-temperature fuel cells will continue to require platinum catalysts in the foreseeable future. However, the most critical issue is to reduce the amount of Pt loadings. An average Pt loading is around 0.5 mg/cm^2 and the total amount of Pt used in current 80 kW class stacks is estimated to be about 100–120 g. From the standpoint of cost and resource conservation, it is not possible to use this amount of Pt in commercial vehicles. By comparison, the amount of Pt used in the exhaust-gas after-treatment system for ICEVs is <10 g. To achieve a realistic cost the amount of Pt used should be reduced to one-tenth of the present level. However, there is currently no technology available for reducing the amount of Pt loading to this extent. Innovative ideas for improving the performance of the electrode catalyst is therefore most desired.

One can expect continuous improvements and developments in conventional power trains. Engine downsizing and the hybridization of the power trains will partly compensate for the low-load efficiency deficit of ICEVs. As electric components have replaced many mechanical parts of the vehicle, hybridization combined with regenerative braking will bring increased electrification of power trains. As many elements of the drivetrain and the refueling infrastructure are new, there are still major improvements needed for the FCVs to become competitive with ICEVs. Today, cost, range, and refueling parameters for FCVs are inferior to gasoline power trains. However, these figures are better than advanced BEVs, and projections show that FCVs can become cost competitive to ICEVs. Until FCVs become competitive in cost and convenience to ICEVs and until significant advancement is made in hydrogen production and storage, FCVs will remain a niche technology (Edwards et al. 2008). To meet the challenges of lowering system costs, R&D programs on fuel cells for automotive applications are desired to develop components with low-cost materials, advanced manufacturing processes, and higher operating efficiency. Ultimately, the major driver for FCVs would be their inherently high efficiency on a well-to-wheel basis, especially when renewable energy sources will become feasible for hydrogen production.

Acknowledgment

Financial support from CSIR, New Delhi through a supra-institutional project under EFYP is gratefully acknowledged.

References

Adamson, K. A., J. Butler, and M. Hugh. 2008. Fuel-cell today industry review 2008. *Fuel cells: Commercialization.* Royston, UK: Fuel-Cell Today.

Adcock, P., A. Kells, and C. Jackson. 2008. PEM Fuel Cells for Road Vehicles. EET-2008 European Ele-Drive Conference. *International Advanced Mobility Forum*, Geneva, Switzerland, March 11–13, 2008.

Ahluwalia, R. K., X. Wang, A. Rousseau, and R. Kumar. 2003. Fuel economy of hydrogen fuel-cell vehicles. *Journal of Power Sources* 130:192–201.

Arico, A. S., P. Bruce, B. Scrosati, J. Tarascon, and W. Schalkwijk. 2005. Nano structured materials for advanced energy conversion and storage devices. *Nature Materials* 4:366–77.

Arita, M. 2002. Technical issues of fuel-cell systems for automotive application. *Fuel Cells* 2:10–14.

Beuscher, U., S. J. C. Cleghorn, and W. B. Johnson. 2005. Challenges for PEM fuel-cell membranes. *International Journal of Energy Research* 29:1103–12.

Brett, D. J. L., A. Atkinson, N. P. Brandon, and S. J. Skinner. 2008. Intermediate temperature solid oxide fuel cells. *Chemical Society Reviews* 37:1568–78.

Britsche, O., and G. Gutmann. 2004. Systems for hybrid cars. *Journal of Power Sources* 127:8–15.

Bruijn, F. D. 2005. The current status of fuel-cell technology for mobile and stationary applications. *Green Chemistry* 7:132–50.

Cropper, M. A. J., S. Geiger, and D. M. Jolli. 2004. Fuel cells: A survey of current developments. *Journal of Power Sources* 131:57–61.

Demirdoven, N., and J. Deutch. 2004. Hybrid cars now, fuel-cell cars later. *Science* 305:974–76.

Edwards, P. P., V. L. Kuznetsov, W. I. F. David, and N. P. Brandon. 2008. Hydrogen and fuel cells: Towards a sustainable energy future. *Energy Policy* 36:4356–62.

Emadi, A., and S. S. Williamson. Fuel-cell vehicles: Opportunities and challenges. *Proceedings of IEEE power engineering society general meeting*, June 2004, Denver, CO, 1640–45.

Folkesson, A., C. Andersson, P. Alvfors, M. Alakula, and L. Overgaard. 2003. Real life testing of a hybrid PEM fuel-cell bus. *Journal of Power Sources* 118:349–57.

Garnier, J., M.-C. Pera, D. Hissel, F. Harel, D. Candusso, N. Gloandut, J.-P. Diard, A. De Bernardinis, J.-M. Kaufmann, and G. Coquery. 2003. Dynamic PEM fuel-cell modelling for automotive applications. *IEEE conference proceedings, vehicular technology conference, IEEE 58th,* October 6–9, 2003, Vol. 5, 3284–88.

Gasteiger, H. A., S. S. Kocha, B. Sompalli, and F. T. Wagner. 2005. Activity benchmarks and requirements for Pt, Pt alloy, and non-Pt oxygen reduction catalysts for PEFCs. *Applied Catalysis B—Environmental* 56:9–35.

Helmolt, R. V., and U. Eberle. 2007. Fuel-cell vehicles: Status. *Journal of Power Sources* 165:833–43.

Hocevar, S., and W. Summers. 2008. *Hydrogen technology,* ed. A. Leon, 15–79. Berlin Heidelberg: Springer-Verlag.

Li, Q., J. O. Jensen, R. He, and H. J. Bjerrum. 2003. *Proceedings of the European hydrogen energy conference*, Grenoble, September 2–5. Paris: Association Francaise de I'Hydrogene.

Loferski, P. J. 2008. Platinum-group metals. In *2007 Minerals year book.* Washington, DC: U.S. Geological Survey. Available at http://minerals.usgs.gov/minerals/pubs/commodity/platinum/myb1-2007-plati.pdf

Mallant, R. K. A. M. 2003. PEMFC systems: The need for high temperature polymers as a consequence of PEMFC water and heat management. *Journal of Power Sources* 118:424–29.

Martin, A. 2002. *In Proceedings of the fuel-cell world*, ed. M. Nurdin, Lucerne, July 1–5. Oberrohrdorf, Switzerland: European Fuel-Cell Forum.

Masten, D. A., and A. D. Bosco. 2003. System design for vehicle applications: GM/Opel. Chapter 53 in *Handbook of fuel cells: Fundamentals, technology and applications,* eds. W. Vielstich, A. Lamm, and H. A. Gasteiger, 714–24. Chichester: Wiley.

Matsuo, S. 2003. Honda fuel-cell activities. Chapter 86 in *Handbook of fuel cells: Fundamentals, technology and applications*, eds. W. Vielstich, A. Lamm, and H. A. Gasteiger, 1180–83. Chichester: Wiley.

McNicol, B. D., D. A. J. Rand, and K. R. Williams. 2001. Fuel cells for road transportation purposes—Yes or no? *Journal of Power Sources* 100:47–59.

Moghbelli, H., K. Ganapavarapu, and R. Langari. 2003. A comparative review of fuel-cell vehicles (FCVs) and hybrid electric vehicles (HEVs)—Part II. In Proceedings of SAE future transportation technology conference, June 2003, Costa Mesa, CA.

Oosterkamp, P., G. Kraaj, P. Laag, E. Stobbe, and D. Wouters. 2006. The development of fuel-cell systems for mobile applications. *WHEC* 16/June 13–16.

Ormerod, R. M. 2003. Solid oxide fuel cells. *Chemical Society Reviews* 32:17–28.

Rodrigues, A., M. Fronk, and B. McCormick. 2003. General Motors/OPEL fuel-cell activities—Driving towards a successful future. Chapter 85 in *Handbook of fuel cells: Fundamentals, technology and applications*, eds. W. Vielstich, A. Lamm, and H. A. Gasteiger, 1173–79. Chichester: Wiley.

Ross, D. K. 2006. Hydrogen storage: The major technological barrier to the development of hydrogen fuel-cell cars. *Vacuum* 80:1084–89.

Savogado, O. 2004. Emerging membranes for electrochemical systems: Part II. High temperature composite membranes for polymer electrolyte fuel-cell (PEFC) applications. *Journal of Power Sources* 127:135–61.

Schafer, A., J. B. Heywood, and M. A. Weiss. 2006. Future fuel-cell and internal combustion engine automobile technologies: A 25-year life cycle and fleet impact assessment. *Energy* 31:2064–87.

Schmid, H. P., and J. Ebner. 2003. Daimler Chrysler fuel-cell activities. Chapter 86 in *Handbook of fuel cells: Fundamentals, technology and applications,* eds. W. Vielstich, A. Lamm, and H. A. Gasteiger, 1168–71. Chichester: Wiley.

Shao, P. Z., and S. M. Haile. 2004. A high-performance cathode for next generation of solid-oxide fuel cells. *Nature* 431:170–73.

Shukla, A. K. 2005. Fuelling future cars. *Journal of the Indian Institute of Science* 85:51–65.

Shukla, A. K. A., S. Aricò, and V. Antonucci. 2001. An appraisal of electric automobile power sources. *Renewable Sustainable Energy Reviews* 5:137–55.

Singhal, S. C., and K. Kendall, eds. 2003. *High temperature solid oxide fuel cells—Fundamentals, design and applications.* Oxford: Elsevier Science.

Steele, B. C. H., and A. Heinzel. 2001. Materials for fuel-cell technologies. *Nature* 414:345–52.

Tawfik, T., Y. Hung, and D. Mahajan. 2007. Metal bipolar plates for PEM fuel cells—A review. *Journal of Power Sources* 163:755–67.

Wee, J. H. 2007. Applications of proton exchange membrane fuel-cell systems. *Renewable and Sustainable Energy Reviews* 11:1720–38.

Wicks, F. E., and D. Marchionne. 1992. Development of a model to predict vehicle performance over a variety of driving conditions. Proceedings 27th intersociety energy conversion engineering conference, August 3–7, 1992, San Diego, CA, *IEEE* 3:151.

Williamson, S. S., and A. Emadi. 2002. Fuel-cell applications in the automotive industry. In Proceedings of Electrical Manufacturing and Coil Winding Expo, October 15, Cincinnati, OH.

Yang, W.-C. 2000. Fuel-cell vehicles: Recent advances and challenges—Review. *International Journal of Automotive Technology* 1:9–16.

Zhang, L., J. Zhang, D. P. Wilkinson, and H. Wang. 2006. Progress in preparation of non-noble electrocatalysts for PEM fuel-cell reactions. *Journal of Power Sources* 156:171–82.

12

Hybrid Vehicles

K. T. Chau

CONTENTS

12.1 Introduction

Electric vehicles (EVs) used to be classified into two groups: the pure EV that was purely fed by batteries and propelled solely by an electric motor, and the hybrid EV (HEV) that was sourced by both batteries and liquid fuel and powered by both the engine and electric motor. With the advent of other energy sources (Chau, Wong, and Chan 1999), namely fuel cells, ultra capacitors, and ultrahigh-speed flywheels, this classification became ill-suited. Sometimes, the HEV was even extended to embrace a vehicle using any two different energy sources such as the battery-ultra capacitor hybrid. In recent years, there is a consensus that EVs refer to vehicles with at least one of the propulsion devices being the electric motor. When the energy source is solely batteries and the propulsion device is only an electric motor, they are termed the Battery EV (BEV) or sometimes loosely called the EV. When the energy source involves fuel cells working together with batteries and the propulsion device is only an electric motor, they are termed the Fuel Cell EV (FCEV) or simply called the fuel cell vehicle. When both batteries and gasoline/diesel fuel are the energy sources as well as both the engine and electric motor are the propulsion devices, they are termed the HEV or simply called the hybrid vehicle. Figure 12.1 depicts the classification of the Internal Combustion Engine Vehicle (ICEV), HEV, BEV, and FCEV based on the energy source and the propulsion device.

In 1834, the first EV, actually a BEV, was built by Thomas Davenport. With the drastic improvements in the ICEV, the BEV almost vanished from the scene by the 1930s. The rekindling of interest in BEVs started at the outbreak of the energy crisis and oil shortage in the 1970s. The actual revival of BEVs was due to the ever increasing concerns on energy conservation and environmental protection throughout the world in the 1990s. Namely, BEVs offer high-energy efficiency, allow diversification of energy resources, enable load equalization of power systems, show zero local and minimal global exhaust emissions, and operate quietly. However, there are two major barriers hindering the popularization of BEVs—short driving range and high initial cost. These barriers cannot be easily solved by the available energy source technologies (including batteries, fuel cells, ultra capacitors, and ultrahigh-speed flywheels) in the near future (Chan and Chau 2001; Chau and Wong 2001).

People may not buy a BEV, no matter how clean, if its range between charges is only 1–200 kilometers. By the same token, people may not buy a FCEV, no matter how clean, if its price is over 10 times the ICEV counterpart. The HEV, incorporating the engine and electric motor, was introduced as an

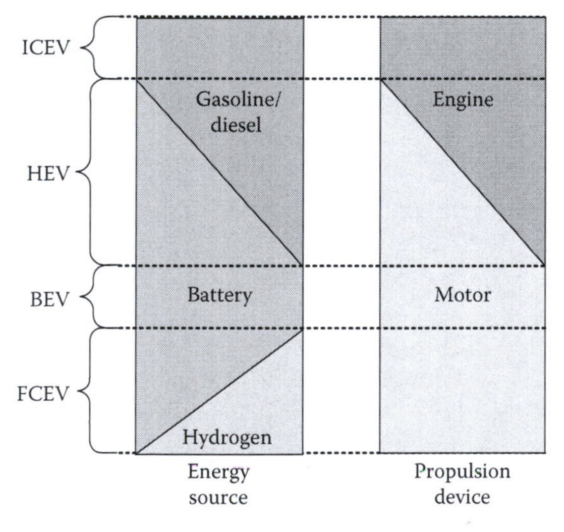

FIGURE 12.1
Classification of vehicles.

interim solution before the full implementation of pure EVs when there is a breakthrough in energy sources (Wakefield 1998). The definite advantages of the HEV are to greatly extend the original BEV driving range by two to four times, and to offer rapid refueling of liquid gasoline or diesel. An important plus is that it requires only little changes in the energy supply infrastructure. The key drawbacks of the HEV are the loss of zero-emission vehicle (ZEV) concept and the increased complexity. Nevertheless, the HEV is vastly less polluting and has less fuel consumption than the ICEV while having the same range. These merits are due to the fact that the engine of the HEV can always operate in its most efficient mode, yielding low emissions and low fuel consumption. Also, the HEV may be purposely operated as a BEV in the zero-emission zone. It is becoming a consensus that the HEV is not only an interim solution for implementation of ZEVs, but also a practical solution for commercialization of super, ultra-low emission vehicles (SULEVs).

The concept of HEVs is nothing new. A patent in 1905 delineated that a battery-powered electric motor was used to boost the acceleration of an ICEV (Wouk 1995). However, over the years, the development of HEVs had been discouraging. One possible reason was due to their complexity, especially on how to coordinate and combine the mechanical driving forces from both the engine and electric motor. The turning point of HEV development was the advent of the Toyota Prius in 1997 (Hermance and Sasaki 1998). Subsequently, the development of HEVs has taken on an accelerated pace.

12.2 Hybrid Vehicle Configurations

Based on the flows of electrical and mechanical powers, HEVs used to have two basic configurations—the series hybrid and parallel hybrid. With the introduction of some HEVs having the features of both the series hybrid and parallel hybrid, the basic configurations have been extended to three kinds—series, parallel, and series-parallel. Additionally, some HEVs with both front-axle and rear-axle drives cannot be represented by these three kinds. Thus, they are further extended to four kinds (Chau and Wong 2002; Ehsani et al. 2005): series hybrid, parallel hybrid, series-parallel hybrid, and complex hybrid.

Figure 12.2 shows the corresponding functional block diagrams, in which the electrical link is bidirectional, the hydraulic link is unidirectional, and the mechanical link (including shafts, brakes, clutches, and gears) is also bidirectional. It can be found that the key feature of the series hybrid is to couple the engine with the generator to produce electricity for pure electric propulsion, whereas the key feature of the parallel hybrid is to couple both the engine and electric motor with the transmission via the same driveline to propel the wheels. The series–parallel hybrid is a combination of both the

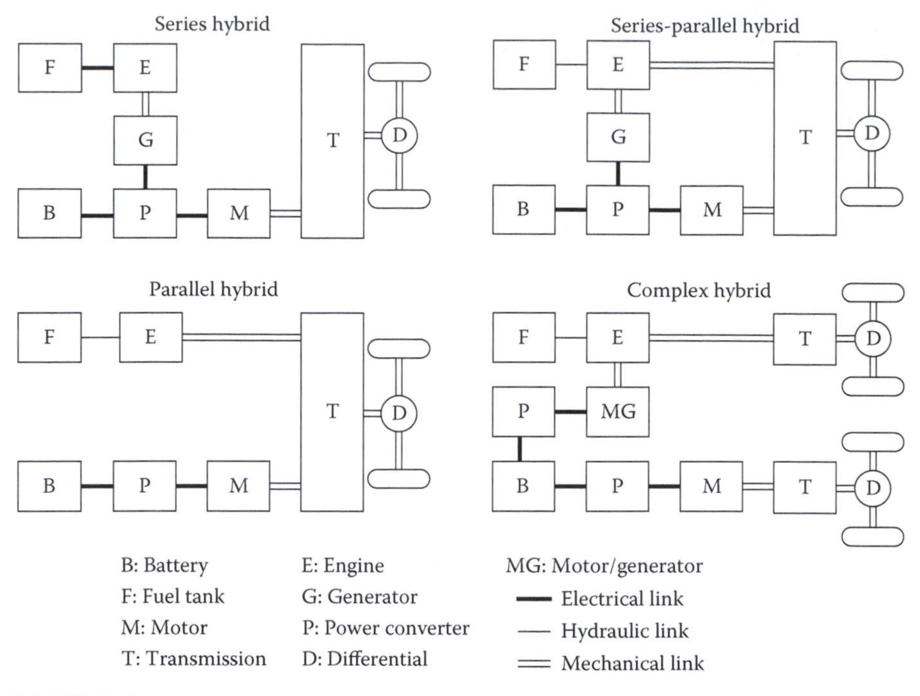

FIGURE 12.2
Basics configurations of HEVs.

series and parallel hybrids. On top of the series-parallel hybrid, the complex hybrid can offer additional and versatile features.

The series hybrid is the simplest kind of HEV. Its engine mechanical output is first converted into electricity by using a generator. The converted electricity can charge the battery and/or feed the electric motor to propel the wheels via the mechanical transmission and differential. Due to the absence of clutches throughout the mechanical link, it has the definite advantage of flexibility for locating the engine–generator set. Although it has the advantage of simplicity of its drivetrain, it needs three propulsion devices—the engine, generator, and electric motor. Another disadvantage is that all these propulsion devices need to be sized for the maximum output power if it is designed to be able to climb a long grade. Unless it is only needed to serve short trips such as commuting to work and shopping, the corresponding engine–generator set can adopt a lower rating.

Different from the series hybrid, the parallel hybrid allows both the engine and electric motor to deliver power in parallel to drive the wheels. Since both the engine and electric motor are generally coupled to the driveline of the wheels via two clutches, the propulsion power may be supplied by the engine alone, by the electric motor alone, or by both. The electric motor can be used as a generator to charge the battery by regenerative braking or absorbing power from the engine when its output power is greater than that required to drive the wheels. Better than the series hybrid, the parallel hybrid needs only two propulsion devices—the engine and electric motor. Another advantage over the series hybrid is that a smaller engine and a smaller electric motor can be sized because they share the maximum output power. Even for long-trip operation, only the engine needs to be rated for the maximum output power while the electric motor may still be about a half.

In the series-parallel hybrid, the configuration incorporates the features of both the series hybrid and parallel hybrid, but involves an additional mechanical link compared with the series hybrid, and also an additional generator compared with the parallel hybrid. Although possessing the advantageous features of both the series hybrid and parallel hybrid, the series–parallel hybrid is relatively more complicated and costly. Nevertheless, with the advances in control and manufacturing technologies, some modern HEVs prefer to adopt this system.

As reflected by its name, the complex hybrid involves a complex configuration that cannot be classified into the above three kinds. It seems to be similar to the series–parallel hybrid, since the generator and electric motor are both electric machinery. However, the key difference is due to the bidirectional power flow of the electric machine in the complex hybrid and the unidirectional power flow of the generator in the series–parallel hybrid. This bidirectional power flow can allow for versatile operating modes, especially the three propulsion power (due to the engine and two electric motors) operating mode that cannot be offered by the series–parallel hybrid. Compared with the series–parallel hybrid, the complex hybrid suffers from higher complexity and cost. Nevertheless, some advanced HEVs adopt this system for dual-axle propulsion.

12.3 Hybrid Vehicle Classification

Based on the hybridization level and operation features between the engine and electric motor, HEVs have been classified as the micro hybrid, mild hybrid, and full hybrid (Ebron and Cregar 2005). Recently, this classification has been further extended to include the latest plug-in hybrid EV (PHEV) and the range-extended EV (REV). Figure 12.3 depicts their classification in terms of the energy source and propulsion device.

For the micro hybrid, the conventional starter motor is eliminated while the conventional generator is replaced by a belt-driven integrated-starter-generator (ISG). This ISG is typically 3–5 kW. Instead of propelling the vehicle, it offers two important hybrid features. One is to shut down the engine when the vehicle is at rest, so-called idle stop feature, hence improving fuel economy for city driving. Another is to recharge the battery primarily during

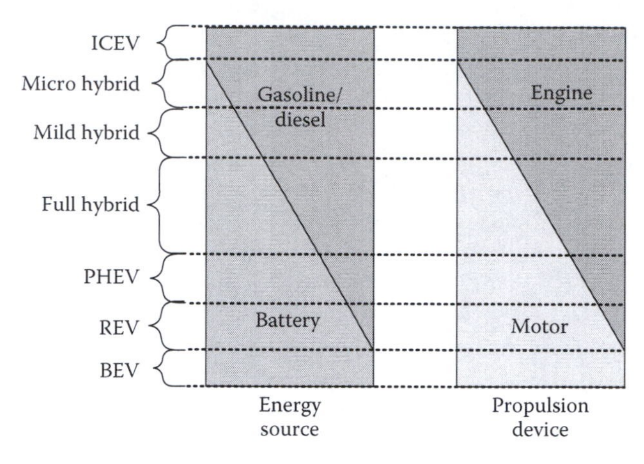

FIGURE 12.3
Classification of HEVs.

vehicle deceleration and braking, hence offering a mild amount of regenerative braking. The battery voltage is generally 12 V.

For the mild hybrid, the ISG is generally placed between the engine and the transmission. This ISG is typically 7–12 kW. It can provide the hybrid features of idle stop and regenerative braking. Also, as the ISG assists the engine to propel the vehicle, it can allow for a downsized engine. However, since the engine and ISG share the same shaft, it cannot offer the electric launch feature, namely accelerating under electric power only. The battery voltage is typically 36–144 V.

For the full hybrid, the key technology is the electronic continuously variable transmission (ECVT) system that functions to perform power splitting and power regulation. So, it can offer all hybrid features, including the electric launch, idle stop, regenerative braking, and a downsized engine. The corresponding electric motor and battery ratings are typically 30–50 kW and 200–500 V, respectively. Instead of downsizing the engine, the electric motor can provide additional torque and hence better acceleration performance than a conventional vehicle with the same size engine.

For the PHEV, it provides all features of the full hybrid, while having an additional feature of being plug-in rechargeable. Since it incorporates a large high-capacity battery bank that can be recharged by plugging it into an external charging port or using the onboard charging capabilities of the full hybrid, it can offer more electric drive and reduce the requirement for refueling. The corresponding electric motor and battery ratings are typically 30–50 kW and 400–500 V, respectively.

For the REV, it provides all features of the plug-in hybrid, but with a small engine connected to a generator to recharge the battery bank when its capacity is lower than a threshold. This avoids the range anxiety that is associated with the BEV. So, it can offer an energy-efficient electric drive in its initial pure electric range and hence significantly reduce the requirement for refueling. The corresponding electric motor and battery ratings are similar to that of the plug-in hybrid, typically 30–50 kW and 400–500 V.

12.4 Hybrid Vehicle Operations

Due to the variations in HEV configurations, different power control strategies are necessary to regulate the power flow to and from different components. These control strategies aim to satisfy a number of goals for HEVs (Beretta 2000; Van Mierlo 2000). There are four key goals: maximum fuel economy, minimum emissions, minimum system costs, and good driving performance. The design of power control strategies for HEVs involves different considerations. Some key considerations are summarized below:

- *Optimal engine operation point:* The optimal operation point on the torque–speed plane of the engine can be based on the maximization of fuel economy, the minimization of emissions, or even a compromise between fuel economy and emissions.
- *Optimal engine operation line:* In case the engine needs to deliver different power demands, the corresponding optimal operation points constitute an optimal operating line.
- *Optimal engine operation region:* The engine has a preferred operation region on the torque–speed plane, in which fuel economy and emissions remain optimum.
- *Proper power distribution:* The distribution of power demand between the engine and electric motor should be properly split up during the driving cycle, aiming to ensure that the battery can provide instantaneous discharge for hill-climbing or acceleration, and allow instantaneous recharge for regenerative braking.
- *Zero-emission policy:* In some areas, the HEV needs to be operated as a ZEV. The changeover between the pure electric mode and hybrid mode should be controlled manually by the driver or automatically by the GPS-based controller.

12.4.1 Series Hybrid Operation

In the series hybrid system, the power flow control can be illustrated by four operating modes shown in Figure 12.4. During launching, normal driving or accelerating, both the engine (via the generator) and battery deliver electrical energy to the power converter that then feeds the electric motor and hence drives the wheels via the transmission. At light load,

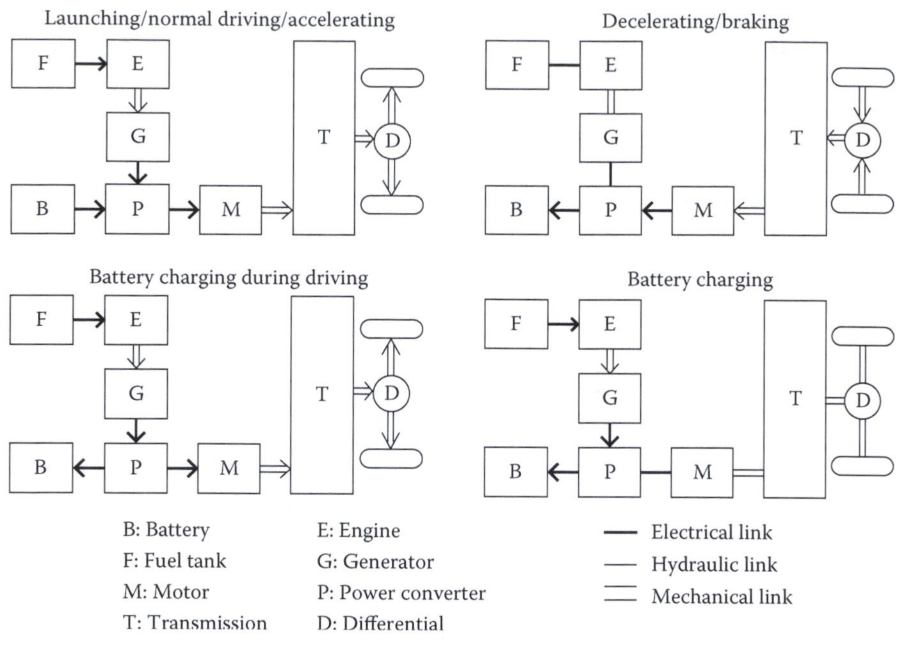

FIGURE 12.4
Series hybrid operating modes.

the engine output power is greater than that required to drive the wheels so that the generated electrical energy is also used to charge the battery until its capacity reaches a proper level. During decelerating or braking, the electric motor acts as a generator that converts the reduction of kinetic energy of the wheels into electricity, hence charging the battery via the power converter. Also, the battery can be charged by the engine via the generator and power converter, even when the vehicle is completely at a standstill.

12.4.2 Parallel Hybrid Operation

Figure 12.5 illustrates the four operating modes of the parallel hybrid. During launching or accelerating, both the engine and electric motor proportionally share the required output power to propel the vehicle. Typically, the relative distribution between the engine and electric motor is 80–20%. During normal driving, the engine solely supplies the necessary power to propel the vehicle while the electric motor remains in the off mode. During decelerating or braking, the electric motor acts as a generator to charge the battery via the power converter. Also, since both the engine and electric motor are coupled to the same drive shaft, the battery can be charged by the engine via the electric motor when the vehicle is at light load.

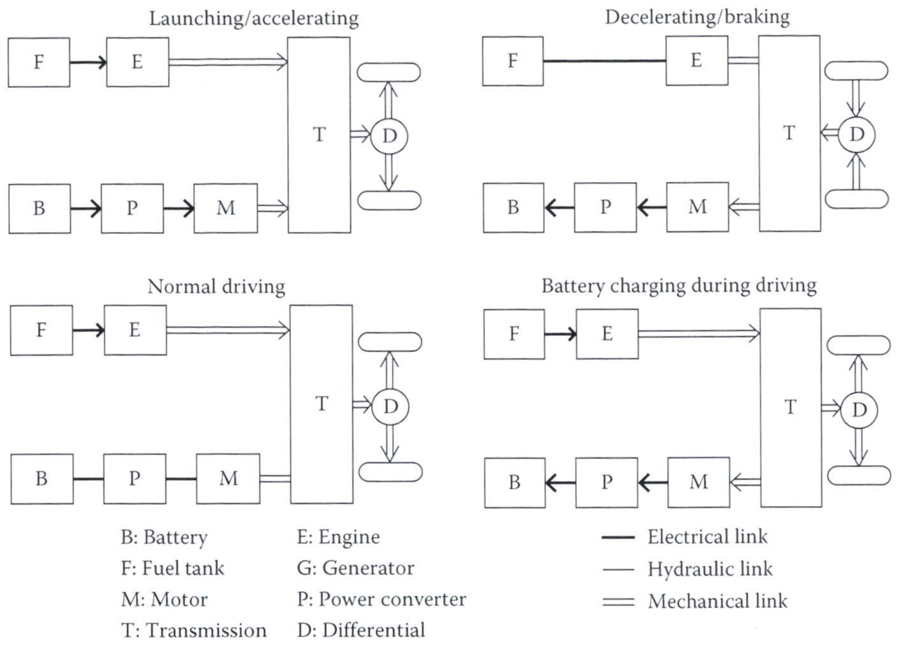

FIGURE 12.5
Parallel hybrid operating modes.

12.4.3 Series–Parallel Hybrid Operation

In the series–parallel hybrid system, it involves the features of a series hybrid and a parallel hybrid. Thus, there are many possible operating modes to carry out its power flow control. Basically, they can be divided into two groups; namely, engine-heavy and motor-heavy. The engine-heavy one denotes that the engine is more active than the electric motor for series–parallel hybrid propulsion, whereas the motor-heavy one indicates that the electric motor plays a more active role.

Figure 12.6 shows an engine-heavy series–parallel hybrid system, in which there are six operating modes. For launching, the battery solely provides the necessary power to feed the electric motor to propel the vehicle while the engine is in the off mode. During accelerating, both the engine and electric motor proportionally share the required power to propel the vehicle. During normal driving, the engine solely provides the necessary power to propel the vehicle while the electric motor remains in the off mode. During decelerating or braking, the electric motor acts as a generator to charge the battery via the power converter. At light load, the engine not only drives the vehicle but also the generator to charge the battery via the power converter. When the vehicle is at a standstill, the engine can individually drive the generator to charge the battery.

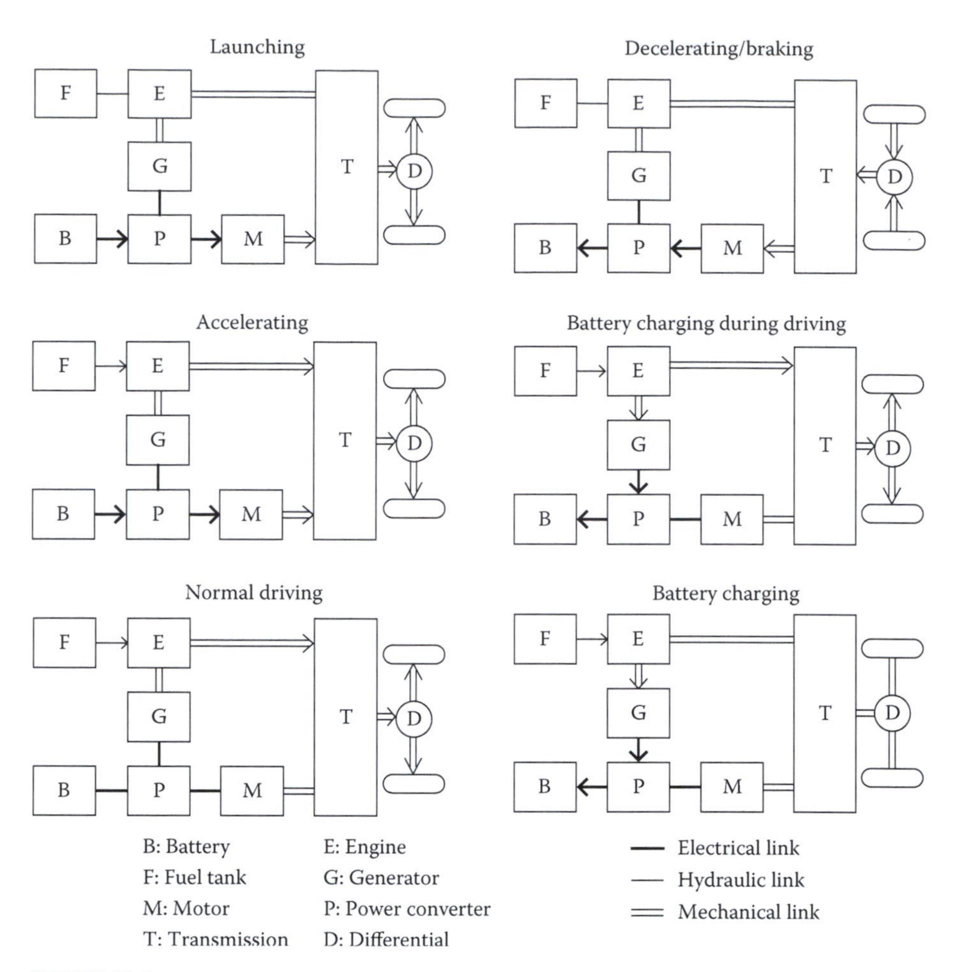

FIGURE 12.6
Engine-heavy series–parallel hybrid operating modes.

Figure 12.7 shows a relatively motor-heavy, series–parallel hybrid system, in which there are six operating modes. For launching, the battery solely feeds the electric motor to propel the vehicle while the engine is in the off mode. For both accelerating and normal driving, both the engine and electric motor work together to propel the vehicle. The key difference is that the electrical energy used for accelerating comes from both the generator and battery whereas that for normal driving it comes solely from the generator driven by the engine. Notice that a planetary gear is usually employed to split up the engine output, hence to propel the vehicle and to drive the generator. During decelerating or braking, the electric motor acts as a generator to charge the battery via the power converter. Also, at light load, the engine

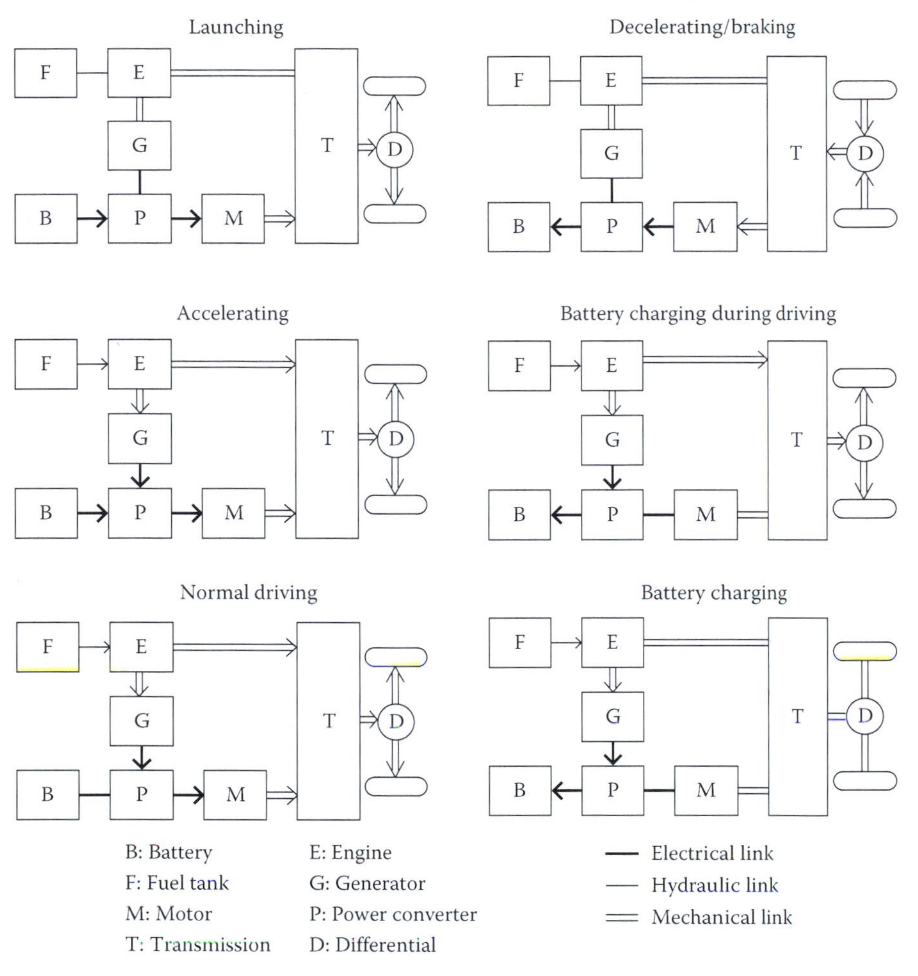

FIGURE 12.7
Motor-heavy series–parallel hybrid operating modes.

not only drives the vehicle but also the generator to charge the battery. When the vehicle is at a standstill, the engine can maintain driving the generator to charge the battery.

12.4.4 Complex Hybrid Operation

The development of complex hybrid control has been focused on the dual-axle propulsion system for HEVs. In this system, the front-wheel axle and rear-wheel axle are separately driven. There is no mechanical shaft or transfer to connect the front and rear axles, so it enables a more lightweight propulsion system and increases the vehicle packaging flexibility. Moreover, regenerative braking on all four wheels can significantly improve vehicle fuel economy.

Figure 12.8 shows a dual-axle complex hybrid system, where the front-wheel axle is propelled by a hybrid drivetrain and the rear-wheel axle is driven by an electric motor. There are six operating modes. For launching, the battery delivers electrical energy to feed both the front and rear electric motors to individually propel the front and rear axles of the vehicle whereas the engine is in the off mode. For accelerating, both the engine and front

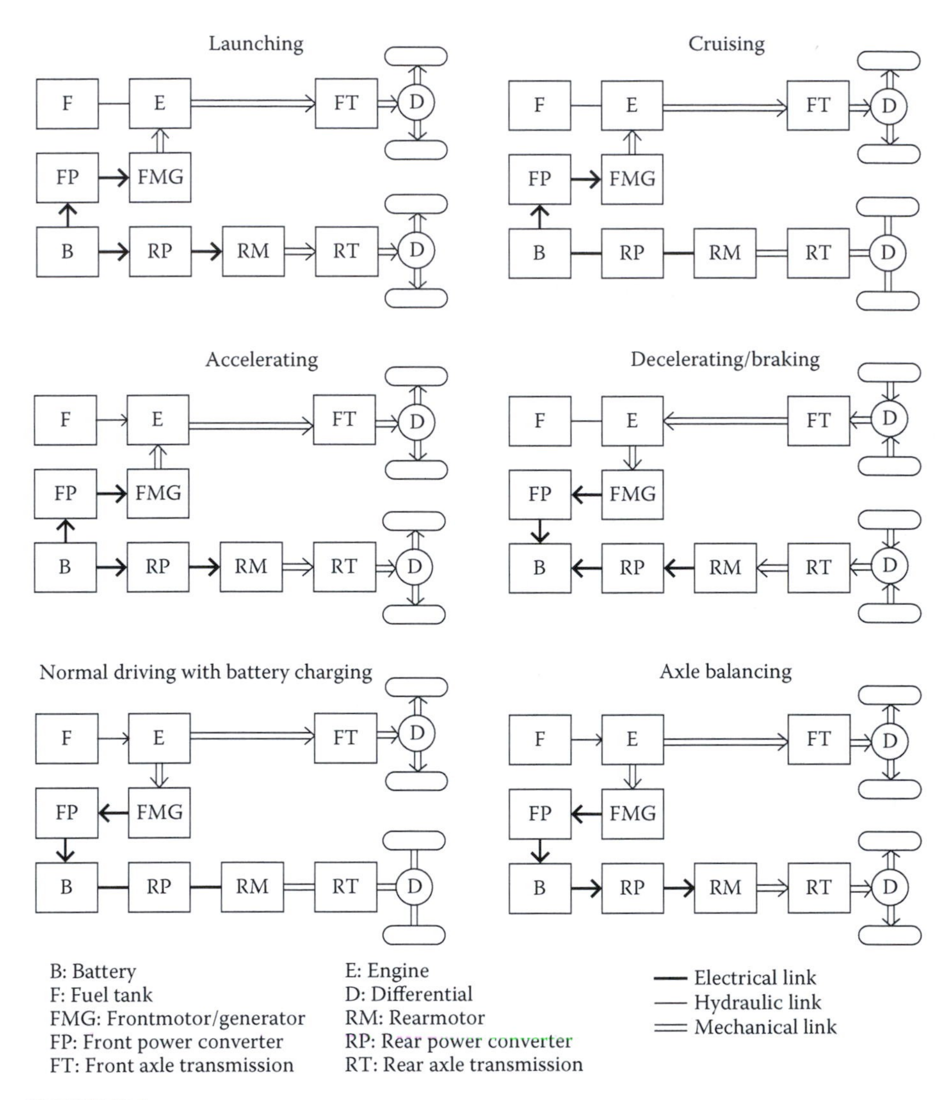

FIGURE 12.8
Dual-axle (front-hybrid rear-electric) complex hybrid operating modes.

electric motor work together to propel the front axle, while the rear electric motor also drives the rear axle. Notice that this operating mode involves three propulsion devices (one engine and two electric motors) to simultaneously propel the vehicle. During normal driving with battery charging, the engine output is split up to propel the front axle and to drive the front generator to charge the battery. The corresponding device to mechanically couple the engine, front electric motor/generator and front axle altogether is usually based on a planetary gear. During cruising, the battery delivers electrical energy to the front electric motor only to drive the front axle whereas both the engine and rear electric motor are off. During decelerating or braking, both the front generator and rear electric motor (which works as a generator) serve to simultaneously charge the battery. A unique feature of this dual-axle system is the capability of axle balancing. In case the front wheels slip, the front generator absorbs the change of engine output power; through the battery, this power difference is then used to feed the rear motor to drive the rear wheels.

Figure 12.9 shows another dual-axle complex hybrid system, where the front-wheel axle is driven by an electric motor and the rear-wheel axle is propelled by a hybrid drivetrain. Focusing on vehicle propulsion, there are six operating modes. For launching, the battery delivers electrical energy only to the front electric motor that in turn drives the front axle of the vehicle whereas both the engine and rear electric motor are off. Once the vehicle moves forward, the battery also delivers electrical energy to the rear electric motor that functions to quickly rise up the engine speed, thus cranking the engine. For accelerating, the front electric motor drives the front axle while both the engine and rear electric motor work together to propel the rear axle. So, there are three propulsion devices (one engine and two electric motors) simultaneously propelling the vehicle. During normal driving, the engine works alone to propel the rear axle of the vehicle. During decelerating or braking, both the front motor (which acts as a generator) and rear generator serve to simultaneously charge the battery. For battery charging during driving, the engine output is split up to propel the rear axle and to drive the rear generator to charge the battery.

12.4.5 ISG System Operation

In conventional automobiles, the starter motor and generator are separately coupled with the engine as shown in Figure 12.10, hence providing high starting torque for cold cranking and generating electricity for battery charging, respectively. This arrangement takes the advantage of simplicity, but suffers from poor utilization of both machines and hence results in heavy weight and bulky size. In order to incorporate both functions in a single unit, the development of ISG systems is accelerating (Walker et al. 2004). As shown in Figure 12.11, the ISG system serves both cold cranking and battery charging, while eliminating the use of transmission belts and flywheels. Essentially,

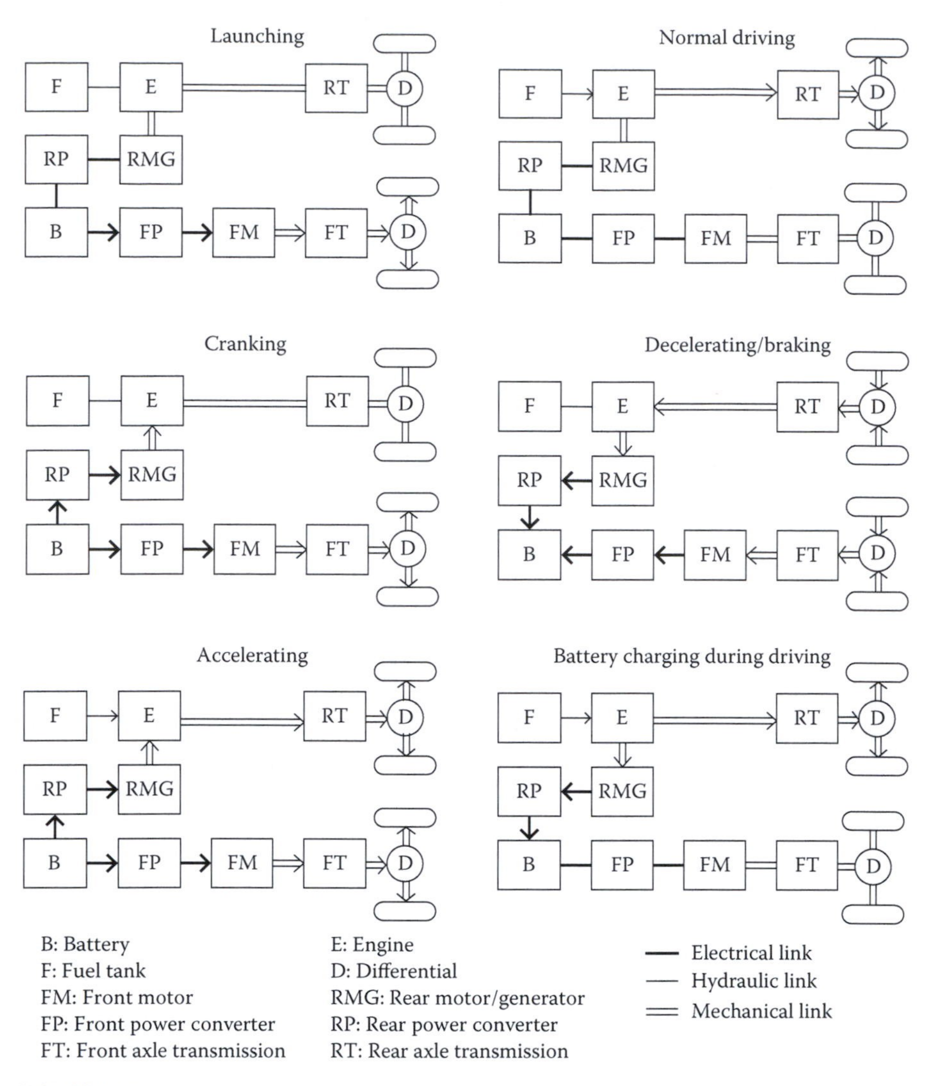

FIGURE 12.9
Dual-axle (front-electric rear-hybrid) complex hybrid operating modes.

the ISG system is fueled by two technologies—the electric machine and power converter.

The electric machine is the core of the ISG system (Cai 2004). Because of the fundamental drawback of wear and tear of carbon brushes, the DC machine is ill-suited. Among those viable brushless machines, the induction machine exhibits the advantages of low cost and high robustness, but needs complicated control to provide wide-speed generation (Chen et al. 2002; Jiang, Chau, and Chan 2003). The switched reluctance (SR) machine takes

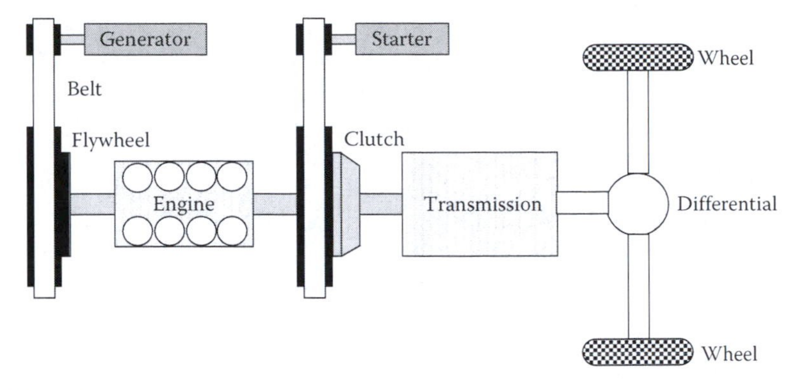

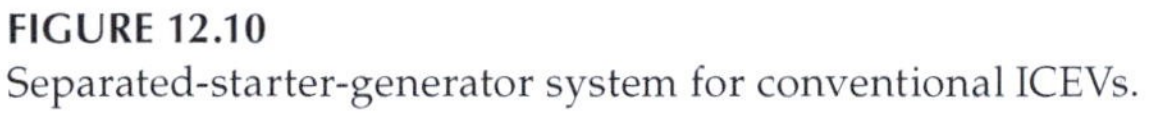

FIGURE 12.10
Separated-starter-generator system for conventional ICEVs.

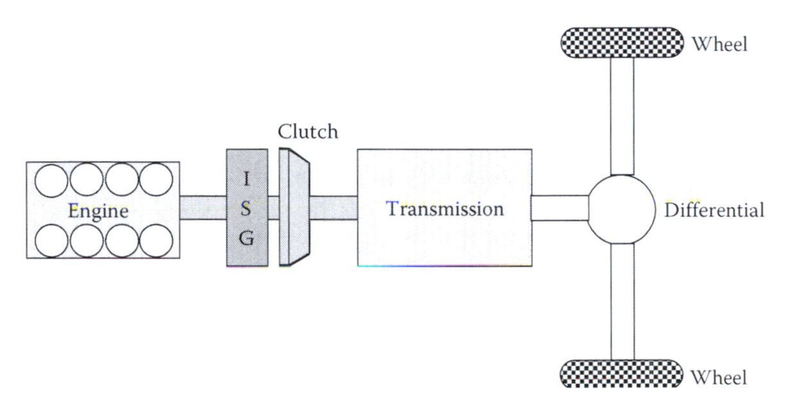

FIGURE 12.11
Integrated-starter-generator system for HEVs.

the definite advantages of high robustness and low inertia, but has the drawbacks of relatively lower efficiency and lower power density (Zhan, Chan, and Chau 1999). The permanent-magnet (PM) brushless machines, including those PM brushless AC ones (Chau et al. 2006a; Cui et al. 2005) and PM brushless DC ones (Gan et al. 2000; Wang et al. 2002), offer the advantages of high efficiency and high-power density. However, they suffer from the incapability of flux control (Bae and Sul 2003). A double-stator PM brushless topology is proposed to tune the flux, but it needs a complicated mechanism to shift the spatial angle between two stators (Zhang et al. 2006). By incorporating the advantageous features of both the PM brushless and SR machines, the doubly salient PM (DSPM) machine offers the merits of high efficiency, high-power density, high robustness, and low inertia (Chau et al. 2005; Cheng et al. 2003). But, it still has the drawback of inflexible flux control. By replacing the PM materials with a DC field winding, a brushless doubly fed doubly salient (BDFDS) machine (or specifically called the stator doubly fed doubly salient, SDFDS machine) can be derived from the DSPM machine (Chau, Cheng, and Chan 2002). This arrangement can solve the problem of uncontrollable PM flux, and can offer the possibility to on-line optimize efficiency. However, it seriously deteriorates an important merit of PM excitation, namely high-energy density. Consequently, by incorporating PMs into the BDFDS machine, the BDF-DSPM machine is created that can offer a very wide constant-power speed range for ISG application (Chau et al. 2006b).

The power converter functions to regulate bidirectional power flow within the ISG system. Conventionally, it is a low-voltage high-current inverter that generally suffers from high-power losses (Liu, Hu, and Xu 2004). By incorporating an additional DC-DC stage, the DC-DC-AC converter can provide a flexible DC-link voltage, hence improving the torque–speed characteristics and system efficiency (Xu and Liu 2004). However, these converters have the drawbacks of high-switching stress and electromagnetic interference (EMI). To alleviate these drawbacks, both the resonant inverter and multilevel inverter are proposed. The resonant inverter can offer the merit of soft switching, but needs a complicated operation of an additional resonant circuitry (Alan and Lipo 2000; Chau, Yao, and Chan 1999). The multilevel inverter can be classified as a high-frequency pulse-width-modulation (PWM) multilevel inverter type, and a fundamental-frequency, multilevel inverter type (Loh, Holmes, and Lipo 2005; Rodriguez, Lai, and Peng 2002). Particularly, the latter type generates staircase AC output that takes advantage of low switching loss and minimum EMI. This includes the diode-clamped, multilevel inverter, capacitor-clamped multilevel inverter, cascaded multilevel inverter, and switched capacitor converter multilevel inverter (Axelrod, Berkovich, and Ioinovici 2005). However, they have the drawbacks that many switches and capacitors are necessary to generate a sufficient level of staircase AC voltage waveform, whereas the maximum number of capacitors is governed by the stipulated output voltage. Recently, these drawbacks have been alleviated by using partial charging; namely, multiple voltage steps per capacitor (Chan and Chau 2007).

12.4.6 ECVT System Operation

The turning point of HEV development was the advent of the Toyota Prius in 1997, which first adopted the ECVT propulsion system. Subsequently, the development of HEVs is one of the major strategic plans for all giant automakers. After the introduction of ECVT that was marketed under the Toyota hybrid system, there have been many derivatives. Those viable ones are the Ford hybrid system input split ECVT, the GM–Allison compound split ECVT, the Timken compound split ECVT, and the Renault compound split ECVT (Miller 2006; Miller and Everett 2005).

As shown in Figure 12.12, a basic ECVT propulsion system is mainly composed of a planetary gear, a motor, and a generator. Figure 12.13 shows this planetary gear that consists of a sun gear, a planet carrier that holds several

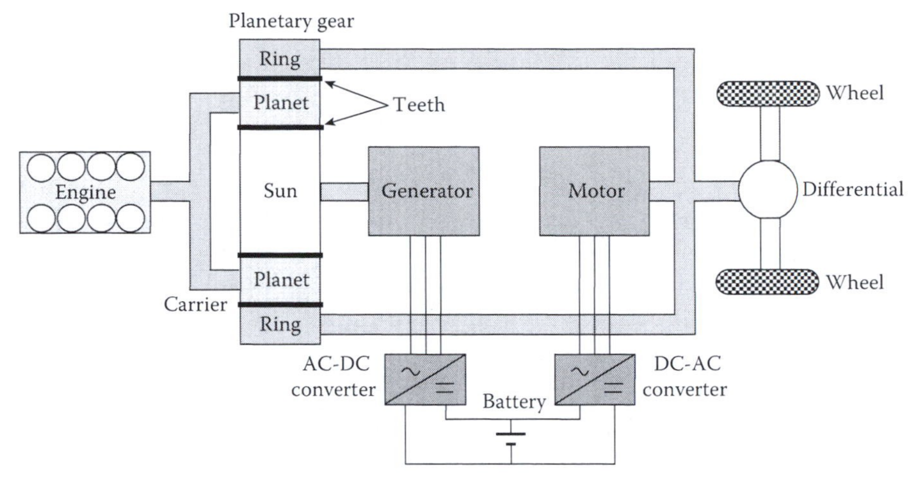

FIGURE 12.12
ECVT using two machines coupled with planetary gearing.

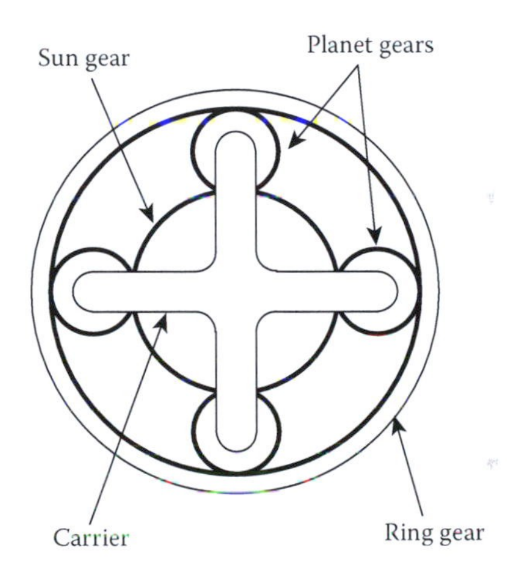

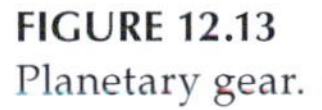
FIGURE 12.13
Planetary gear.

planet gears meshed with the sun gear, and a ring gear with inward-facing teeth that mesh with the planet gears. The engine is attached to the planet carrier, the motor is coupled with the driveline shaft so that both are attached to the ring gear, and the generator is mounted to the sun gear. Their relationships are given by (Miller and Everett 2005):

$$T_d = T_m + \left(\frac{1}{1+\rho}\right)T_e,$$

$$\omega_d = (1+\rho)\ \omega_e - \rho\omega_g,$$

$$\rho = \frac{N_s}{N_r},$$

where T_d is the driveline torque, T_m is the motor torque, T_e is the engine torque, ω_d is the driveline speed, ω_e is the engine speed, ω_g is the generator speed, ρ is the planetary gear ratio, N_S is the number of sun gear teeth, and N_r is the number of ring gear teeth. By controlling the power taken by the generator and then feeding back into the motor, ω_e can be maintained constant when ω_d is varying. Hence, a continuously variable ratio between engine speed and wheel speed can be achieved.

The advantages of this ECVT propulsion system are summarized below:

- Because of the absence of clutches, torque converters or shifting gears, it can significantly improve the transmission efficiency and reduce the overall size, hence increasing both the energy efficiency and power density. Also, since it is mechanically simple, it offers high reliability.
- In the presence of a continuously variable ratio between engine speed and wheel speed, the engine can always operate at its most energy-efficient or optimal operation line (OOL) as shown in Figure 12.14, hence resulting in a considerable reduction of fuel consumption.
- The system can fully enable the idle stop feature (the engine is completely shut down when the vehicle is stopped) and the electric launch feature (the motor provides all necessary torque required to put the vehicle in motion). These two features are particularly essential to improve the energy efficiency of the full hybrid.
- The system can fully enable regenerative braking when the vehicle is slowing or downhill coasting, and full-throttle acceleration where both the engine and electric motor operate simultaneously to provide the required power demand.

The key disadvantage of all available ECVT propulsion systems is the reliance on planetary gearing. This planetary gear is a mechanical device in which transmission loss, gear noise, and regular lubrication are inevitable. In recent years, active researches are being conducted, aiming to get rid of this mechanical planetary gear while retaining ECVT propulsion.

Rather than using a planetary gear, the combination of two concentrically arranged machines can realize the power split for the full hybrid. Figure 12.15 shows the system concept in which the two machines are mechanically coupled and electrically connected (Eriksson and Sadarangani 2002; Hoeijmakers and Ferreira 2006). The first machine is a double-rotor

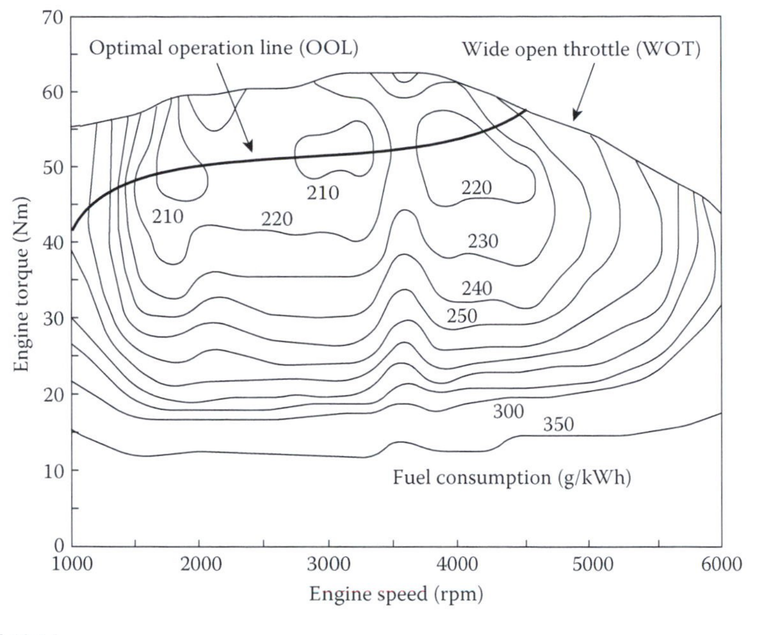

FIGURE 12.14
Optimal operation line of engine.

machine in which its inner rotor is directly coupled with the engine, whereas its outer rotor is coupled with the rotor of the second machine. The inner rotor of the first machine has a three-phase winding, which is connected to a power converter via slip rings and carbon brushes. The rotor of the second machine is also mechanically connected to the driveline shaft. Its stator has a three-phase winding that is fed by another power converter. The two power converters are electrically connected via a DC link. Thus, there are two paths of power flow from the engine shaft to the driveline shaft. One is the mechanical power directly passing from the outer rotor of the first machine to the rotor of the second machine. Another path is the electrical power flow through the slip rings and two power converters. The corresponding relationships are given by

$$P_e = P_{\text{mech}} + P_{\text{elec}},$$

$$P_{\text{elec}} = (\omega_e - \omega_d)\, T_e,$$

where P_e is the engine power, P_{mech} is the mechanical power transfer, P_{elec} is the electrical power transfer, ω_e is the engine speed, ω_d is the driveline speed, and T_e is the engine torque. It is obvious that P_{elec} linearly decreases with the increase of ω_d. By controlling P_{elec}, ω_e can be maintained constant when ω_d

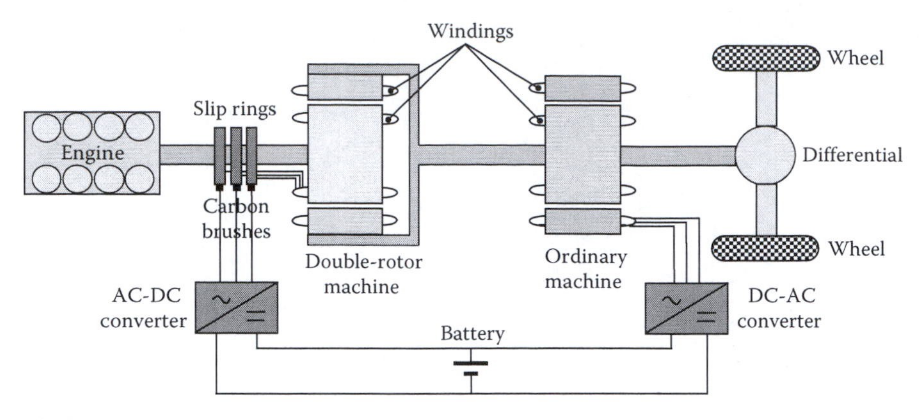

FIGURE 12.15
ECVT using two concentrically arranged double-rotor machines.

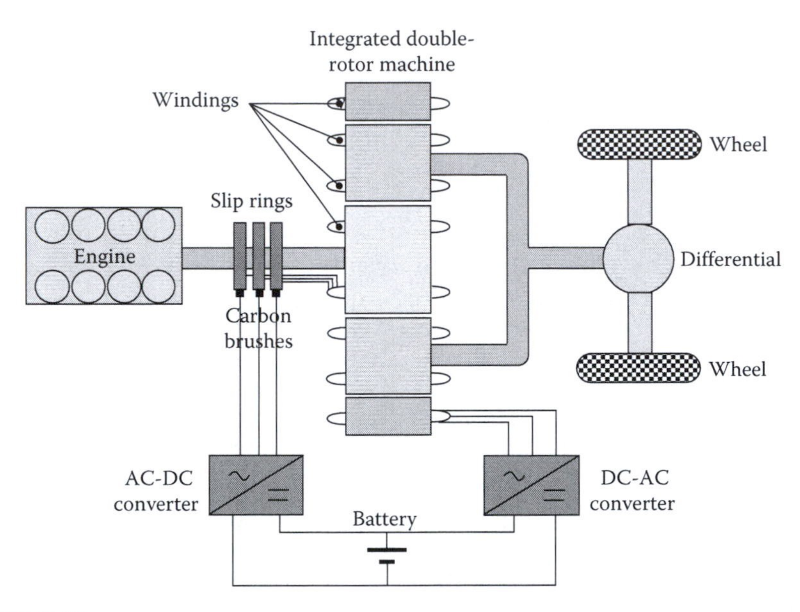

FIGURE 12.16
ECVT using integrated double-rotor induction machine.

is varying. Hence, the desired continuously variable ratio between engine speed and wheel speed can be attained.

In order to reduce the system weight and size, the two machines can be integrated into a single machine. The key is to share the outer rotor of the first machine with the rotor of the second machine so that the stator is placed concentrically around the outer rotor. Figure 12.16 shows the configuration of an integrated double-rotor induction machine (Hoeijmakers and Ferreira 2006). It has the definite advantage of being highly compact and lightweight. The corresponding principle of operation is the same as the two concentrically arranged double-rotor machines. In order to further improve the efficiency and power density, the integrated double-rotor machine can adopt the PM brushless topologies. Figure 12.17 and 12.18 show the corresponding radial-field and axial-field topologies, respectively (Eriksson and

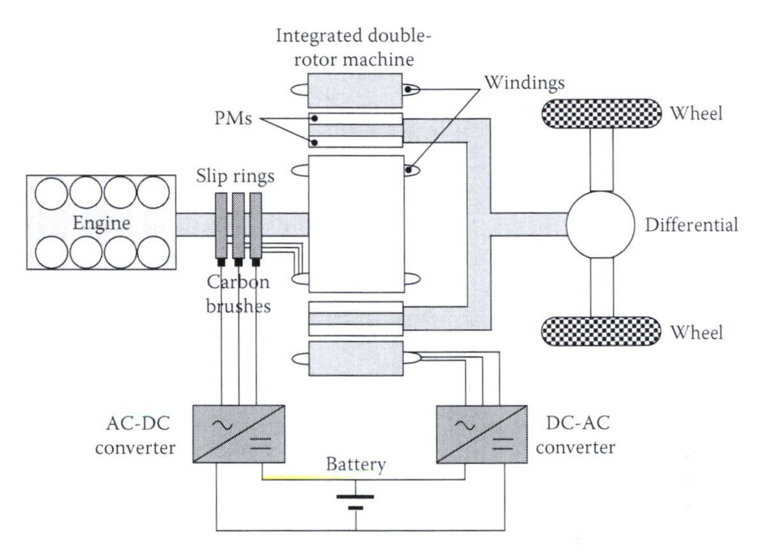

FIGURE 12.17
ECVT using integrated radial-field double-rotor PM machine.

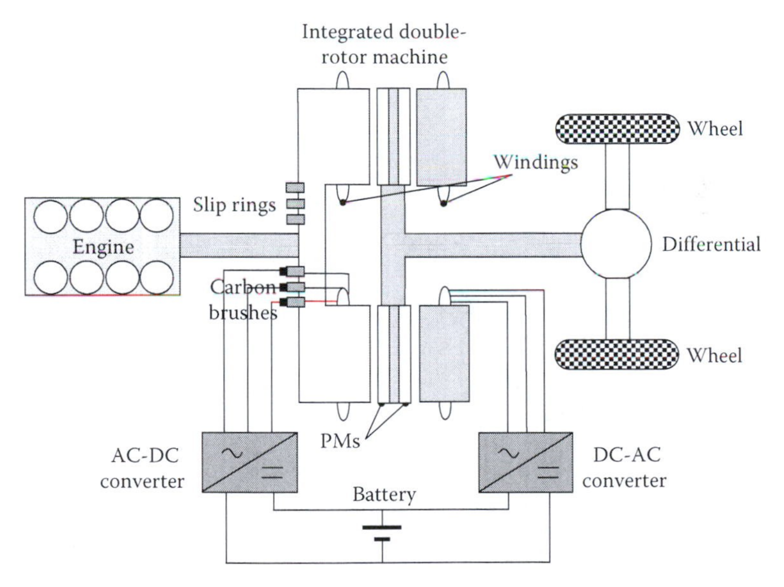

FIGURE 12.18
ECVT using integrated axial-field double-rotor PM machine.

Sadarangani 2002). Different topologies have different space and thermal requirements.

Although the above integrated double-rotor machines can perform ECVT operations without using a planetary gear, they still need carbon brushes and slip rings to withdraw power from the primary rotor. With no doubt, this will deteriorate the reliability of the whole system.

Recently, a high-performance coaxial magnetic gear has been proposed (Atallah and Howe 2001; Jian and Chau 2009). It can provide noncontact torque transmission and speed variation using the modulation effect of PM fields. Hence, the drawbacks of low-transmission efficiency, significant mechanical friction, and annoying audible noise brought by the mechanical planetary gear can be avoided. Figure 12.19 shows the configuration of the coaxial magnetic gear. It employs high-energy rare-earth PMs on both the inner rotor and outer rotor, and has the modulating ring sandwiched between the two rotors. The modulating ring takes charge of modulating the magnetic fields in the two air-gaps that interface with the inner rotor and the outer rotor. In order to provide magnetic paths with high permeability as well as to reduce eddy current loss, the modulating ring is built of thin sheets of laminated ferromagnetic material. Also, epoxy is filled in its slots to reinforce structural strength for high-torque transmission. The coaxial magnetic gear can achieve stable torque transmission if it satisfies the following relationship:

$$n_s = p_1 + p_2,$$

where p_1 and p_2 are the pole-pair numbers of the inner rotor and outer rotor respectively, and n_s is the number of ferromagnetic segments on the

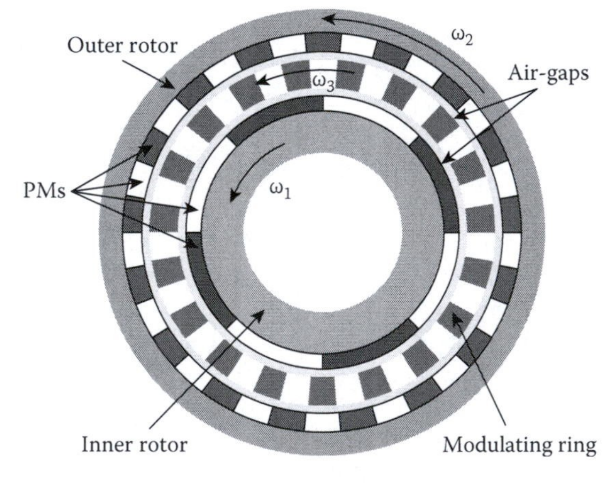

FIGURE 12.19
Coaxial magnetic gear.

modulation ring. When the modulating ring is fabricated as a stationary body, it offers fixed-ratio variable speed transmission as governed by

$$\omega_1 = -\frac{p_2}{p_1}\omega_2 = -G_r\omega_2,$$

where ω_1, ω_2 are the rotational speeds of the outer rotor and inner rotor respectively, G_r is the so-called gear ratio, and the minus sign indicates that the two rotors rotate in opposite directions.

In order to achieve the power split, the modulating ring is designed as another rotational body. Thus, it can offer continuously variable speed transmission, and the corresponding speed relationship is governed by

$$\omega_1 + G_r\omega_2 - (1 + G_r)\omega_3 = 0,$$

where ω_3 is the rotational speed of the modulating ring. Without considering the power losses occurred at the coaxial magnetic gear, it yields

$$T_1\omega_1 + T_2\omega_2 + T_3\omega_3 = 0,$$

where T_1, T_2, and T_3 are the developed torques on the inner rotor, outer rotor, and modulating ring, respectively. Consequently, instead of using the mechanical planetary gear, the coaxial magnetic gear-based ECVT propulsion system is shown in Figure 12.20. The corresponding inner rotor is attached to the rotor of the generator, the modulating ring is coupled with the engine shaft, and the outer rotor is connected with the rotor of the motor

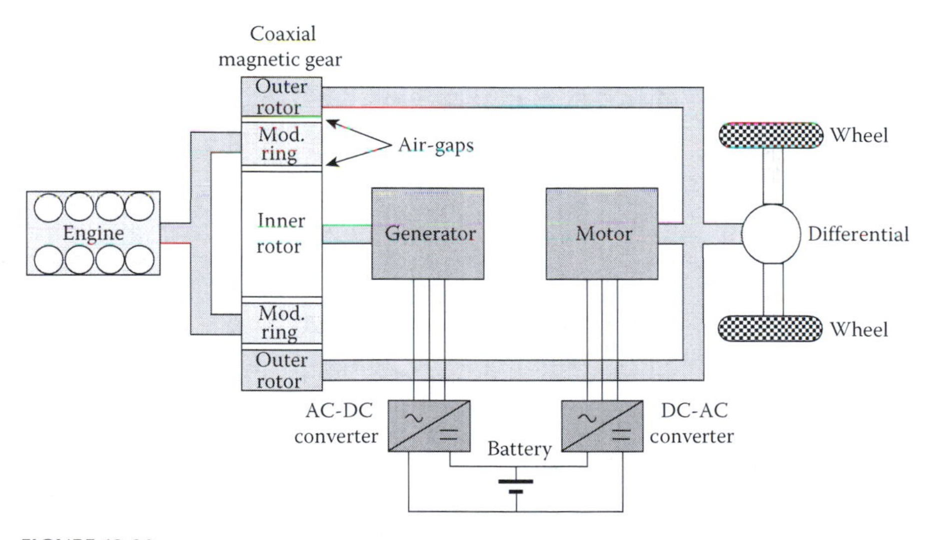

FIGURE 12.20
ECVT using two machines coupled with coaxial magnetic gearing.

and the driveline shaft. The principle of operation is similar to that of the planetary gear-based ECVT propulsion system.

12.5 Hybrid Vehicle Benefits

The use of EVs, no matter if they are BEV, HEV, or FCEV can have two major benefits. Namely, the energy benefit resulting from better energy diversification and higher energy efficiency, as well as the environmental benefit resulting from better air quality and lower noise pollution.

12.5.1 Energy Benefit

Derived from oil, gasoline and diesel are the major fuels for ICEVs. Although the development of biofuels has taken on an accelerated pace in recent years, our present transportation means are still heavily dependent on crude oil. EVs are an excellent solution to regulate this unhealthy dependence because electricity can be generated by almost any kind of energy resource in the world. Figure 12.21 illustrates the merit of energy diversification for two types of HEVs (namely the PHEV and REV) in which electrical energy can be obtained from the power grid via thermal, solar, nuclear, hydropower, wind, geothermal, oceanic, and biomass power generation, as well as from a generator coupled with an engine.

Conversion processes for various energy resources, namely oil (both conventional and nonconventional), natural gas, coal, renewables (including biomass, hydro, wind, solar, geothermal, and oceanic), and nuclear, as well as

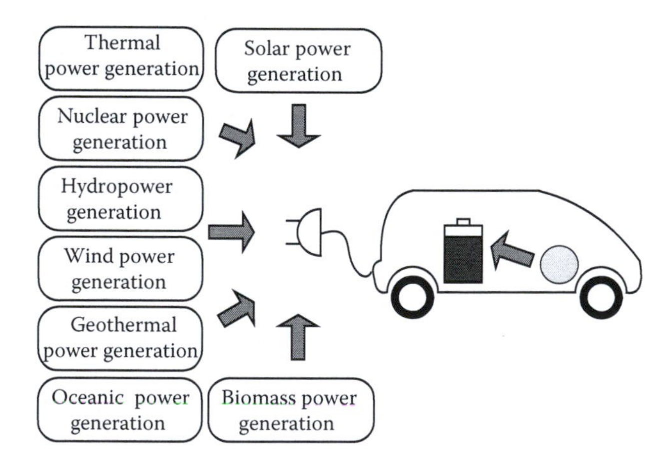

FIGURE 12.21
Power generations of electricity for two types of HEVs—PHEV and REV.

various types of EVs, namely the ICEV, HEV (micro, mild, full), PHEV, REV, BEV, and FCEV, are depicted in Figure 12.22. It can be found that electricity is the most convenient energy carrier between various energy resources and various EVs, while liquid fuels (including gasoline, diesel, LPG, and biofuels) are the major energy carrier for the ICEV and various HEVs. Consequently, it can be concluded that the latest two types of HEVs; namely the PHEV and REV, can provide the greatest energy diversification by accepting both liquid fuels and electricity.

Besides the definite merit of energy diversification resulting from the use of EVs, another advantage is the high-energy efficiency offered by EVs. In order to compare the overall energy efficiency of the BEV with the ICEV, their energy conversion flows from crude oil to road load are illustrated in Figure 12.23, where the numerical data are for indicative purposes only. By taking crude oil as 100%, the overall energy-efficiencies for the BEV and ICEV are 18 and 13%, respectively. Therefore, even when all electricity is generated by oil-fueled power plants, the BEV is more energy-efficient than the ICEV by about 38%. For the HEVs, the corresponding energy efficiencies are between the BEV and ICEV. Typically, they are 20–30% higher in energy efficiency or fuel economy than the ICEV. It should be noted that since the ICEV presently represents over 60% of oil demand use in advanced countries, the use of HEVs can significantly save on the consumption of oil, hence savings in both energy and money.

Moreover, all EVs possess one distinct advantage over the ICEV in energy utilization, namely regenerative braking. As shown in Figure 12.24, HEVs

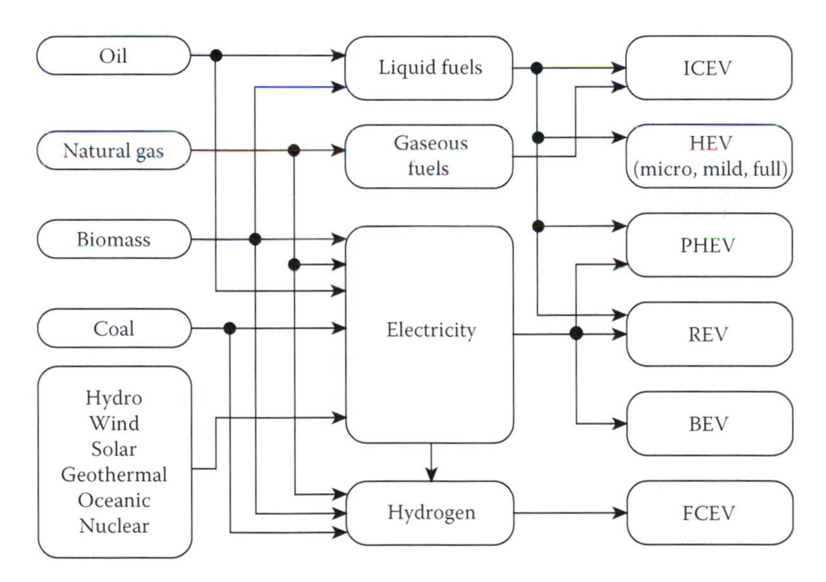

FIGURE 12.22
Energy conversions for various EVs.

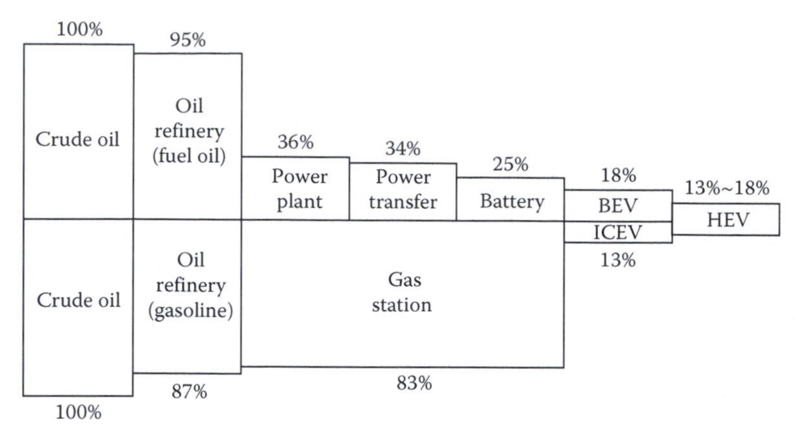

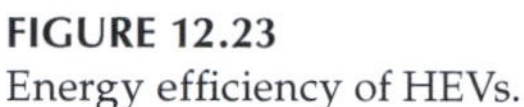

FIGURE 12.23
Energy efficiency of HEVs.

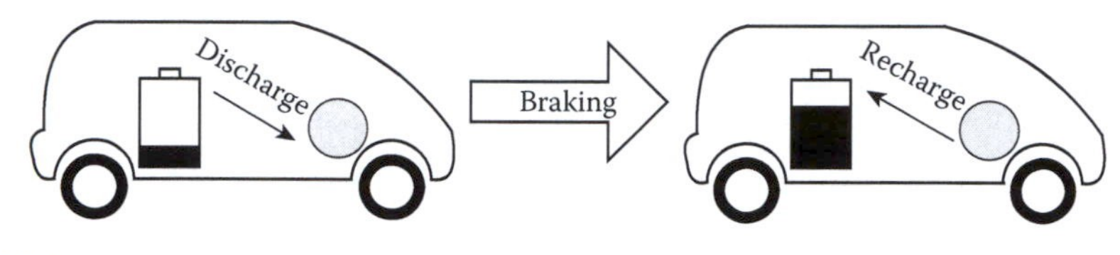

FIGURE 12.24
Energy savings by regenerative braking.

can recover kinetic energy during braking and utilize it for battery recharging, whereas the ICEV wastefully dissipates this kinetic energy as heat in the brake discs and drums. With this technology, the energy efficiency of HEVs shows further value.

12.5.2 Environmental Benefit

Currently, in many metropolises, ICEVs are responsible for over 50% of hazardous air pollutants and smog-forming compounds. Although the engine of ICEVs is continually improved to reduce the emitted pollutants, the increase in the number of ICEVs is much faster than the reduction of emissions per vehicle. Hence, the total emitted pollutants due to ICEVs, including carbon monoxide (CO), hydrocarbons (HC), nitrogen oxides (NO_x), sulfur oxides (SO_x), particulate matter (dust), and nonmethane organic gases (NMOG), continue to grow in a worrying way.

In order to alleviate or at least control the growth of air pollution due to road transportation, the use of various EVs is the most viable choice (Chan and Chau 2001). Figure 12.25 shows an indicative comparison of harmful emissions locally generated by the ICEV and the BEV, while those of the HEVs lie between them. As expected, the BEV offers zero local emissions. Taking into

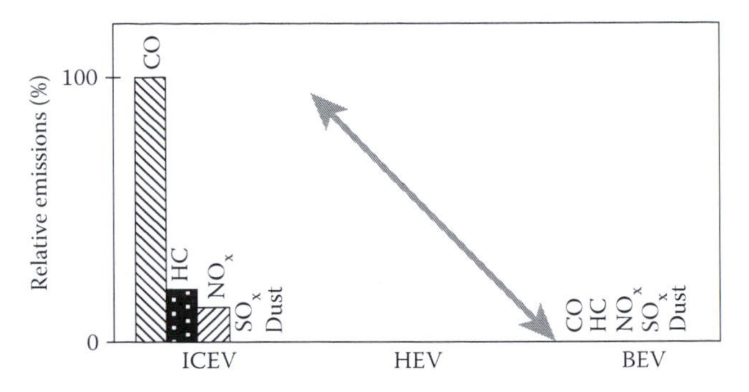

FIGURE 12.25
Comparison of local harmful emissions.

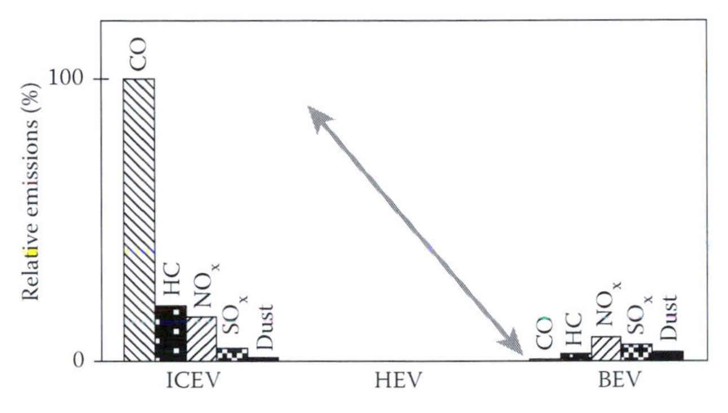

FIGURE 12.26
Comparison of global harmful emissions.

account the emissions generated by refineries to produce gasoline and diesel for the ICEV as well as the emissions by power plants to generate electricity for the BEV, an indicative comparison of global harmful emissions is shown in Figure 12.26. It can be found that the global harmful emissions of the BEV are still much lower than those of the ICEV. For the HEVs, the corresponding global harmful emissions are also between the BEV and ICEV.

Currently, many automakers produce HEVs, which are not only commercially available, but also economically sustainable. The latest flagships include the Chevrolet C15 Silverado hybrid, Ford Fusion hybrid, Honda Civic hybrid, Mercedes Benz S400 hybrid, Nissan Altima hybrid, Saturn VUE hybrid, and Toyota Prius hybrid. In general, these HEVs exhibit significant reductions in exhaust emissions over their ICEV counterparts. Nevertheless, different HEV models have different emission levels of different pollutants. For example, based on the data listed by the Air Resources Board of the California Environmental Protection Agency (Air Resources Board 2009), the Toyota Prius can offer a very low level of CO emission while the Mercedes-Benz S400 hybrid can provide a very small content of NMOG emission.

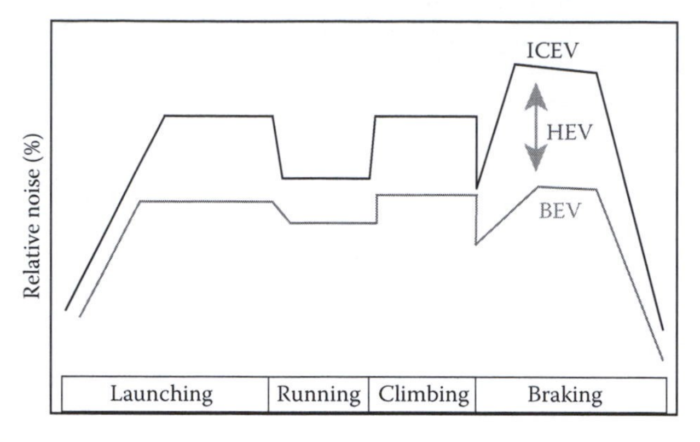

FIGURE 12.27
Comparison of vehicle noises.

EVs have another definite advantage over ICEVs, namely the suppression of noise pollution. Contrary to ICEVs with their combustion engine and complicated mechanical transmission that produce severe noise problems to our surroundings, the BEV is powered by an electric motor operating with very low-acoustic noise. Moreover, it can offer either gearless or single-speed mechanical transmissions so that the corresponding annoying noise is minimal. Figure 12.27 gives an indicative comparison of noise created by the ICEV and BEV during launching, running, climbing, and braking, while that of the HEVs lie between them.

12.6 Hybrid Vehicle Trends

Currently, many automobile companies throughout the world are accelerating the development of EVs for the coming huge market. It is anticipated that the HEV will be a practical and sustainable solution for SULEVs, while the BEV and FCEV will share the market of ZEVs. In view of the fact that the micro hybrid is virtually a motor-assisted ICEV while the REV is essentially an engine-assisted BEV, the HEVs will be dominant in the automotive market in the near future.

Focusing on the HEVs, the development trends will have two research directions. On the one hand, both the electric motor as well as battery and fuel cell technologies will be further developed so that electric propulsion will play a more active role to further increase fuel economy and reduce exhaust emissions. On the other hand, both the advanced engine and biofuel technologies will be actively developed so that the fuel efficiency and exhaust emissions can be improved. Definitely, these two development trends will make our future HEVs cleaner and greener.

Acknowledgments

I would like to express sincere thanks to my research student Mr. Wenlong Li for his contribution to the preparation of figures in this chapter. Especially, I must express my indebtedness to my wife, Joan Wai Yi, and my son, Aten Man Ho, for their patients and support all along the way.

References

Air Resources Board. 2009. *On-road new vehicle and engine certification program*. Sacramento, CA: California Environmental Protection Agency.

Alan, I., and T. A. Lipo. 2000. Starter/generator employing resonant-converter-fed induction machine. I. Analysis. *IEEE Transactions on Aerospace and Electronic Systems* 36:1309–28.

Atallah, K., and D. Howe. 2001. A novel high performance magnetic gear. *IEEE Transactions on Magnetics* 37:2844–46.

Axelrod, B., Y. Berkovich, and A. Ioinovici. 2005. A cascade boost-switched-capacitor-converter-two level inverter with an optimized multilevel output waveform. *IEEE Transactions on Circuits and Systems I* 52:2763–70.

Bae, B. H., and S. K. Sul. 2003. Practical design criteria of interior permanent magnet synchronous motor for 42V integrated starter-generator. Proceedings of IEEE International Conference on Electric Machines and Drives June:656–62.

Beretta, J. 2000. New tools for energy efficiency evaluation on hybrid system architecture. Proceedings of International Electric Vehicle Symposium (CD-ROM).

Cai, W. 2004. Comparison and review of electric machines for integrated starter alternator applications. Proceedings of IEEE Industry Applications Society Annual Meeting October:386–93.

Chan, C. C., and K. T. Chau. 2001. *Modern electric vehicle technology*. Oxford: Oxford University Press.

Chan, M. S. W., and K. T. Chau. 2007. A switched-capacitor boost-multilevel inverter using partial charging. *IEEE Transactions on Circuits and Systems II* 54:1145–49.

Chau, K. T., J. M. Yao, and C. C. Chan. 1999. A new soft-switching vector control approach for resonant snubber inverters. *International Journal of Electronics* 86:101–5.

Chau, K. T., M. Cheng, and C. C. Chan. 2002. Nonlinear magnetic circuit analysis for a novel stator-doubly-fed doubly-salient machine. *IEEE Transactions on Magnetics* 38:2382–84.

Chau, K. T., Q. Sun, Y. Fan, and M. Cheng. 2005. Torque ripple minimization of doubly salient permanent magnet motors. *IEEE Transactions on Energy Conversion* 20:352–58.

Chau, K. T., W. Cui, J. Z. Jiang, and Z. Wang. 2006a. Design of permanent magnet synchronous motors with asymmetric air gap for electric vehicles. *Journal of Applied Physics* 99:80R322:1–3.

Chau, K. T., Y. B. Li, J. Z. Jiang, and C. Liu. 2006b. Design and analysis of a stator doubly fed doubly salient permanent magnet machine for automotive engines. *IEEE Transactions on Magnetics* 42:3470–72.

Chau, K. T., and Y. S. Wong. 2001. Hybridization of energy sources for electric vehicles. *Energy Conversion and Management* 42:1059–69.

Chau, K. T., and Y. S. Wong. 2002. Overview of power management in hybrid electric vehicles. *Energy Conversion and Management* 43:1953–68.

Chau, K. T., Y. S. Wong, and C. C. Chan. 1999. An overview of energy sources for electric vehicles. *Energy Conversion and Management* 40:1021–39.

Chen, S., B. Lequesne, R. R. Henry, Y. Xue, and J. J. Ronning. 2002. Design and testing of a belt-driven induction starter-generator. *IEEE Transactions on Industry Applications* 38:1525–33.

Cheng, M., K. T. Chau, C. C. Chan, and Q. Sun. 2003. Control and operation of a new 8/6-pole doubly salient permanent magnet motor drive. *IEEE Transactions on Industry Applications* 39:1363–71.

Cui, W., K. T. Chau, J. Z. Jiang, and Y. Fan. 2005. Design of a novel phase-decoupling permanent magnet brushless AC motor. *Journal of Applied Physics* 97:10Q515:1–3.

Ebron, A., and R. Cregar. 2005. Introducing hybrid technology. *National Alternative Fuels Training Consortium eNews* July:1–5.

Ehsani, M., Y. Gao, S. E. Gay, and A. Emadi. 2005. *Modern electric, hybrid electric, and fuel cell vehicles: Fundamentals, theory, and design*. Boca Raton, FL: CRC Press.

Eriksson, S., and C. Sadarangani. 2002. A four-quadrant HEV drive system. *Proceedings of IEEE Vehicular Technology Conference* September:1510–14.

Gan, J., K. T. Chau, C. C. Chan, and J. Z. Jiang. 2000. A new surface-inset, permanent-magnet, brushless DC motor drive for electric vehicles. *IEEE Transactions on Magnetics* 36:3810–18.

Hermance, D., and S. Sasaki. 1998. Hybrid electric vehicles take to the streets. *IEEE Spectrum* 35:48–52.

Hoeijmakers, M. J., and J. A. Ferreira. 2006. The electric variable transmission. *IEEE Transactions on Industry Applications* 42:1092–1100.

Jian, L., and K. T. Chau. 2009. Analytical calculation of magnetic field distribution in coaxial magnetic gears. *Progress in Electromagnetics Research* 92:1–16.

Jiang, S. Z., K. T. Chau, and C. C. Chan. 2003. Spectral analysis of a new six-phase pole-changing induction motor drive for electric vehicles. *IEEE Transactions on Industrial Electronics* 50:123–31.

Liu, J., J. Hu, and L. Xu. 2004. Design and control of a kilo-amp DC/AC inverter for integrated starter-generator (ISG) applications. *Proceedings of IEEE Industry Applications Society Annual Meeting* October:2754–61.

Loh, P. C., D. G. Holmes, and T. A. Lipo. 2005. Implementation and control of distributed PWM cascaded multilevel inverters with minimal harmonic distortion and common-mode voltage. *IEEE Transactions on Power Electronics* 20:90–99.

Miller, J. M. 2006. Hybrid electric vehicle propulsion system architectures of the e-CVT type. *IEEE Transactions on Power Electronics* 21:756–67.

Miller, J. M., and M. Everett. 2005. An assessment of ultracapacitors as the power cache in Toyota THS-II, GM-Allison AHS-2 and Ford FHS hybrid propulsion systems. Proceedings of IEEE Applied Power Electronics Conference and Exhibition March:481–90.

Rodriguez, J., J. S. Lai, and F. Z. Peng. 2002. Multilevel inverters: A survey of topologies, controls, and applications. *IEEE Transactions on Industrial Electronics* 49:724–38.

Van Mierlo, J. 2000. Views on hybrid drivetrain power management. Proceedings of International Electric Vehicle Symposium (CD-ROM).

Wakefield, E. H. 1998. *History of the electric automobile: Hybrid electric vehicles*. Warrendale, PA: Society of Automotive Engineers.

Walker, A., P. Anpalahan, P. Coles, M. Lamperth, and D. Rodgert. 2004. Automotive integrated starter generator. Proceedings of International Conference on Power Electronics, Machines and Drives March:46–48.

Wang, Y., K. T. Chau, C. C. Chan, and J. Z. Jiang. 2002. Design and analysis of a new multiphase polygonal-winding permanent-magnet brushless DC machine. *IEEE Transactions on Magnetics* 38:3258–60.

Wouk, V. 1995. Hybrids: Then and now. *IEEE Spectrum* 32:16–21.

Xu, L., and J. Liu. 2004. Comparison study of DC-DC-AC combined converters for integrated starter generator applications. Proceedings of International Conference on Power Electronics and Motion Control August:1130–35.

Zhan, Y. J., C. C. Chan, and K. T. Chau. 1999. A novel sliding mode observer for indirect position sensing of switched reluctance motor drives. *IEEE Transactions on Industrial Electronics* 46:390–97.

Zhang, D., K. T. Chau, S. Niu, and J. Z. Jiang. 2006. Design and analysis of a double-stator cup-rotor PM integrated-starter-generator. Proceedings of IEEE Industry Applications Society Annual Meeting October:20–26.

13

Future Fuels

Arumugam Sakunthalai Ramadhas

CONTENTS

13.1 Introduction

Energy is the prime mover of economic growth of any country and is vital to the sustenance of modern economies. Future economic growth crucially depends on the long-term availability of energy from sources that are affordable, accessible, and environmentally friendly as well. Most developing countries are net importers of energy, more than 25% of primary energy needs are being met through imports, mainly in the form of crude oil and natural gas. Petroleum products are not only extensively used for the transport sector but also are being used by industries to meet peak demand and domestic purposes.

Renewable energy is a basic ingredient for overall sustainable development. Renewable fuel sources have the potential to provide energy supplies for an indefinite period of time. A lot of research and development activities on alternative fuels are being investigated around the world. However, biomass derived fuels can contribute significantly toward the problems related to the fuel crisis because these can be produced from locally available resources. Biomass origin fuels include alcohols, vegetable oils, and synthesis gases, and these can be formulated to bring their properties closer to that of conventional petroleum fuels.

Use of biofuels had been initiated at the time of the invention of the internal combustion engine. However, due to the abundant availability of petroleum distillates at that time, the use of biofuels for engine applications did not materialize. Because of recent concern over the environment and the energy crisis, vegetable oils and ethanol usage is again picking up. Use of biodiesel in diesel engines and ethanol in gasoline engines has become a commercial success in recent years. Most of the countries in the world have made their own national energy policies to blend biofuel with petroleum fuels to improve energy security and reduce environmental pollution. The biofuel produced from vegetable oils like sunflower, soybean, and palm, and ethanol produced from sugar cane, are called first-generation biofuels (Ramadhas 2009).

13.2 First-Generation Biofuels

While analyzing the commercial viability of any new fuel, particularly biofuels, the following are to be considered:

- Availability of land for future expansion on commercialization
- Economics of the project—feedstock, production process, and application

- Environmental impact of the new fuel
- Compatibility with in-use vehicles and fuel distribution infrastructure
- Other uses of the particular biomass source

First generation biofuels have good fuel properties and their production processes are commercially well developed and successfully implemented in automotive vehicles in many countries. However, the source of first-generation biofuel is to be considered for its success in the scenario. First-generation biofuel technologies are hindered by the fact of limited available land and risk of competition with food crops. The feedstock for first-generation biofuels are vegetable oils (sunflower, soybean, palm, and jatropha), corn, wheat, and sugar cane. If we go on increasing the percentage share of biofuel in petroleum products, it is necessary to increase the cultivation of biomass crops. But, most of the vegetable oils (except jatropha) are used for food purposes. Moreover, sugar is a most desirable food product and not intended for fuel purposes. Hence, after a few years it will be shown that we cannot produce enough biofuels without threatening food supplies and biodiversity.

The development of a new fuel will be commercially successful if the price of fuel is low. The overall production cost could be reduced if the by-products in the process also have some commercial value. The first-generation biofuels have a medium energy content and greenhouse effect except sugarcane ethanol. The first-generation biofuels are not very highly successful except for ethanol in Brazil. First generation biofuels are not cost competitive with existing fossil fuels. Moreover, considering emissions from production and transport of first-generation biofuels, life-cycle emissions from first-generation biofuels frequently exceed those of traditional fossil fuels. For the past few years, research and development efforts have sought to alleviate these limitations by exploring new pathways to convert little-used plant feedstocks to biofuels with better efficiencies. Second-generation biofuels can help solve these problems and can supply a larger proportion of our fuel supply sustainably, affordably, and with greater environmental benefits (www.wikipedia.com).

13.3 Second Generation Biofuels

Second generation biofuels are made from nonedible and lignocellulosic materials. For the production of second-generation biofuels, some specific types of plants are grown and converted into fuel. These are sustainable energies and have environmental benefits. Second-generation biofuels refer to biomass-to-liquid (BTL) technologies, namely, biological conversion; that is, cellulosic ethanol and thermochemical conversion that includes Fischer–Tropsch fuel, methanol, dimethyl ether (DME), and biohydrogen, and so on.

These technologies are in the development stage and have yet to be commercialized. The goal of second-generation biofuel processes is to extend the amount of biofuel that can be produced sustainably by using biomass consisting of the residual nonfood parts of current crops as well as other crops not used for food purposes. These includes stems, leaves, and husks that are left behind once the food crop has been extracted from the current crops and other nonfood crops such as switch grass and jatropha, and also industry waste such as wood chips, skins, and pulp. There are varieties of conversion methods to produce second-generation biofuels from lignocelluloses. However, two commonly followed pathways are as follows:

- Biological method: Enzymes and other microorganisms are used to convert cellulose and hemicellulose components of feedstocks to sugars prior to their fermentation to produce ethanol
    1. Bioethanol (cellulose, sorghum, cassava)
- Thermochemical method: Pyrolysis/gasification technologies produce a synthesis gas ($CO + H_2$) from which a wide range of long-carbon chain biofuels can be obtained and are reformed to the requirements of:
    1. Fischer–Tropsch liquids
    2. Biohydrogen
    3. Dimethyl ether

Thermal conversion processes convert all biomass feedstocks into useful energy forms and have a high degree of feedstock flexibility. In the biological conversion process, only a limited fraction of biomass can be converted with the known enzymatic technology to date. Biological processes may require limited feedstock flexibility and hence microorganisms must be tailored to the feedstock. Further research and development are needed to improve the conversion rate and economics.

13.4 Biological Conversion

13.4.1 Bioethanol

Ethanol is an alternative that can be blended with gasoline to improve the octane number and also for energy conservation reasons. Ethanol is a high octane fuel. It can be blended with gasoline to improve the octane number and also improve the fuel properties of gasoline. Ethanol blended gasoline improves the engine power and emission characteristics. In India, 5% ethanol blending with gasoline is permitted. In some countries, flexi-fuel vehicles (FFV) fueled with E85 (85% ethanol and 15% gasoline) are available.

First generation ethanol is produced by fermentation of sugar to ethanol, which is similar to the wine-making process. Ethanol can also be produced from starch, sucrose, and lignocellulosic biomass. The cost of feedstock plays a major role in production of ethanol. Lignocellulosic materials such as agri-residue and wood are attractive materials for ethanol fuel production as it is abundantly available.

Lignocellulose materials comprising of lignin, hemicellulose, and cellulose. These are complex carbohydrates (molecules based on sugar). Second-generation ethanols are derived from organic plant materials (such as wood, grasses, corn stover, wheat straw, etc.). The composition of various types of cellulose biomass materials is given in Table 13.1 (Deimbras 2005).

Hydrolysis breaks down the cellulose chains into sugar molecules that are then fermented to produce ethanol in the same way as first-generation bioethanol production. Hydrolysis breaks the hydrogen bonds in the hemicelluloses and cellulose into their sugar components; that is, pentoses and hexoses. These sugars can be fermented into ethanol. The hydrolysis process may require a pretreatment step to make the biomass more accessible for reaction. There are two ways for hydrolysis; namely, chemical and enzymetic. The outline of the methods are described in the following sections (www.plantoils.in).

13.4.1.1 Chemical Hydrolysis

In the chemical method, pretreatment and hydrolysis takes place in a single step. Diluted acid process; that is, 1% sulfuric acid under high temperature

TABLE 13.1

Properties of Cellulose Biomass

Materials	Cellulose (%)	Hemicellulose (%)	Lignin (%)	Ash (%)	Others (%)
Algae	20–40	20–50	—	—	—
Grasses	25–40	25–50	10–30	—	—
Hardwoods	43–47	25–35	16–24	0.4–0.8	2–8
Softwoods	40–44	25–29	25–31	0.4–0.6	1–5
Cornstalks	39–47	26–31	3–5	12–16	1–3
Wheat straw	37–41	27–32	13–15	11–14	5–9
Newspaper	40–55	25–40	18–30	—	—
Chemical pulps	60–80	20–30	2–10	—	—

Source: Deimbras, A., *Energy Sources*, 27, 327–37, 2005. Reprinted with permission from Taylor & Francis.

and pressure for few minutes in a continuous reactor provides a yield of about 50%. The concentrated acid hydrolysis process uses relatively mild temperatures and pressures. This process has a much longer reaction time than the diluted acid hydrolysis process. This process provides high sugar recovery and is cost effective as it recovers acid during recycling. However, acid hydrolysis results in the formation of by-products such as formic acid that have adverse effects on yeast growth during fermentation.

13.4.1.2 Enzymatic Hydrolysis

Biomass is cooked at higher temperatures and large amounts of amylolytic enzymes are added to hydrolyze the starchy/celluliosic biomass prior to fermentation. Different kinds of celluloses are used to break the cellulose and hemicelluloses. These enzymes attack the cellulose chains to produce polysaccharides of shorter length. Enzymes commonly used are endoglucanases, exoglucanases, and cellobiohydrolases. The pretreatment process is necessary for enzymatic hydrolysis; that is, pretreatment processes remove hemicellulose and breaks down the crystalline structure of cellulose or removes the lignin to expose hemicellulose and cellulose molecules. Figure 13.1 shows the bioethanol production process.

13.4.1.3 Bioethanol Properties

The properties of bioethanol are depicted in Table 13.2. The octane number of ethanol is higher than that for petrol; hence ethanol has better antiknock characteristics. This increases the fuel efficiency of the engine. The oxygen content of ethanol also leads to a higher efficiency, which results in a cleaner combustion process at relatively low temperatures.

The Reid vapor pressure, a measure for the volatility of a fuel, is very low for ethanol. This indicates a slow evaporation, which has the advantage of lower evaporative emissions and reduces the risk of explosions. Without aids, engines using ethanol cannot be started at temperatures below 20°C. The most cost-effective aid is the blending of ethanol with a small proportion of a volatile fuel such as gasoline. Thus, various mixtures of bioethanol

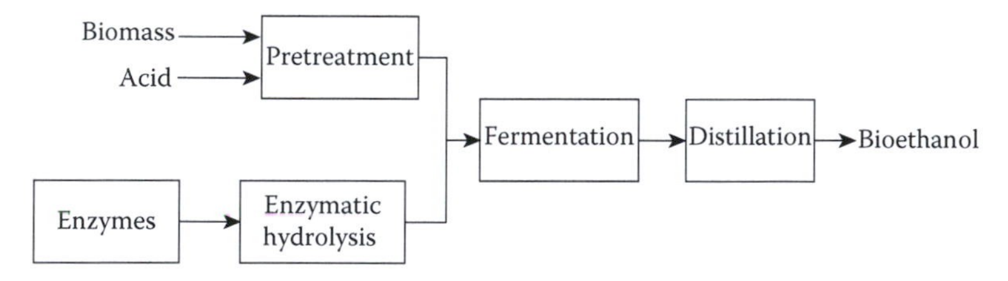

FIGURE 13.1
Bioethanol production process.

TABLE 13.2

Bioethanol Properties

Fuel Properties	Gasoline	Bioethanol
Molecular weight (kg/kmol)	111	46
Density (kg/l) at 15°C	0.75	0.80–0.82
Oxygen content (wt-%)	—	34.8
Lower calorific value (MJ/kg) at 15°C	41.3	26.4
Lower calorific value (MJ/l) at 15°C	31.0	21.2
Octane number (RON)	97	109
Octane number (MON)	86	92
Cetane number	8	11
Stoichiometric air/fuel ratio (kg air/kg fuel)	14.7	9.0
Boiling temperature (°C)	30–190	78
Reid Vapor Pressure (kPa) at 15°C	75	16.5

Source: www.eubia.org/212.0.html.

with gasoline or diesel fuels have been used. Bioethanol has been extensively tested in light duty FFV as E85G. Ethyl tertiary butyl ether (ETBE) is also used in blends of 10–15% with gasoline to enhance its octane rating and reduce emissions. Blends of gasoline with up to 22% ethanol can be used in spark ignition engines without any material or operating problems. Blends of diesel with up to 15% ethanol do not introduce any technical engine problems and require no ignition improver (www.eubia.org/212.0.html).

13.5 Thermochemical Conversion

There is a wide range of processes available for converting biomass into more valuable fuels. Coal/biomass can be converted to liquid hydrocarbon suitable for engine operation by coal to liquid (CTL) or BTL conversion processes. The liquefaction process can be classified into two processes, namely, direct liquefaction and indirect liquefaction.

Direct liquefaction process:

- Coal/biomass + catalyst + Hydrogen → hydrocarbons.

Indirect liquefaction process:

- Gasification

  Coal/biomass + oxygen + steam → CO + H_2 (synthesis gas)
- F-T synthesis

  Synthesis gas + catalyst → hydrocarbons

13.5.1 Gasification

Biomass gasification is one of the biomass conversion technologies developed in order to produce a combustible gas mixture (called producer gas) using agroresidues. Gasification is the process of converting biomass into a gaseous fuel by means of partial oxidation at high temperature. Gasification involves the devolatilization and conversion of biomass in an atmosphere of steam and/or air to produce a medium or low calorific value gas. The ratio of oxygen to biomass used is typically around 0.3. Gasification is a form of pyrolysis, carried out at high temperatures in order to optimize gas production. The producer gas is a mixture of carbon monoxide, hydrogen, and methane, together with carbon dioxide and nitrogen. It is one of the important conversion technologies that can be effectively utilized for decentralized power generation and thermal applications. The major portion of agricultural wastes undergoes natural decomposition resulting in production of various greenhouse gases and other environmental problems (Wagner 2007). The gasification process is suitable for a wide range of biomass resources ranging from agricultural crops, biomass residues, and organic wastes.

The major classification of gasifiers include a fixed bed, a fluidized bed, and an entrained bed. These gasifiers differ in size, type of feedstock, feed, production rate, residence time, and reaction temperature.

Figure 13.2 shows the schematic of the production method of second-generation biofuels by the gasification process.

13.5.1.1 Fixed-Bed Gasifiers

Fixed-bed gasifiers can be broadly classified into three according to the gas flow direction: updraft gasifier, downdraft gasifier, and crossdraft gasifier. In updraft gasifiers the air enters beneath the combustion zone. The

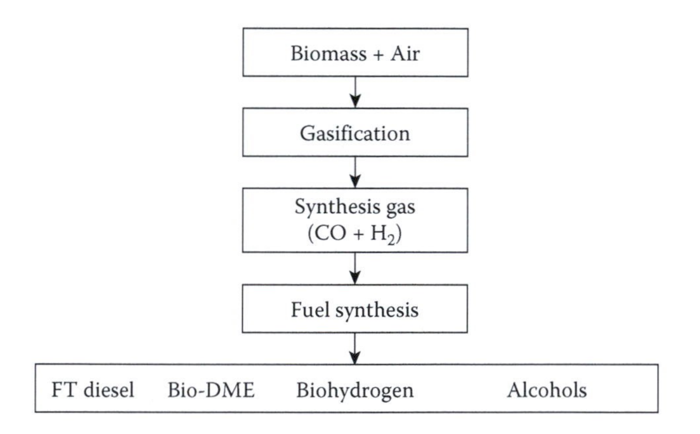

FIGURE 13.2
Schematic of production of second-generation biofuels.

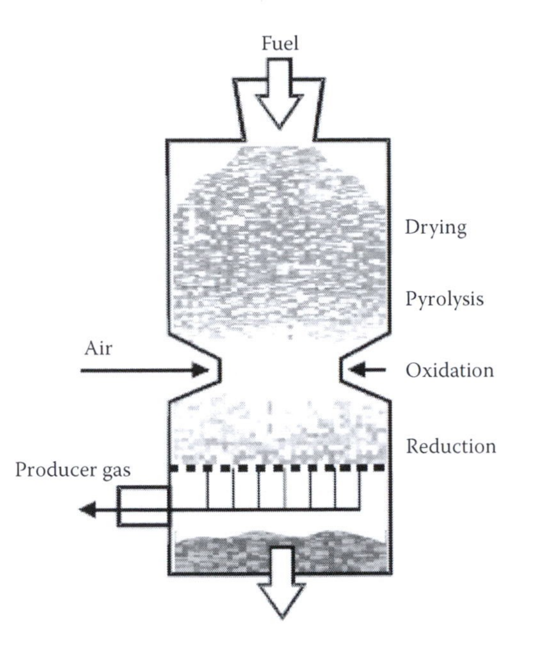

FIGURE 13.3
Schematic of downdraft gasifier.

gas formed will be drawn out from the top of the gasifier. In this type of gasifier, since the gas formed passes through the unburned biomass, the gas contains tar and water vapor. So updraft gasifiers are more suitable for tar-free biomass.

In downdraft gasifiers air enters at the combustion zone and is drawn out from the bottom of the gasifier. The schematic of down draft gasifier is shown in Figure 13.3. The main disadvantage of this type of gasifier is that the gas contains fewer amounts of tar and volatiles. This is because the gas is passing through the high temperature zone where the unburned tars can be cracked into gaseous hydrocarbons, thus producing relatively cleaner gas. These gasifiers are most suitable for fuels like wood and agricultural wastes.

In crossdraft gasifiers the gas inlet and outlet are arranged transverse to the fuel bed. The gas produced passes upward in the annular space around the gasifier that is filled with charcoal. The charcoal acts as an insulator and duct filter. Since the path length is short in the crossdraft gasifier, it can respond to the rapid changes in gas requirements.

Drying: biomass fuel consists of moisture ranging from 5 to 35%. At temperatures above 100°C, water is removed and converted into steam. In the drying process, fuels do not experience any kind of decomposition.

Pyrolysis: pyrolysis is the thermal decomposition of biomass fuels in the absence of oxygen. Pyrolysis involves the release of three kinds of products: solid, liquid, and gas. The ratio of products is influenced by the chemical composition of biomass fuels and the operating conditions. The heating value of the gas produced during the pyrolysis process is low (3.5–8.9 MJ/m^3).

Oxidation: introduced air in the oxidation zone contains, besides oxygen and water vapors, inert gases such as nitrogen and argon. These inert gases are considered to be nonreactive with fuel constituents. Oxidation takes place at the temperature range of 700–2000°C. Hydrogen in the biomass reacts with oxygen in the air blast and produces steam.

$$C + O_2 \Leftrightarrow CO_2 + 406\ \text{MJ/kmol}$$

$$H_2 + \frac{1}{2}O_2 \Leftrightarrow H_2O + 242\ \text{MJ/kmol.}$$

Reduction: in the reduction zone, a number of high temperature chemical reactions take place in the absence of oxygen. The principal reactions that take place in reduction are as follows:

$$\text{Boudouard reaction: } CO_2 + C \Leftrightarrow 2CO - 172.6\ \text{MJ/kmol}$$

$$\text{Water-gas reaction: } C + H_2O \Leftrightarrow CO + H_2 - 131.4\ \text{MJ/kmol}$$

$$\text{Water-shift reaction: } CO_2 + H_2 \Leftrightarrow CO + H_2O + 41.2\ \text{MJ/kmol}$$

$$\text{Methane production reaction:} C + 2H_2 \Leftrightarrow CH_4 + 75\ \text{MJ/kmol.}$$

Fixed bed gasifiers operate at 26 bar and closely resemble a blast furnace. The residence time of fixed bed gasifiers is between 30 and 60 minutes. Coal size distribution must be controlled to ensure good bed permeability. To ensure stable fluid bed operation, gasification temperatures are kept below the ash fusion temperature of the coal. Above this temperature, particles become sticky and excessive levels of agglomeration will occur, resulting in bed defluidization (http://www.ccsd.biz/factsheets/igcc.cfm).

13.5.1.2 Fluidized-Bed Gasifiers

A fluidized bed reactor is a vessel in which fine solids are kept in suspension by a gas such that the whole bed exhibits a fluid like behavior. This type of reacting system is characterized by high heat and mass transfer rates between the solid and gas. In a fluidized bed gasifier, air, oxygen, or steam is passed through distributor plates at the bottom of the gasifier. Limestone is mixed with coal to remove the sulfur present in the coal. Char/ash taken along with producer gas are pyrolyzed in the hot agglomerating zone. Any ash particles that stick together will fall and are cooled and removed at the bottom of the gasifier. In fluidized bed gasification, rising oxygen enriched gas reacts with suspended coal at a temperature of 950–1100°C and a pressure of 20–30 bar. High levels of back mixing result in fluid bed gasifiers having a uniform temperature distribution. A typical diagram of a fluidized bed gasifier is shown in Figure 13.4.

The major advantages of fluidized bed gasifiers stem from their feedstock flexibility, resulting from easy control of temperatures and their ability to deal with fluffy and fine grained materials (sawdust, etc.) without the need of preprocessing. The pressurized operation has an advantage of higher methane formation (higher calorific value), and increased heat from methanation reactions that reduces the amount of oxygen needed. Moreover, the reduced heat losses through the wall increase the efficiency of the system. Other drawbacks of the fluidized bed gasifier lie in the rather high tar content of the product gas (up to 500 mg/m^3 gas), the incomplete carbon burnout, and poor response to load changes.

13.5.1.3 Entrained Gasifiers

Entrained flow is the most aggressive form of gasification, with pulverized coal and oxidized gas flowing cocurrently. High reaction intensity is provided by a high pressure (20–30 atm), high temperature (>1400°C) environment. A typical composition of syngas obtained from biomass gasification on a volumetric basis is shown in Table 13.3.

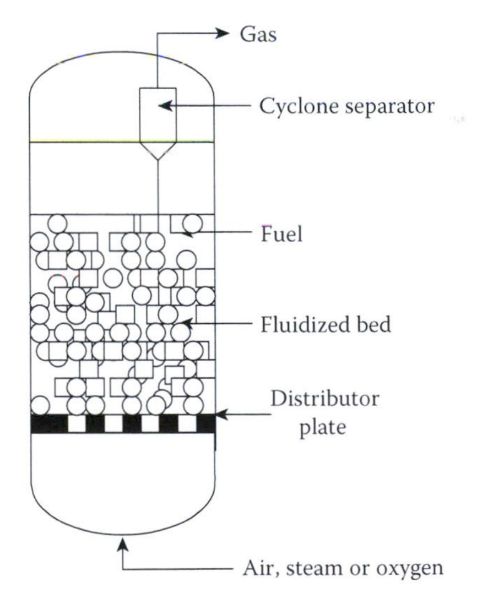

FIGURE 13.4
Schematic of fluidized bed gasifier.

TABLE 13.3

Typical Composition of Syngas

Composition	Percentage
Carbon monoxide	18–22%
Hydrogen	15–19%
Methane	1–5%
Hydrocarbons	0.2–0.4%
Nitrogen	45–55%
Water vapor	4%

TABLE 13.4

Properties of Various Biomass

Biomass	Bulk Density (kg/m³)	Ash Content (%)	C (%)	H (%)	N (%)	O (%)	Calorific Value (MJ /kg)
Bagasse	74	4.00	47.00	6.50	0.0	42.50	17.5
Coir pith	47	13.60	41.27	4.02	1.51	39.60	16.7
Ground nut shell	165	3.10	33.90	1.97	1.10	59.93	18.8
Saw dust	177	1.20	52.28	5.20	0.47	40.85	18.4
Straw	80	15.50	35.97	5.28	0.17	43.08	15.5
Wood	330	1.50	52.30	5.20	0.50	42.00	18.4

Table 13.4 describes the properties of various biomass sources used for gasification. The ash content is generally higher for powdered biomass whereas calorific value is almost the same for all the biomass fuels. The important properties related to the gasification process are moisture content, elemental composition, bulk density, and volatile matter content. It is preferred to use the biomass dry as it produces a high quality gas of high calorific value and low tar. The chemical composition of ash affects the melting behavior of the ash. Ash melting may cause slagging and channel formation of the gasifier. Biomass gasification generally produces low sulfur and low nitrogen content gases. It is preferred to use a biomass of high bulk density to get a good profit margin in the gasification process because low bulk density biomass tends to decrease the capacity of the reactor. Moreover, it increases the transportation, handling, and storage expense. Preparation of biomass suitable for the gasification process depends upon the type of the gasifier used. Downdraft gasifiers require uniform biomass size more than updraft gasifiers. Gasifiers are equipped with sizing arrangements so that it allows the uniform or required size biomass. A briquette method is used to make pellets of a specified size suitable for the gasifier.

The gasification of biomass yields CO and H_2 as major constituents. In the synthesis process, presence of catalysts yield and quality of the product can be varied. Based on the gasification technology and synthesis process F-T diesel, DME, and biohydrogen can be produced. These are the potential alternative fuels for use in internal combustion engines.

13.5.2 Biohydrogen

Hydrogen is considered the "energy of the future" since it is a clean energy source with high energy content as compared to hydrocarbon fuels and not readily available in nature as fossil fuels. Hydrogen is a clean fuel, as on combustion it emits only carbon dioxide and water. It is nonpolluting of the atmosphere and an environmentally friendly fuel. Hydrogen is produced using nonrenewable technologies such as steam reformation of natural gas (~50% of global H_2 supply), petroleum refining (~30%), or the gasification of coal (~20%). Hydrogen has an approximate three-fold higher energy content than gasoline (119 MJ/kg for hydrogen vs. 44.2 MJ/kg for gasoline), while bioethanol has 50% less energy content (22.1 MJ/kg for ethanol) compared to gasoline and 84% less energy content compared to hydrogen. Hydrogen could be up to three times more costly than gasoline but produce three times the energy output, while ethanol would require a cost per gallon of 50% less than gasoline to produce the same energy outputs (www.biohydrogen.co.uk).

13.5.2.1 Gasification Process

Hydrogen, produced from water and biomass feedstock, is called biohydrogen. Current technologies for the production of hydrogen are much more expensive than that of petroleum fuel. Hydrogen can be produced from biomass by gasification and fast pyrolysis. The gasification process converts biomass into synthesis gas; that is, carbon monoxide and hydrogen, which depends upon the process applied, oxidation medium, and fuel/oxygen ratio. Low temperature gasification processes (less than 1000°C) yield products containing significant amounts of hydrogen as compared to higher temperature processes. Fixed bed and fluidized bed gasification technologies are employed to produce syngas with good conversion efficiencies.

Hydrogen can also be produced from pyrolysis of lignocelluloses biomass (second-generation biofuel). Pyrolysis is carried out at temperatures in the range of 650–800 K and at atmospheric pressure or slightly higher pressure and converts biomass into bio-oils. This is a complex mixture of compounds including acids, alcohols, aldehydes, esters, chemicals, and aromatics. Bio-oils are used as boiler fuel for stationary power plants and heating applications. However, this fuel can be upgraded to use as automobile engine fuel.

This pyrolysis can be classified into: slow pyrolysis and fast pyrolysis. Slow pyrolysis yields high charcoal and is not considered for hydrogen production.

Fast pyrolysis takes place at higher temperatures; a rapid heating process in which water vapor is produced that is condensed to a dark brownish liquid bio-oil. The yield of bio-oil depends upon proper choice of reactor configuration; biomass feedstock particle size, and heat and mass transfer rate. Fast pyrolysis produced bio-oils can be easily transported. Moreover, small pyrolysis can be installed where sufficient amounts of biomass are available for fast pyrolysis. This bio-oil is steam reformed to hydrogen and carbon monoxide. The water–gas shift reaction is used to convert the reformed gas into hydrogen. The pressure swing adsorption is used to purify the product. The yield from the gasification process is higher than the pyrolysis process. Moreover, the yield increases with steam to sample ration in the reaction. Jong (2008) assumed the typical reaction for biohydrogen production is given as:

$$CH_{1.98}O_{0.76} + 1.24H_2O \rightarrow CO_2 + 2.23H_2.$$

13.5.2.2 Supercritical Water Gasification (SCWG)

Biohydrogen can also be produced from the Supercritical Water Gasification (SCWG) method. It is a relatively novel gasification method, in which biomass is transformed into a hydrogen-rich gas by introducing it in supercritical water. Supercritical water is obtained at pressure above 221 bar and temperatures above 374°C. By treatment of biomass in supercritical water—but in the absence of added oxidants—organics are converted into fuel gases and are easily separated from the water phase by cooling to ambient temperature. The produced high pressure gas is very rich in hydrogen (www.news.mongabay.com/bioenergy/2008/02/researchers-find-large-potential-for.html).

SCWG process consists of operations such as feed pumping, heat exchanging, reactors, gas-liquid separators, and product upgrading. The reactor operating temperature is typically between 600°C and 650°C; the operating pressure is around 300 bar. A residence time of 0.5–2 minutes is required to achieve complete carbon conversion depending on the feedstock. This process is in particular suitable for the conversion of wet organic materials (renewable or nonrenewable) with moisture content in the range of 70–95%. This process is very cost effective as it uses wet feedstock also reducing the pretreatment processes. Selectivity toward methane, hydrogen, or syngas, can be controlled with temperature, pressure, and using the proper catalysts.

13.5.2.3 Fermentation Process

The main source of hydrogen during a biological, fermentative process is carbohydrates, which are very common in plant tissues, either in the form of oligosaccharides or as polymers, cellulose, hemicellulose, and starch. Thus, the biomass of certain plants with a high content of carbohydrates could be considered as a very promising substrate for biohydrogen production. The maximum theoretical hydrogen yield is 4 moles per mole of utilized carbohydrates, expressed as glucose equivalents when carbohydrates are used as substrates (Nandi and Sengupta 1998). Sweet sorghum stalks are rich in sugars, mainly in sucrose that amounts up to 55% of dry matter and in glucose (3.2% of dry matter). They also contain cellulose (12.4%) and hemicelluloses (10.2%; Billa et al. 1997). Sweet sorghum biomass is rich in readily fermentable sugars and thus it can be considered as an excellent raw material for fermentative hydrogen production.

13.5.2.4 Applications

Biohydrogen has properties similar to that of conventional hydrogen.

- Hydrogen, a colorless and odorless gas, is a clean fuel suitable for automobiles at normal atmospheric conditions.
- Hydrogen can be stored in metal hydrates and compressed gas cylinders for vehicle application purposes.
- It is stable and coexists harmlessly with free oxygen until an input of energy drives the exothermic (heat releasing) reaction.
- Fuel cells use hydrogen as a fuel that converts the chemical energy contained in the hydrogen molecule into electrical energy.

13.5.3 Fischer–Tropsch Diesel

Synthetic fuels are often referred to as Fischer–Tropsch (F–T)diesel fuel or gas-to-liquid fuel. The F–T process was named after the two German scientists who developed it in the 1920s—Frans Fischer and Hans Tropsch. The Fischer-Tropsch process (F–T process) is the synthesis of long-chain hydrocarbons from CO and H_2 gas mixtures using a metal catalyst. Natural gas is the major feedstock for the F–T process; the production comes under the gas-to-liquid conversion process. F–T synthesis using coal is referred to as the indirect liquefaction of coal. F–T manufactured from fossil fuels brings no distinct greenhouse gas benefits in comparison with petroleum products. A greenhouse benefit can only be achieved when F–T fuels are produced from biomass.

Cleaning of synthesis gas is required before F–T processing and the process yields mixtures of different hydrocarbons. H_2S removal is necessary to protect synthesis catalysts as well as limit SO_2 emissions. In order to maintain the catalyst for the maximum lifetime, sulfur levels in syngases entering the synthesis reactor must be below 1 ppm by volume. These levels are achieved through one of two methods—physical absorption by organic fluids or chemical reactions between amines and sulfur compounds. The cleaning synthesis process includes removal of tar, dust separation in the cyclone separator, bag filter and carbonyl sulfide hydrolysis in a scrubber at 100–250°C, tar condensation, and ammonia and hydrogen cyanide scrubbing with sulfuric acid (Speight 2008; http://blogs.princeton.edu/chm333/f2005/group2/03_chemistry/01_current_chemistry/03_sulfur_and_carbon_dioxide_removal/).

Fischer–Tropsch fuels may be produced in three steps:

1. Synthesis gas generation: production of synthesis gas ($CO + H_2$) using the gasification process. If the synthesis gas with an H_2/CO ratio less than two is used, the composition is not stoichiometric for the F–T reactions. Hence, the water–gas shift reaction is used to modify the H_2/CO ratio to two.
2. Hydrocarbon synthesis: conversion of syngas into a mixture of liquid hydrocarbon and wax.
3. Fuel upgrading: syngas upgraded through hydrocracking and isomerization and fractionated middle distillate fuels (liquid fuels).

The F–T synthesis equation can be described as:

$$nCO + (2n + 1)H_2 \rightarrow C_nH_{2n+2} + nH_2O \text{ (Exothermic).}$$

There are three different reactor designs for the F–T process. These include

- Fixed-bed (gas phase);
- Fluidized-bed (gas phase); and
- Liquid phase (slurry reactor).

Catalysts used for the synthesis process are based on iron and cobalt. As the reactions are exothermic in order to avoid an increase in temperature, which results in lighter hydrocarbons, it is necessary to have sufficient cooling for stable reaction conditions. Low temperature Fischer–Tropsch (200–240°C) reactions use an iron catalyst and high temperature Fischer–Tropsch (300–350°C) reactions use an iron or cobalt catalyst. Moreover, nickel-based and ruthenium-based catalysts are used for commercial processes. Cobalt catalysts are of higher activity and longer life as compared to iron catalysts, but cobalt catalysts are much more expensive than iron catalysts. In the synthesis reactor, cleaned gases come in contact with the catalyst that transforms the H_2 and CO into long-chain hydrocarbon molecules.

F–T diesel is composed of only straight chain hydrocarbons with nil sulfur and aromatics. However, it may contain traces of NH_3, H_2S, dust, and so on. These components poison the catalyst, and hence these need to be removed. The process output depends upon the catalyst, reaction temperature, reaction pressure, and reaction duration. Indirect liquefaction F–T power plants generate carbon dioxide in concentrated forms allowing capture and sequestration. F–T power plants with CO_2 capture have better life-cycle greenhouse emission profiles in comparison with petroleum products production in refineries.

F–T processes yield diesel as the main product and gasoline, kerosene, and gases as by-products. Synthetic diesel fuel provides benefits in terms of PM and NO_x. The comparison of properties of commercial diesel with two typical synthetic fuels is given in Table 13.5.

From Table 13.5, it has been seen that properties of F–T diesel are much better than petroleum diesel. F–T diesel can be used as a blending component in diesel. F–T diesel contains parafins and olefins, and nil sulfur and aromatics. Particulate matter is reduced, which is essential in this emission scenario. Moreover, it has a high cetane number that improves combustion characteristics. It is colorless and odorless in appearance. The energy density of F–T diesel is similar to diesel. However, F–T diesel has poor lubricity and cold flow properties. Lubricity properties can be improved by adding suitable lubricity improvers. Blending biodiesel with F–T diesel remarkably

TABLE 13.5

Characteristics of the Fischer–Tropsch Diesel

Characteristic	Diesel	Aliphatic Diesel A	Aliphatic Diesel B
Density at 15°C (kg/m^3)	832.0	775.7	775.7
Viscosity at 40°C (mm^2/s)	2.78	2.39	2.68
Cetane number	52.8	71.3	74.8
Sulfur content (mg/kg)	206	<1	<1
Water content (mg/kg)	—	23	16
Cold filter plugging point (°C)	–27	–28	–6
Distillation curve (°C)			
Initial point	188	160.0	189.2
5%	210	186.2	206.6
10%	225	199.0	213.2
20%	—	219.2	227.4
30%	—	236.2	241.7
50%	271	264.0	271.6
70%	—	290.7	302.9
80%	—	305.0	318.9
90%	329	322.8	337.3
95%	345	335.2	349.3
Final point	356	344.3	359.9
Residue, %	—	1	1
Monoaromatics (%v)	—	0.2	<0.1
Bioaromatics (%v)	—	<0.1	<0.1
Total aromatics (%v)	—	0.2	<0.1
H/C	1.92	2.18	2.24

Source: Reprinted with permission from Zervas, E., *Fuel*, 87, 1141–47, 2008.

TABLE 13.6

Exhaust Emissions of F–T Diesel in Comparison with Conventional Diesel

Fuel	HC	CO	NO_X	PM	CO_2
Conventional U.S. #2 diesel	0.346	1.584	5.373	0.120	643.75
California #2 diesel	0.274	1.091	4.893	0.109	615.85
F-T diesel	0.198	0.968	4.607	0.104	611.49

Source: www.environment.gov.au/settlements/transport/comparison/pubs/ 2ch3.pdf (accessed on October 25, 2009).

improves lubricity properties and reduces engine exhaust emissions. F–T fuels can provide 40–60% emission benefits in light duty vehicles and 5% in heavy duty vehicles for CO, HC, and PM (Majewski and Khair 2006).

Table 13.6 shows the exhaust emissions of an engine in comparison with conventional emissions as per hot-start FTP engine tests in g/bhp-h. From Table 13.6 it is seen that F–T diesel reduces exhaust emission very well.

The Fischer–Tropsch process has been receiving so much attention because

- Synthetic fuels are compatible with diesel, have comparable energy density, and can use existing infrastructure facilities.
- Sulfur content is nil, hence, it will be compatible with a wide range of after treatment devices that are NO_x absorbers.
- It can be added to diesel to improve the fuel properties of diesel.

13.5.4 Dimethyl Ether

Dimethyl ether is a diesel engine fuel with a high cetane number and has an interesting emission potential. In the chemical industry, dimethyl ether is produced from pure methanol by means of a process called catalytic dehydration, which chemically separates water from methanol. This methanol can be produced from coal, natural gas, or biomass. It is possible to use black liquor, a residual product from the production of paper pulp (from forest produce), as a feedstock for manufacturing DME. Often the production of methanol and DME is combined in one process.

Dimethyl ether is one of the most promising second-generation biofuels. DME produced from biomass is characterized by high overall energy efficiency and low greenhouse gas emissions from well-to-wheel. DME distribution networks require an infrastructure for storage and delivery; that is, bulk storage at the production industry, transportation to bottling plants, and cylinder distribution. Existing liquefied petroleum gas (LPG) distribution networks can be utilized for DME production.

It is nontoxic and produces very low exhaust emissions, but it requires a new kind of filling station. Bio-DME can also be produced directly from synthesis gas. DME (CH3–O–CH3) is an LPG-like synthetic fuel that is produced through gasification of various renewable substances or fossil fuels. The synthetic gas is then catalyzed to produce DME in a single step at lower costs. The optimum H_2/CO ratio for DME synthesis is lower than that for methanol and ideally should be around one. The synthesis of DME from producer gas/syngas is a highly exothermic reaction. DME can be produced as a single product or in combination with methanol.

13.5.4.1 DME from Methanol

Single step DME process possesses both methanol synthesis and methanol dehydration activity. There are two types of catalyst systems, namely, dual catalyst systems and single catalyst systems. In dual catalyst systems, the catalyst is a mixture of methanol synthesis and a dehydration catalyst. In general, methanol catalysts are copper based and dehydration catalysts are from solid acid materials such as zeolites and γ-alumina. The catalyst used in the reactors may be arranged in mixed forms, one layer or in several layers. In single catalyst systems, the two functions can be achieved by the precipitated methanol synthesis components onto high surface area solid acid materials. The DME production process occurs either in a gaseous phase in a fixed bed reactor or a liquid phase in a slurry bed reactor.

DME production through methanol dehydration involves the following main reactions:
Water–gas shift:

$$CO + H_2O \rightarrow CO_2 + H_2 \qquad \Delta H_r = -40.9 \text{ kJ/mol.}$$

Methanol synthesis (Zeolite):

$$CO + 2H_2 \rightarrow CH_3OH \qquad \Delta H_r = -90.7 \text{ kJ/mol}$$

$$2CH_3OH \rightarrow CH_3OCH_3 + H_2O \qquad \Delta H_r = -23.4 \text{ kJ/mol.}$$

Net reaction:

$$3H_2 + 3CO \rightarrow CH_3OCH_3 + CO_2 \qquad \Delta H_r = -245.3 \text{ kJ/mol.}$$

This methanol dehydration is exothermic and compares with methanol synthesis. The vaporized methanol is fed into the reactor. The synthesis pressure ranges from 10 to 20 bar and reactor inlet temperatures 220–250°C and the outlet temperature is 300–350°C. The conversion efficiency of 70–85% could be achieved. The unconverted methanol and produced water are separated at the bottom and the methanol is returned back to the reactor after purification. The DME is collected at the top of the reactor. For the production of 1 ton of DME, 1.4 tons of methanol are required (Smith, Asaro, and Naqvi 2008).

13.5.4.2 DME from Coal/Biomass

Gasification processes that produce synthesis gases can use a high sulfur content coal or biomass feedstock. A typical schematic of production of DME by the gasification process is shown in Figure 13.5.

Gasification produced CO and H_2 are cooled and the H_2/CO ratio is adjusted as per the requirements of DME synthesis by water–shift gas reaction. The sulfur and unburned carbon are removed by a separator. Purified gases are synthesized in the reactor to produce DME. Oil and gas companies are focusing on DME as a way to efficiently use their natural gas resources. Moreover, research is going on to efficiently use DME as a substitute for LPG and diesel.

13.5.4.3 Properties of DME

DME is a clean burning (completely sootless) synthetic fuel that can be substituted for conventional diesel, LPG, or be reformed into hydrogen for fuel cells. It can be transported as a pressurized liquid similar to LPG. DME is gaseous at ambient temperature, but it is a liquid if the pressure is above 5 bar or the temperature is below –25°C. It is likely to be used as a liquid at 5–10 bar. Transport, storage, and distribution of DME are similar to that of LPG. The main challenges for further development of bio-DME are similar to those of Fischer–Tropsch liquids and biomethanol.

The DME has a higher cetane number (55–60) that is the desirable property of diesel engine fuel. There are no major modifications of engines required. However, new fuel storage, handling, and injection systems are required. The volumetric energy content is much lower, approximately half that of diesel. Retrofitting diesel engines for the use of DME is relatively simple. Although DME does not corrode metals, unlike bioethanol and biomethanol, it may affect certain kinds of plastics, elastomers, and rubbers after some time. Engine calibration has to be modified to lower the NO_x emissions and

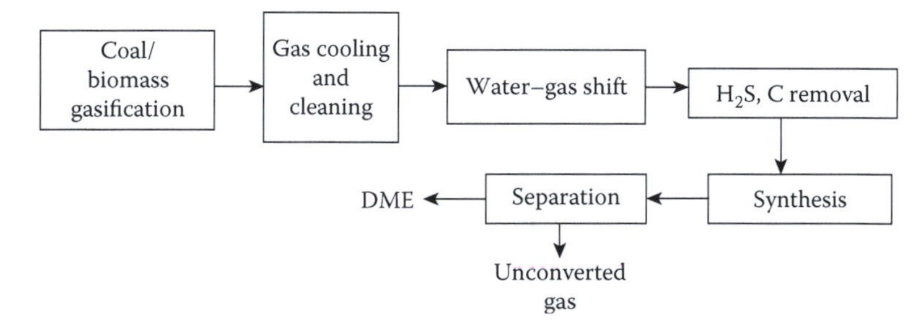

FIGURE 13.5
Biodimethyl ether production process.

TABLE 13.7

Properties of Dimethyl Ether

Property	DME
Boiling point, °C	–25.1
Liquid density, g/cm^3 at 20°C	0.67
Cetane number	55–60
Explosion limits, %	3.4–17
Ignition temperature at 1 atm, °C	235–350
Lower heat value, kJ/kg	28,870
Specific density, gas	1.59
Water, wt%	3.2 ± 0.3
Methanol, wt%	7.5 ± 1.0
DME, wt%	88.9 ± 0.9
Oxygenates, ethers and gases, wt%	0.4 ± 0.1

Source: Smith, R., Asaro, M., and Naqvi, S., Fuels of the future. Menlo Park, CA: SRI Consulting, 2008. www.dmeforum.jp

PM emissions. The thermal efficiency and emissions with DME is almost the same as that of natural gas. The important properties of DME are given in Table 13.7 (www.dmeforum.jp; Smith et al. 2008).

Dimethyl ether is a safe, environmentally benign energy source for fuel cells or diesel engines. It is gaseous at ambient conditions but can be liquefied at moderate pressure. With a high cetane number, DME has very attractive characteristics as an alternative fuel for diesel engines. More details on engine testing and emissions are available in Chapter 7.

DME can also be used as a clean burning substitute for diesel in transportation vehicles and as a clean fuel for power generation. DME is an excellent alternative fuel for diesel engines where ultra low emissions are required. DME can be blended up to 20% with LPG and used for household cooking and heating, without any modifications to equipment or used as a replacement. Compared to LPG, DME has a higher oxygen content that permits lower CO, smoke, HC emissions, and has rapid vaporization as it has a lower boiling point. While storing DME in household application cylinders, DME would weigh 18 kg compared to an LPG weight of 14 kg and contain 18% less energy (Smith, Asaro, and Naqvi 2008).

The advantages of bio-DME include

- DME can be used as a diesel fuel additive or blend
- Extremely low exhaust emissions
- Almost CO_2 neutral if produced from biomass
- Best well-to-wheel energy efficiency from biosources
- Produced from many resources and can be used for several applications

The disadvantages of bio-DME include

- DME has poor lubricating properties whereas diesel has good lubricating properties. Hence, lubricating additives need to be added to DME to avoid problems in fuel injectors and other components of the engine. Research is going on to develop DME injectors for better performance.
- DME has a low boiling point (–23°C) and hence at atmosphere it is in a gaseous state and its liquid also has a very low density (approximately 0.67 g/cm^3).

13.6 Algae Biodiesel

Biodiesel is generally produced from vegetable oils and animal fats. Biodiesel produced from algae is known as algae biodiesel. Algae grows naturally all over the world. Under optimal conditions, it can be grown in massive, almost limitless amounts. Half of algae's composition, by weight, is lipid oil. For its growth (photosynthesis), algae consumes carbon dioxide from the atmosphere replacing it with oxygen. Recycling carbon dioxide in the atmosphere reduces environmental pollution and greenhouse gas effects (Wagner 2007).

Algae have emerged as one of the most promising sources especially for biodiesel production because of the following (www.castoroil.in):

- Yields of oil from algae are orders of magnitude higher than those for traditional oilseeds.
- Algae can grow in places away from farmlands and forests, thus minimizing damages caused to the ecological and food chain systems.
- Algae can be grown in sewages and next to power plant smoke stacks where they digest pollutants and give us oil.

Biodiesel makers claim that they will be able to produce more than 100,000 gallons of algae oil per acre per year depending on the type of algae being used, the way the algae is grown, and the method of oil extraction. It has been estimated that 100-acre algae biodiesel plants could produce 10 million gallons of biodiesel in a year and 140 billion gallons of algae biodiesel to replace petroleum-based products each year. Since algae can be grown anywhere indoors, it is a promising alternative fuel source.

13.6.1 Cultivation of Algae

Algae can grow in open ponds or closed, controlled atmospheres. Like plants, algae require primarily three components to grow: sunlight, carbon dioxide, and water. Photosynthesis is an important biochemical process in which plants, algae, and some bacteria convert the energy of sunlight to chemical energy. In open ponds unassisted growth is slow, owing to the lower concentration of carbon dioxide; where carbon dioxide concentrations are increased artificially, higher growth rates can be achieved in open ponds as well. Algae could be grown in closed structures called photobioreactors, where the environment is better controlled than in open ponds. The cost of setting up and operating a photobioreactor would be higher than for those for open ponds. But, the oil yields from these photobioreactors could be significantly higher than open ponds. This offsets the initial investment in the medium/long run (Shay 1993 and www.oilgae.com).

Some of the challenges in cultivation of algae include: too much sunlight kills algae growth, overcrowding reduces the growth of algae, excess/waste produced in the ponds is required to be removed continuously, temperature needs to be steadily maintained, and open ponds, which are subjected to evaporation and rain fall, may cause an imbalance in pH value (Wagner 2007).

Algae are cultivated in closed containers and fed sugar to promote growth. This method eliminates problems since it allows growers to control all environmental factors. The benefit of this process is that it allows the algae biodiesel to be produced anywhere as the environmental conditions are controlled. Photobioreactor equipment is used to harvest algae. Photobioreactor set up can be continuous or batch process. A batch photobioreactor is set up with nutrients and algal seed and allowed to grow until the batch is harvested. A photobioreactor is continually harvested daily or at frequent intervals. Photobioreactors can be of glass or plastic tubes, tanks, plastic sleeves, or bags. A typical algae biodiesel production process is described in Figure 13.6.

Algae are made up of eukaryotic cells; that is, with nuclei and organelles. All algae have plastids, the bodies with chlorophyll that carry out photosynthesis. But the various lines of algae can have different combinations of chlorophyll molecules; some have just Chlorophyll A, some A and B, and other lines, A and C. All algae is primarily comprised of proteins, carbohydrates, fats, and nucleic acids in varying proportions. Algae contain anywhere between 2 and 40% of lipids/oils by weight. This fatty acid (oil) can be extracted and converted into biodiesel (Shay 1993). Algae range from small, single-celled organisms to multicellular organisms, some with fairly complex and differentiated forms. Microalgae are an organism capable of photosynthesis that is less than 2 mm in diameter. Microalgae have much more oil than macroalgae and it

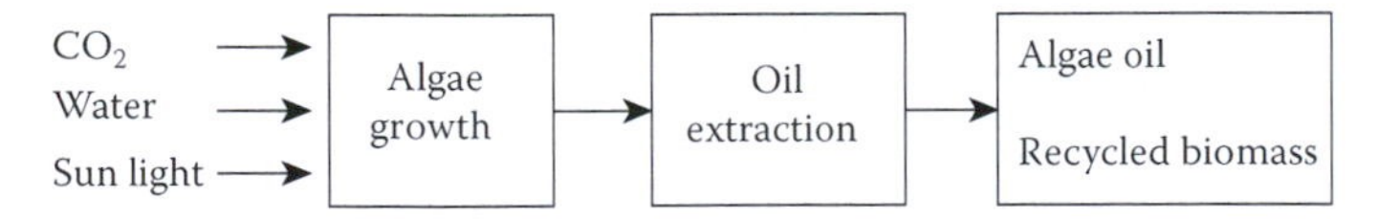

FIGURE 13.6
Algae biodiesel production processes.

TABLE 13.8

Oil Production Rate Per Hectare

Oil Seed	Liters
Castor	1413
Sunflower	952
Safflower	779
Palm	5950
Soy	446
Coconut	2689
Algae	100,000

Source: www.oilgae.com.

is much faster and easier to grow (www.oilgae.com). Algae can yield biodiesel about 15–250 times more than that of vegetable oils. Table 13.8 shows the quantity of oil produced per hectare of land (www.oilgae.com).

Pressing of algae extracts 75% of oil, but the hexane solvent method extracts up to 95% of the oil from algae. In this process, after removing the oil by pressing, the balance is mixed with hexane, then filtered and purified to remove unwanted chemicals.

The supercritical fluids method extracts up to 100% of the oil from algae. In this method carbon dioxide acts as the supercritical fluid. The algae is pressurized and heated to change its composition into a liquid as well as a gas. At this point, carbon dioxide is mixed with the algae and the carbon dioxide turns the algae completely into oil.

13.6.2 Commercialization

Many companies are pursuing the development of algae bioreactors for biodiesel production and CO capturing. Research into algae for the mass-production of oil is mainly focused on microalgae due largely to its less complex structure, fast growth rate, and high oil content. Interest in linking up commercial power plants and sewage treatment plants with an algae production plant is under way.

Infinifuel Biodiesel LLC Company built a geothermal powered and heated biodiesel plant in 2006 and hence is combining geothermal energy and renewable biodiesel production from algae oil. Aqua flow Bionomic Corporation in Marlborough, New Zealand has produced its first sample of biodiesel fuel made from algae found in sewage ponds. Enhanced Biofuels & Technologies developed an algae process that combines a bioreactor with an open pond, using waste CO_2 from coal-fired power plant flue gases as a food for the algae. Valcent Products has developed a high density vertical bioreactor for the mass production of oil bearing algae while removing large quantities of CO_2 from the atmosphere (www.castoroil.in).

13.6.3 Advantages of Algae Biofuel

- Higher yield per hectare and hence lower cost.
- Algae can grow in any place where there is enough sunshine and water.
- Algae captures CO_2 for its growth. If the algae cultivation set up is established near thermal power plants/industry it absorbs the CO_2 emitted by them and hence reduces the greenhouse effect.
- Algae biofuel contains no sulfur, is nontoxic, and biodegradable.

13.7 Comparison between First and Second-Generation Biofuels

A comparison of first and second generation biofuels are given in Table 13.9.

13.8 Challenges for Future Biofuels

There is a wide scope for second generation biofuels in the coming years, however, some hindrances are in commercial production and marketing.

1. First generation biofuels are generally derived from crop based feedstocks, while second-generation biofuels can be derived from a broader range of feedstocks, such as grasses, wood, landfill waste, and potentially algae.
2. New technologies for second-generation biofuels are being developed, but very few are yet viable in commercial scale productions.

TABLE 13.9

Comparison of First and Second-Generation Biofuels

Issue	First Generation	Second Generation
Capital cost	Low/ton	High/ton
Feedstock cost	High/ton	Low/ton
Operating cost	Low/ton	High/ton
Energy yield	High	Low
Logistics	Low cost and practical	High cost and impractical

3. Some second-generation biofuels that are derived from feedstocks with a low energy density may face logistical issues and land use efficiency questions and these may present as big a barrier to their uptake as the development of the core technology.

However, these drawbacks can be overcome with policies and the development of economical production methods.

References

Billa, E., D. P. Koullas, B. Monties, and E. G. Koukios. 1997. Structure and composition of sweet sorghum stalk components. *Industrial Crops and Products* 6:297–302.

Deimbras, A. 2005. Bioethanol from cellulosic materials: A renewable motor fuel from biomass. *Energy Sources* 27:327–37.

Jong, W. D. 2008. Sustainable hydrogen production by thermo chemical biomass processing. *In Hydrogen fuel-production, transportation and storage*, ed. R. B. Gupta. Boca Raton, FL: CRC Press 185–226.

Majewski, A. W., and M. K. Khair. 2006. *Diesel emissions and their control.* Warrendale, PA: SAE International.

Nandi, R., and S. Sengupta. 1998. Microbial production of hydrogen: An overview. *Critical Review on Microbiology* 24:61–84.

Ramadhas, A. S. 2009. New generation biofuels. *In Workshop on optimum energy utilization in thermal systems*. Eds. C. Muraleedharan and S. Jayaraj Calicut, India: National Institute of Technology 38–48.

Shay, E. G. 1993. Diesel fuel from vegetable oils: Status and opportunities. *Biomass Bioenergy* 4:227–42.

Smith, R., M. Asaro, and S. Naqvi. 2008. *Fuels of the future*. Menlo Park, CA: SRI Consulting.

Wagner, L. 2007. *Biodiesel from algae.* Research report. London: Mora Associates.

Speight, J. G. 2008. *Synthetic fuels hand book: Properties, Process and Performance.* USA: McGraw-Hill Hand books.

Appendix: Fuel Properties

APPENDIX 1

Fuel Properties

Property	Gasoline	No. 2 Diesel Fuel	Methanol	Ethanol	MTBE	Propane	CNG	Hydrogen	Biodiesel
Chemical formula	C_4–C_{12}	C_3–C_{25}	CH_3OH	C_2H_5OH	$(CH_3)_3COCH_3$	C_3H_8	CH_4 (83–99%) C_2H_6 (1–13%)	H_2	C_{12}–C_{22} FAME
Molecular weight	100–105	200	32.04	46.07	88.15	44.1	16.04	2.02	~292
Composition, weight %									
Carbon	85–88	87	37.5	52.2	68.1	82	75	0	77
Hydrogen	12–15	30	12.6	13.1	13.7	18	25	100	12
Oxygen	0	0	49.9	34.7	18.2	—		0	11
Specific gravity, 15.5°C/15.5°C	0.72–0.78	0.85	0.796	0.794	0.744	0.508	0.424	0.07	0.88
Density, kg/m³ @ 15.5°C	719–779	848	795	735	742	506	128	—	878
Boiling temperature, °C	27–225	180–340	65	78	55	–42	–164 to –88	–253	315–350
Reid vapor pressure, bar	0.55–1.03	<0.01	0.32	0.16	0.54	14.34	166	—	<0.01
Research octane number.	88–98	—	107	108	116	112	—	130+	—
Motor octane number.	80–88	—	92	92	101	97	—	—	—
(R + M)/2	86–94	—	100	100	108	104	120+		
Cetane number.	—	40–55		0–54					48–65

Water solubility, @ 21°C									
Fuel in water, volume %	Negligible	Negligible	100	100	4.8	—	—	—	—
Water in fuel, volume %	Negligible	Negligible	100	100	1.5	—	—	—	—
Freezing point, °C	–40	–40 to –1 (pour point)	–97.5	–114	–108.8	–187.7	–182.2	–259.4	–3.3 to 18.8 (cloud point)
Viscosity									
Centipoise @ 20°C	0.5–0.6	2.8–5.0	0.74	1.50	0.47	—	—	—	—
Flash point, closed cup, °C	–43	60–80	11	13	–26	–104	–184	—	100–170
Autoignition temperature, °C	257	316	464	423	435	450	482–632	500	—
Flammability limits, volume %									
Lower	1.4	1	7.3	4.3	1.6	2.2	5.3	4.1	—
Higher	7.6	6	36	19	8.4	9.5	15	74	—
Latent heat of vaporization									
kJ/m^3 @ 15.5°C	251	195	931	663	241	216	—	—	—
kJ/kg @ 15.5°C	348.9	232.6	1177	921.1	321.0	449.2	509.4	446.8	—
kJ/kg air for stoichiometric mixture 15.5°C	23.26	18.61	182.4	102.3	27.5	—	—	—	—
Heating value									
Higher (liquid fuel–liquid water) MJ/kg @ 15.5°C	46.53	45.76	22.88	29.84	37.95	50.23	52.22	139.11	40.16

(*Continued*)

APPENDIX 1 (CONTINUED)

Fuel Properties

Property	Gasoline	No. 2 Diesel Fuel	Methanol	Ethanol	MTBE	Propane	CNG	Hydrogen	Biodiesel
Lower (liquid fuel–water vapor) MJ/kg @ 15.5°C	43.44	42.78	20.09	26.95	35.10	46.29	47.13	121.46	37.52
Specific heat, J/kg °K	2008.3	1799.1	2510.4	2384.9	2092.0	—	—	—	—
Stoichiometric air/fuel, weight	14.7	14.7	6.45	9	11.7	15.7	17.2	34.3	13.8
Volume % fuel in vaporized stoichiometric mixture	2	—	12.3	6.5	2.7	—	—	—	—

Source: www.afdc.energy.gov/afdc/pdfs/fueltable.pdf and www.methanol.org/pdf/fuelproperties.pdf

Index

D

M

P

R

S

W

X

Z